Adriaan C. Zaanen

Introduction to Operator Theory in Riesz Spaces

Springer

Berlin
Heidelberg
New York
Barcelona
Budapest
Hong Kong
London
Milan
Paris
Santa Clara
Singapore
Tokyo

Adriaan C. Zaanen

Introduction to Operator Theory in Riesz Spaces

Springer

Adriaan C. Zaanen
Nassaulaan 15
NL-2628 GA Delft
The Netherlands

Department of Mathematics and Computer Science
University of Leiden
Niels Bohrweg 1
2333 CA Leiden
The Netherlands

Mathematics Subject Classification (1991): 47A, 46

Cataloging-in-Publication Data applied for

Die Deutsche Bibliothek - CIP-Einheitsaufnahme
Zaanen, Adriaan C.:
Introduction to operator theory in riesz spaces / Adriaan C. Zaanen
Berlin; Heidelberg; New York; Barcelona; Budapest; Hong Kong; London;
Milan; Paris; Santa Clara; Singapore; Tokyo: Springer, 1997
ISBN-13:978-3-642-64487-0   e-ISBN-13:978-3-642-60637-3
DOI: 10.1007/978-3-642-60637-3

ISBN-13:978-3-642-64487-0

Production: PRODUserv Springer Produktions-Gesellschaft, Berlin
Cover-Design: Erich Kirchner, Heidelberg
Typesetting in TeX by author using a Springer TeX macro-package

SPIN: 10536964   41/3020 Printed on acid-free paper

Preface

Since the beginning of the thirties a considerable number of books on functional analysis has been published. Among the first ones were those by M.H. Stone on Hilbert spaces and by S. Banach on linear operators, both from 1932. The amount of material in the field of functional analysis (including operator theory) has grown to such an extent that it has become impossible now to include all of it in one book. This holds even more for textbooks. Therefore, authors of textbooks usually restrict themselves to normed spaces (or even to Hilbert space exclusively) and linear operators in these spaces. In more advanced texts Banach algebras and (or) topological vector spaces are sometimes included. It is only rarely, however, that the notion of order (partial order) is explicitly mentioned (even in more advanced expositions), although order structures occur in a natural manner in many examples (spaces of real continuous functions or spaces of measurable functions). This situation is somewhat surprising since there exist important and illuminating results for partially ordered vector spaces, in particular for the case that the space is lattice ordered. Lattice ordered vector spaces are called vector lattices or Riesz spaces. The first results go back to F. Riesz (1929 and 1936), L. Kantorovitch (1935) and H. Freudenthal (1936). Any Riesz space provided with a norm such that norm and order are connected in a certain appropriate manner is called a normed Riesz space. If the space is then norm complete (i.e., every norm Cauchy sequence has a norm limit), the space is called a Banach lattice. Important standard examples are the space of continuous functions on an interval in the real line (or, more generally, on a compact Hausdorff space) and certain spaces of measurable functions (the L_p–spaces). There are several books available dealing with Riesz spaces and operators in these spaces, but nearly all of these books are written for experts and not very suitable for use as a textbook.

In the present work an effort is made to facilitate the situation for the reader who is not already familiar with ordered spaces. It is sufficient that the reader has some knowledge about the simplest notions in functional analysis and basic measure theory (Lebesgue integral for one or two variables is enough for most purposes). In the examples dealing with continuous functions on a Hausdorff space the reader may restrict himself (herself) to continuous functions on an interval in the real line. The only exception is in the last two chapters where a little more knowledge about compact sets and compact operators is required, although everything used without proof is first explained. The axiom of choice (in the form of Zorn's lemma) occurs in one chapter only, where the Hahn-Banach theorem and several of its consequences are discussed. Proofs depending on so-called representation theorems do not occur at all. Direct proofs, although sometimes a little longer than proofs by means of representation, often reveal more about the situation under discussion. Some more expert readers may regret the absence of detailed (prime) ideal theory or the absence of the Dedekind completion. There is, however, a limit to the size of what can be done in an introductory work. The decision to include the last two chapters was taken because in these chapters it is shown clearly how relatively recent results about the spectra of positive operators in Banach lattices are related to classical results on matrices with non-negative entries which go back to the beginning of this century.

It has been the intention while preparing the text to adhere thoughout to a patient expository style with attention for the details. Supplementary material has been included in the form of exercises, many with a hint for the proof.

A brief survey of the contents follows. The first six chapters are devoted to the algebraic theory of Riesz spaces. In the first chapter some simple, but fundamental facts on partially ordered sets, lattices and Boolean algebras are collected, with proofs. Then, in the next chapters, Riesz spaces are defined (with examples), ideals and bands in these spaces are introduced, order completeness properties, order convergence properties and projection properties are discussed and, finally, complex Riesz spaces are defined. In the next two chapters we deal with normed Riesz spaces. Spaces having an order continuous norm (such as L_p–spaces for $1 \leq p < \infty$) receive special attention. From chapter 9 on linear operators in Riesz spaces play a major part, in particular those operators that can be written as a difference of two positive operators. Order continuous operators get special attention. The order dual of a Riesz space is defined and (in the normed case) is compared with the norm dual space. Similarly, order adjoint operators are defined and compared with norm adjoint operators. Among the results proved are Freudenthal's spectral theorem, the Radon-Nikodym theorem and Synnatschke's theorem on adjoint operators. In chapter 18 it is shown that in some Riesz spaces it is possible to define a multiplication (with the usual properties one expects from a multiplication), in chapter 19 the operator theory is complexified and in chapter 20 the Hahn-Banach theorem together with some of its consequences is proved. Finally, in chapters 21 and 22, we discuss the spectral properties of operators in a Banach lattice that are positive, compact and irreducible. At the end of the last chapter the reader will find a short list of other books on the subject of Riesz spaces.

The end of each proof is indicated by a small black square (∎).

My sincere thanks to my son Maarten who transformed the original document into a TeX document.

Delft, July 1996   Adriaan C. Zaanen

Contents

CHAPTER 1
Lattices and Boolean Algebras

1. Partially Ordered Sets

As observed before, order structures are, together with algebraic and topological structures, of fundamental importance in large parts of mathematics. In the present section we introduce some elementary notions concerning order.

Let X be a non-empty set (i.e., $X \neq \emptyset$), the points of X will be denoted by $x, y, \ldots$. The set of all ordered pairs (x, y) with x and y members of X is called the *Cartesian product* of X and X itself; notation $X \times X$. Note that we deal with ordered pairs, so for $x \neq y$ the points (x, y) and (y, x) of $X \times X$ are different in general. By definition, a *relation* in X is a non-empty subset of $X \times X$; we shall write xRy if (x, y) is an element of the subset defining the relation R. We mention two examples, the first one of which is well-known.

The relation R is called an *equivalence relation* in X if

(i) xRx for every $x \in X$ (R is *reflexive*),
(ii) xRy and yRz implies xRz (R is *transitive*),
(iii) xRy implies yRx (R is *symmetric*).

If the subset of $X \times X$ defining the relation R consists only of all points (x, x), then R is the *relation of equality*. Obviously, this is an equivalence relation.

Definition 1.1. The relation R in X is called a *partial ordering* if

(i) xRx for every $x \in X$ (R is reflexive),
(ii) xRy and yRz implies xRz (R is transitive),
(iii) xRy and yRx implies $x = y$ (R is *anti-symmetric*).

The set X, equipped with a partial ordering, is called a *partially ordered set*.

If R is a relation of partial ordering, we write $x \leq y$ (or, equivalently, $y \geq x$) instead of xRy. If x and y are points of X such that $x \leq y$ or $x \geq y$, we say that x and y are *comparable*. If neither $x \leq y$ nor $x \geq y$, then x and y are *incomparable*. If every pair of points x and y of X is a comparable pair, the partial ordering is called a *linear ordering* or a *strict ordering*. The other extreme occurs if every pair of points x, y satisfying $x \neq y$ is an incomparable pair, i.e., $x \leq y$ if and only if $x = y$. In this case we have again the equality relation.

If X is partially ordered and Y is a non-empty subset of X, then the partial ordering in X induces a partial ordering in Y, i.e., if x and y are points of Y, then $x \leq y$ in Y whenever $x \leq y$ in X. If the induced partial ordering in Y is a linear ordering, then Y is called a *chain* in X.

Example 1.2. (i) $X = \mathbb{R}$ (the set of real numbers) with the familiar linear ordering.

(ii) $X = \mathbb{R}^2$ ($\mathbb{R}^2$ is short for $\mathbb{R} \times \mathbb{R}$) with coordinatewise ordering, i.e., if $x = (x_1, x_2)$ and $y = (y_1, y_2)$, then $x \leq y$ if and only if $x_1 \leq y_1$ and $x_2 \leq y_2$ hold in $\mathbb{R}$. There are many incomparable points, such as for example (0,1) and (1,0). Note that every straight line through two different comparable points is a chain.

(iii) Let S be a non-empty set and let X consist of all subsets of S, partially ordered by inclusion, i.e., if A and B are points of X (in other words, A and B are subsets of S), then $A \leq B$ if and only if $A \subseteq B$. In general, X contains many incomparable pairs of points, certainly if S is an infinite set. Note that the empty subset of S is a point of X.

If X is partially ordered, Y is a non-empty subset of X and the point $x_0 \in X$ satisfies $y \leq x_0$ for all $y \in Y$, then x_0 is called an *upper bound* of Y. The subset Y is now said to be *bounded above*. If x_0 is an upper bound of Y such that $x_0 \leq x^\sim$ for any other upper bound $x^\sim$ of Y, then x_0 is called a *least upper bound* or *supremum* of Y. If Y has such a supremum, then the supremum is uniquely determined. Indeed, assuming that x_0 and $x^\sim$ are suprema of Y, we have $x_0 \leq x^\sim$ and $x^\sim \leq x_0$ by the definition of a supremum. Hence $x_0 = x^\sim$ in view of the anti-symmetry of the partial ordering. We shall use the notation $x_0 = \sup Y$ or $x_0 = \sup(y : y \in Y)$ for the supremum of Y, if it exists. The definitions of *lower bound* and *greatest lower bound infimum*) are analogous.

Assume again that X is partially ordered. The point $x_0 \in X$ is called a *maximal element* of X if it follows from $x_0 \leq x \in X$ that $x_0 = x$ (i.e., there does not exist an element strictly larger than x_0). If there exists an element $x_0 \in X$ such that $x_0 \geq x$ for all $x \in X$, then x_0 is called the *largest element* of X. In this case, x_0 is also a maximal element of X. In fact, the largest element, if it exists, is the only maximal element of X. Conversely, if x_0 is the only maximal element of X, then x_0 is not necessarily the largest element of X. All this is illustrated by the following examples.

(i) Let X be the closed unit disc $\{(x, y) : x^2 + y^2 \leq 1\}$ in $\mathbb{R}^2$ with coordinatewise partial ordering. All points (x, y) with $x \geq 0, y \geq 0, x^2 + y^2 = 1$ are maximal elements of X and X does not have a largest element.

(ii) Let X be the union of the open unit disc $\{(x, y) : x^2 + y^2 < 1\}$ and the point $(1,0)$, again with coordinatewise ordering. The point $(1,0)$ is the only maximal element, but it is not the largest element.

The definitions of *minimal element* and *smallest element* are analogous.

We proceed with some further definitions.

Definition 1.3. Let X be a partially ordered set.

(i) X is called *order complete* (or simply *complete*) if every non-empty subset of X has a supremum and and an infimum.

(ii) X is called *Dedekind complete* if every non-empty subset of X that is bounded above (bounded below) has a supremum (infimum).

(iii) X is called *Dedekind σ-complete* (or *countably Dedekind complete*) if every non-empty finite or countable subset of X that is bounded above (bounded below) has a supremum (infimum).

(iv) X is called a *lattice* if every subset consisting of two points has a supremum and an infimum.

Remark. A lattice is called "un treillis" or "un ensemble réticulé" in French, "ein Verband" in German and "een rooster" or "een tralie" in Dutch.

For Dedekind completeness a one-sided condition is already sufficient, as shown by the following theorem.

Theorem 1.4. *The partially ordered set X is Dedekind complete if and only if every non-empty subset which is bounded above has a supremum.*

Proof. Assume that every non-empty subset of X which is bounded above has a supremum, and let Y be a non-empty subset which is bounded below. We have to prove that $\inf Y$ exists. For this purpose, observe that the set $L(Y)$ of all lower bounds of Y is non-empty and bounded above, so $p = \sup L(Y)$ exists by hypothesis. Since every $y \in Y$ is an upper bound of $L(Y)$, it follows that $p \le y$ for every $y \in Y$ (by the definition of a least upper bound). This shows that p is a lower bound of Y, i.e., p is a member of $L(Y)$. Summing up, we have proved now that p is a lower bound of Y and $q \le p$ holds for any arbitrary lower bound q of Y. In other words, $p = \inf Y$.   ∎

Remark. If the partially ordered set X has a smallest as well as a largest element, then X is order complete if and only if X is Dedekind complete. Conversely, if X is order complete, then X has a largest and a smallest element.

Example 1.5. (i) $\mathbb{R}$ and $\mathbb{R}^2$ (with coordinatewise ordering) are not order complete, but they are Dedekind complete and they are lattices.

(ii) The closed unit disc in $\mathbb{R}^2$ with coordinatewise ordering is Dedekind complete, but not order complete and it is not a lattice.

(iii) Let S be a non-empty set and let X consist of all subsets of S partially ordered by inclusion. Then $\emptyset$ and S are the smallest and largest elements of X. Furthermore, X is a lattice with $\sup(A, B) = A \cup B$ and $\inf(A, B) = A \cap B$. The lattice X is order complete: $\sup(A_\alpha : \alpha \in \{\alpha\}) = \cup A_\alpha$ and $\inf(A_\alpha : \alpha \in \{\alpha\}) = \cap A_\alpha$ for any arbitrary "index set" $\{\alpha\}$.

(iv) We exhibit a lattice X which is not Dedekind σ-complete (and so X is not Dedekind complete). Let X be the set of all real continuous functions on the closed interval [0,2], partially ordered by defining that $f \leq g$ in X whenever $f(x) \leq g(x)$ for all $x \in [0, 2]$. The set X is a lattice; $\sup(f, g)$ and $\inf(f, g)$ are the functions obtained by taking the pointwise maximum and minimum of $f(x)$ and $g(x)$ respectively. The reader who is asking whether these functions are continuous (because otherwise these functions would not be members of X) should first observe that if f is continuous, then so is the function $| f |$. Hence, if f and g are continuous, then so are $f + g$ and $| f - g |$. Note now that

$$\sup(f, g) = \frac{1}{2}(f + g) + \frac{1}{2} | f - g |,$$

$$\inf(f, g) = \frac{1}{2}(f + g) - \frac{1}{2} | f - g | .$$

For the proof that X is not Dedekind complete, let $(f_n : n = 1, 2, 3, \ldots)$ be the sequence of functions in X, defined by $f_n(x) = 0$ for all x such that $0 \leq x \leq 1 - n^{-1}$, $f_n(x) = 1$ for $1 \leq x \leq 2$ and f_n linear on $[1 - n^{-1}, 1]$, i.e., the graph of f_n on this interval is a linesegment connecting $(1 - n^{-1}, 0)$ and $(1,1)$. The sequence (f_n) is a countable set in X, bounded below. Assume that $f_0 = \inf f_n$ exists in X. It is not difficult to see that in this case $f_0(x) = 0$ for $0 \leq x < 1$ and $f_0(x) = 1$ for $1 < x \leq 2$. But then f_0 is not continuous, i.e., f_0 is not a member of X. Contradiction. Hence, $\inf f_n$ does not exist in X.

2. Lattices

We present some simple definitions and theorems on lattices. First some notations. Let X be a lattice. For x and y in X, it is customary to write sometimes $x \vee y$ instead of $\sup(x, y)$ and $x \wedge y$ instead of $\inf(x, y)$. This is the "cap and cup" notation. By induction it is immediately evident that every finite subset of X has a supremum and an infimum. The notations are $\sup(x_1, \ldots, x_n)$ or $x_1 \vee \cdots \vee x_n$ or $\vee_{i=1}^n x_i$ for the supremum. Similarly $\inf(x_1, \ldots, x_n)$ or $x_1 \wedge \cdots \wedge x_n$ or $\wedge_{i=1}^n x_i$ for the infimum.

Observe now that $x \le w$ and $y \le w$ implies $x \vee y \le w$. Hence, since both $x \wedge y$ and $x \wedge z$ are majorized by $x \wedge (y \vee z)$, it follows that

$$(x \wedge y) \vee (x \wedge z) \le x \wedge (y \vee z). \tag{1}$$

Lattices with the property that there is equality in (1) for all x, y, z in X deserve a special name. This is done in the following definition.

Definition 2.1. The lattice X is called *distributive* if there is equality in (1), i.e., if

$$x \wedge (y \vee z) = (x \wedge y) \vee (x \wedge z) \tag{2}$$

for all x, y, z in X.

Remark. The operation of taking suprema in a lattice shows some resemblance with the operation of addition in the system of real numbers. Similarly, taking infima resembles multiplication. Keeping this in mind, we see that the equality (2) resembles the equality $x(y + z) = xy + xz$ for ordinary numbers. This explains the use of the word "distributive" In a certain sense distributive lattices are even more distributive than the number system, because (as will be shown in the next theorem) in the formula (2) it is permitted to interchange $\vee$ and $\wedge$, whereas interchanging addition and multiplication in $x(y + z) = xy + xz$ is certainly not permitted.

Theorem 2.2. *The lattice X is distributive if and only if*

$$x \vee (y \wedge z) = (x \vee y) \wedge (x \vee z) \tag{3}$$

for all x, y, z in X.

Proof. Assume that X is distributive, i.e., (2) holds. Denote the left hand side and the right hand side of (3) by l and r for a moment, and let $w = x \vee y$. Then

$$r = w \wedge (x \vee z) = (w \wedge x) \vee (w \wedge z) =$$

$$\{(x \vee y) \wedge x\} \vee \{(x \vee y) \wedge z\} = (x \wedge x) \vee (y \wedge x) \vee (x \wedge z) \vee (y \wedge z) =$$

$$x \vee (y \wedge x) \vee (x \wedge z) \vee (y \wedge z).$$

Note now that $y \wedge x \le x$ and $x \wedge z \le x$, so

$$x \vee (y \wedge x) \vee (x \wedge z) = x$$

Therefore, $r = x \vee (y \wedge z) = l$. The proof that (3) implies (2) is similar.   ∎

Exercise 2.3. Let X be a lattice. Show that X is distributive if and only if it follows from $x \wedge y \leq w$ and $x \wedge z \leq w$ that $x \wedge (y \vee z) \leq w$, and also if it follows from $x \vee y \geq w$ and $x \vee z \geq w$ that $x \vee (y \wedge z) \geq w$.

If the lattice X has a smallest element, this element is called the *zero element* and denoted by θ. If X has a largest element, this is called the *unit element* and denoted by e. Every finite lattice has a zero element and a largest element. The lattice consists of one element only if and only if $\theta = e$. If X is a lattice with θ and e and if the elements x and x' satisfy $x \wedge x' = \theta$ and $x \vee x' = e$, then x' is called a *complement* of x. Of course, x is then a complement of x'. In a distributive lattice with θ and e complements are unique.

Theorem 2.4. *Let X be a distributive lattice with θ and e. If the element x has a complement x', then x' is uniquely determined. Hence, any $x \in X$ has at most one complement.*

Proof. Assume that x' and $x^\sim$ are complements of x. Then

$$x^\sim = x^\sim \vee \theta = x^\sim \vee (x \wedge x') = (x^\sim \vee x) \wedge (x^\sim \vee x') =$$

$$e \wedge (x^\sim \vee x') = x^\sim \vee x'.$$

Similarly $x' = x' \vee x^\sim$. Hence $x^\sim = x'$. ■

We mention some further notions that are useful in lattices with a zero element. Let X be a lattice with zero θ. Elements x, y in X satisfying $x \wedge y = \theta$ are said to be *disjoint*. If Y is a non-empty subset of X, the set

$$Y^d = (x \in X : x \text{ disjoint to all } y \in Y)$$

is called the *disjoint complement* of Y. A further important notion in lattice theory is that of an ideal.

Definition 2.5. Let X be a lattice with zero element θ. The non-empty subset I of X is called an *ideal* in X if

(i) $x \in I$ and $y \in I$ implies $x \vee y \in I$,
(ii) $y \leq x \in I$ implies $y \in I$.

It is evident that every ideal contains θ. Also, $\{\theta\}$ is the smallest and X is the largest ideal in X. It is easy to see that condition (ii) in the definition of an ideal is equivalent to the condition that $x \in I$ implies $x \wedge y \in I$ for every $y \in X$. This clearly shows the analogy with the definition of an ideal in the theory of commutative rings. Finally, it is also not difficult to prove that if Y is a non-empty subset of the distributive lattice X with zero, then the disjoint complement Y^d is an ideal in X.

Much more can be said about lattices, in particular about distributive lattices. The analogy between ideal theory (in particular prime ideal theory) in distributive lattices with zero and ideal theory in commutative rings goes much farther than a coincidental resemblance between some definitions. We refrain from saying more because it would lead us far beyond our present purposes. It is appropriate to observe here that the modern theory of lattices goes back to the years between 1930 and 1940. Important contributions were made by G.Birkhoff and M.H.Stone. In the next section we briefly look at a special type of lattices, the Boolean algebras.

3. Boolean Algebras

Distributive lattices with zero θ and unit e, having the property that every element has a complement, are called *Boolean algebras*. As before, the complement of the element x will be denoted by x'.

Theorem 3.1. *Let X be a Boolean algebra.*

(i)   *For any $x \in X$ the complement x' is the largest element disjoint to x.*
(ii)   *$x \leq y$ implies $x' \geq y'$.*
(iii) *$(x \vee y)' = x' \wedge y'$ and $(x \wedge y)' = x' \vee y'$.*
(iv)   *If $x = \sup(x_\alpha : \alpha \in \{\alpha\})$, then $x' = \inf x'_\alpha$.*

Proof. (i) We have $x \vee x' = e$ and $x \wedge x' = \theta$. Assume now that the element $x^\sim$ satisfies $x \wedge x^\sim = \theta$. We have to prove that $x^\sim \leq x'$, i.e., prove that $x' \vee x^\sim = x'$. For the proof, let $y = x' \vee x^\sim$. Then

$$x \vee y = x \vee x' \vee x^\sim = e,$$

$$x \wedge y = x \wedge (x' \vee x^\sim) = (x \wedge x') \vee (x \wedge x^\sim) = \theta \vee \theta = \theta.$$

Hence, y is a complement of x, so that (by the uniqueness of the complement) $x' = y = x' \vee x^\sim$.

(ii) Let $x \leq y$. Then $y' \wedge x \leq y' \wedge y = \theta$, so $y' \wedge x = \theta$, i.e., y' is disjoint to x. It follows from (i) that $y' \leq x'$.

(iii) It follows from $x \vee y \geq x$ that $(x \vee y)' \leq x'$. Similarly $(x \vee y)' \leq y'$. Hence

$$(x \vee y)' \leq x' \wedge y'. \tag{1}$$

Also ·

$$(x' \wedge y') \wedge (x \vee y) = (x' \wedge y' \wedge x) \vee (x' \wedge y' \wedge y) = \theta \vee \theta = \theta,$$

so $x' \wedge y'$ is disjoint to $x \vee y$. It follows from (i) again that

$$(x \vee y)' \geq x' \wedge y' \tag{2}$$

Comparing (1) and (2), we see that the equality sign holds in (1). The proof for $(x \wedge y)' = x' \vee y'$ is similar.

(iv) $x = \sup x_\alpha$ implies that $x' \leq x$ for every α, so x' is a lower bound of the set of all x'_α. Assume that y is also a lower bound, i.e., $y \leq x'_\alpha$ for all α. We have to prove that $y \leq x'$. From $y \leq x'_\alpha$ for all α it follows that $y' \geq x''_\alpha = x_\alpha$ for all α, so $y' \geq \sup x_\alpha = x$. But then $y = y'' \leq x'$. ■

If the elements x, y of the partially ordered set X satisfy $x \leq y$, then the non-empty set $(z : z \in X, x \leq z \leq y)$ is called an *order interval* in X. The set is denoted by $[x, y]$, similarly as for an interval in the real numbers. We shall prove now that every order interval in a Boolean algebra is a Boolean algebra on its own.

Theorem 3.2. *Every order interval $[x, y]$ in the Boolean algebra X is a Boolean algebra on its own with x as zero element and y as unit element.*

Proof. As before, we denote zero and unit in X by θ and e respectively. Let $[x, y]$ be an order interval in X. It is evident that $[x, y]$ with the induced ordering is a distributive lattice with x as zero and y as unit. All that we have to prove is that every $z \in [x, y]$ has a complement in $[x, y]$. For this purpose, write $z^* = (z' \wedge y) \vee x$, where z' is the complement of z in X. Then

$$z \wedge z^* = z \wedge \{(z' \wedge y) \vee x\} = \{z \wedge (z' \wedge y)\} \vee (z \wedge x) = \theta \vee x = x,$$

$$z \vee z^* = z \vee (z' \wedge y) \vee x = z \vee (z' \wedge y) = (z \vee z') \wedge (z \vee y) = z \vee y = y.$$

This shows that z^* is the complement of z in $[x, y]$. ■

Corollary 3.3. *Let X be a distributive lattice with zero θ such that every order interval $[\theta, y]$ in X is a Boolean algebra. Then every order interval $[x, y]$ in X is a Boolean algebra.*

It can be proved (although we shall not do so because it would lead us too far) that in a lattice as described in the last corollary it is possible to define an addition and a multiplication such that X becomes a commutative ring in the usual algebraic sense. The product xy is simply defined as $x \wedge y$, but the definition of $x + y$ is not so simple. For this it is convenient to denote, if $y \leq x$, the complement of y in $[\theta, x]$ by $x \ominus y$. Addition is now defined by $x + y = (x \vee y) \ominus (x \wedge y)$. To prove the distributive and associative laws is elementary but not wholly trivial. Any X of this kind is called a *Boolean ring*. Note that $x^2 = x$ and $x + x = \theta$ for every $x \in X$. Obviously, the Boolean ring X has a unit element e in the algebraic sense (i.e., $x = ex$ for every $x \in X$) if and only if X has e as its largest element, i.e., if and only if X is a Boolean algebra.

Example 3.4. (i) Let Γ be a non-empty collection of subsets of the non-empty point set X. We recall that Γ is said to be a *ring of subsets* if $A \in \Gamma, B \in \Gamma$ implies $A \cup B \in \Gamma$ and $A \backslash B \in \Gamma$ (where $A \backslash B$ is the set of points in X belonging to A but not to B). Then it follows that also $A \cap B \in \Gamma$, because $A \cap B = A \backslash (A \backslash B)$. Evidently, finite unions and finite intersections of sets of Γ are sets of Γ. The ring Γ is partially ordered by inclusion; $A \leq B$ if and only if $A \subseteq B$. This makes Γ into a distributive lattice with the empty set $\emptyset$ as zero element. The ring Γ is called an *algebra of subsets* of X whenever the set X itself is an element of Γ. In this case Γ is a Boolean algebra having X as unit element; for $A \in \Gamma$ the set $X \backslash A$ is the complement of A. Note that A and B are disjoint in the lattice (i.e., $A \wedge B = \theta$) whenever $A \cap B = \emptyset$, i.e., whenever A and B are disjoint point sets in the usual sense. It is convenient for further use to mention here still another Boolean algebra, closely connected with the one consisting of all sets in the algebra Γ. We recall the notion of the *characteristic function* of a subset A of X. This is a realvalued function χ_A, defined for all $x \in X$ and such that $\chi_A(x) = 1$ for every $x \in A$ and $\chi_A(x) = 0$ for every $x \in X \backslash A$. Let Γ_1 be the collection of all characteristic functions of sets in the algebra Γ, partially ordered by defining for A and B in Γ that $\chi_A \leq \chi_B$ whenever $A \subseteq B$. Then $\chi_A \vee \chi_B = \chi_{A \cup B}$ and $\chi_A \wedge \chi_B = \chi_{A \cap B}$. Evidently, Γ_1 is a Boolean algebra and the one-one mapping $A \Leftrightarrow \chi_A$ of Γ onto Γ_1 preserves order, i.e., $A \subseteq B$ if and only if $\chi_A \leq \chi_B$. The Boolean algebras Γ and Γ_1 are called *isomorphic* (more precisely, *lattice isomorphic*). Note that the set $(A \backslash B) \cup (B \backslash A)$, called the *symmetric difference* of A and B and sometimes denoted by $\Delta(A, B)$, has the characteristic function $|\chi_A - \chi_B|$.

(ii) The reader is assumed to be familiar with the basic notions of measure theory. Let μ be Lebesgue measure in n-dimensional space $\mathbb{R}^n$ and let Γ be the collection of the corresponding μ-measurable sets in $\mathbb{R}^n$. Hence, $(\mathbb{R}^n, \Gamma, \mu)$ is an example of what is called a *measure space*. The collection Γ is an algebra of subsets of $\mathbb{R}^n$; it is even a σ-algebra, i.e., any countable union of sets in Γ is again a set in Γ. Any set in Γ having measure zero is called a *null set*. Let Γ_0 be the collection of all these null sets. The collection Γ is a Boolean algebra and Γ_0 is an ideal in Γ. To see this, observe that $A \in \Gamma_0, B \in \Gamma_0$ implies $A \cup B \in \Gamma_0$ and $C \subseteq A \in \Gamma_0$ implies $C \in \Gamma_0$.

We introduce a relation $\approx$ in Γ by defining that $A \approx B$ holds whenever the symmetric difference $\Delta(A, B) = (A \backslash B) \cup (B \backslash A)$ is a null set (equivalently, whenever the characteristic function $|\chi_A - \chi_B|$ differs from zero only on a null set). The relation $\approx$ is an equivalence relation. To see that $A \approx B, B \approx C$ implies $A \approx C$, note that $|\chi_A - \chi_C| \leq |\chi_A - \chi_B| + |\chi_B - \chi_C|$. The collection of all $B \in \Gamma$ satisfying $B \approx A$ for a given $A \in \Gamma$ will be denoted by $[A]$ and (as customary with an equivalence relation) $[A]$ is called an *equivalence class*. It is clear that $[A] = [B]$ if and only if $A \approx B$. The collection Γ_0 of all null sets is exactly the equivalence class $[\emptyset]$. The collection Γ/Γ_0 of all equivalence classes is partially ordered by defining that $[A] \leq [B]$ if and only if $A \backslash B$ is a null set, i.e., A is contained in B except for a null set. Prove that this is indeed a partial ordering, i.e., $[A] \leq [B]$ and $[B] \leq [C]$ implies $[A] \leq [C]$ and also prove that the definition does not depend on the choice of A and B in $[A]$ and $[B]$. With respect to this ordering Γ/Γ_0 is now a Boolean algebra (with $[\emptyset]$ as zero and $[\mathbb{R}^n]$ as unit). The Boolean algebra Γ/Γ_0 is called the *measure algebra* corresponding to the measure space $(\mathbb{R}^n, \Gamma, \mu)$. In the terminology concerning these notions there is often a certain "abuse of terminology", that is to say, sets in Γ belonging to the same equivalence class are identified. The collection Γ/Γ_0 is then again denoted by Γ and an element in the "new Γ" is treated as if it were a subset of $\mathbb{R}^n$ instead of an equivalence class of subsets. We shall see soon that something similar occurs in the discussion of measurable functions. As a final remark we observe that $(\mathbb{R}^n, \Gamma, \mu)$ may be replaced by a more general measure space.

(iii) Let Γ be the collection of all open sets in the topological space X (if desired, the reader may think of the familiar open sets in $\mathbb{R}, \mathbb{R}^2$ or $\mathbb{R}^n$). If Γ is partially ordered by inclusion, then Γ becomes a lattice such that $A \vee B = A \cup B$ and $A \wedge B = A \cap B$. The lattice is distributive with $\emptyset$ as zero element and X as largest element. In general Γ is not a Boolean algebra. This occurs as soon as there exists at least one closed set F which is not at the same time an open set. Indeed, in this case the open set $O = X \backslash F$ fails to have a complement in Γ, because every open O^* disjoint to O has to be a subset of F but cannot be equal to F itself, and hence $O \cup O^* \neq X$. Similar arguments hold for the collection of all closed sets in X.

Exercise 3.5. (i) If X is a lattice with the property that, for every subset $(y_\alpha : \alpha \in \{\alpha\})$ for which $\sup y_\alpha$ exists and for every $x \in X$, the element $\sup(x \wedge y_\alpha)$ exists and satisfies

$$\sup(x \wedge y_\alpha) = x \wedge (\sup y_\alpha), \tag{3}$$

then we say that the *infinite distributive law* (3) holds in X. Similarly, if for every subset $(y_\alpha : \alpha \in \{\alpha\})$ for which $\inf y_\alpha$ exists and for every $x \in X$, the element $\inf(x \vee y_\alpha)$ exists and satisfies

$$\inf(x \vee y_\alpha) = x \vee (\inf y_\alpha). \tag{4}$$

then we say that the infinite distributive law (4) holds.

In $\mathbb{R}$, let X be the collection of all open subsets of the open interval $E = (x : -1 < x < 1)$. Show that X, partially ordered by inclusion, is a distributive lattice with $\emptyset$ as zero and E as unit. For any subset $(O_\alpha : \alpha \in \{\alpha\})$ of X we have $\sup O_\alpha = \cup_\alpha O_\alpha$ and $\inf O_\alpha$ is the interior of $\cap_\alpha O_\alpha$. It follows easily that the infinite distributive law (3) holds. Show that the law (4) fails to hold, even for sequences. For the proof, let for example $O_n = (x : -n^{-1} < x < 1)$ for $n = 1, 2, \ldots$ and take the union of $\inf O_n$ and $O = (x : -1 < x < 0)$.

(ii) Let X be a Boolean algebra. Show that if $x \wedge y_\alpha = \theta$ for all $\alpha \in \{\alpha\}$ and $y_0 = \sup y_\alpha$ exists, then $x \wedge y_0 = \theta$.

Hint: Denote the complements of y_α and y_0 by y'_α and y'_0 respectively. Use Theorem 3.1 to see that $y'_0 = \inf y'_\alpha$ and $x \leq y'_\alpha$ for every α. Then $x \leq y'_0$, so $x \wedge y_0 = \theta$.

(iii) Show that in a Boolean algebra the two infinite distributive laws hold.

Hint: For the proof of $\sup(x \wedge y_\alpha) = x \wedge (\sup y_\alpha)$, set $y_0 = \sup y_\alpha$. It is evident that $x \wedge y_0$ is an upper bound of the set of all $x \wedge y_\alpha$. If $x \wedge y_0$ is not the supremum, there exists an upper bound w such that $w \leq x \wedge y_0$ and $w \neq x \wedge y_0$. The complement z of w with respect to $x \wedge y_0$ satisfies now $z \neq \theta$ with $z \leq x \wedge y_0$ and $z \wedge (x \wedge y_\alpha) = \theta$ for all α. Hence, by part (ii) above, $z \wedge x \wedge y_0 = \theta$. But $z \leq x \wedge y_0$, so $z \wedge z = \theta$, i.e., $z = \theta$. Contradiction. The second infinite distributive law follows by taking complements.

CHAPTER 2
Riesz Spaces

4. Riesz Spaces

We assume the reader to be familiar with the basic definitions and simplest properties of real and complex vector spaces. In the present section and the next ones we restrict ourselves to real vector spaces. Elements in the vector spaces will usually be denoted by $f, g, h, \ldots$ and the real numbers which act as scalar multipliers by $\alpha, \beta, \ldots$ (this choice for the notation is related to the fact that in many examples the space consists of realvalued functions). The null element (zero element, neutral element) with respect to addition will be denoted by 0; it will always be clear whether we speak about the null element or about the number zero.

Definition 4.1. The real vector space E (with elements $f, g, \ldots$) is called an *ordered vector space* if E is partially ordered in such a manner that the vector space structure and the order structure are *compatible*, that is to say,

(i) $f \leq g$ implies $f + h \leq g + h$ for every $h \in E$,
(ii) $f \geq 0$ implies $\alpha f \geq 0$ for every $\alpha \geq 0$ in $\mathbb{R}$.

If, in addition, E is a lattice with respect to the partial ordering, then E is called a *Riesz space* or also a *vector lattice*.

Observe that condition (ii) in the definition is equivalent to the condition that $f \leq g$ implies $\alpha f \leq \alpha g$ for every $\alpha \geq 0$ in $\mathbb{R}$. We immediately present some examples of Riesz spaces.

Example 4.2. (1) The n-dimensional space $\mathbb{R}^n$ with the familiar coordinatewise addition and scalar multiplication is a real vector space. If we define the ordering likewise *coordinatewise*, i.e., for $x = (x_1, \ldots, x_n)$ and $y = (y_1, \ldots, y_n)$ we define $x \leq y$ whenever $x_k \leq y_k$ for $k = 1, \ldots, n$, then $\mathbb{R}^n$ is a Riesz space. Of course, we use here that in $\mathbb{R}$ itself (i.e.,for $n = 1$) the order structure and the vector space structure are compatible.

(2) The ordering in (for example) $\mathbb{R}^2$ can be defined differently. By definition, let $(x_1, x_2) \leq (y_1, y_2)$ whenever either $x_1 < y_1$ or $x_1 = y_1, x_2 \leq y_2$. This makes $\mathbb{R}^2$ a Riesz space. The ordering is called a *lexicographical ordering*.

(3) Let E be the vector space of all real functions on the non-empty point set X with addition and scalar multiplication pointwise, i.e., $(f_1 + f_2)(x) = f_1(x) + f_2(x)$ and $(\alpha f)(x) = \alpha f(x)$ for all $x \in X$. The ordering is also defined *pointwise*, i.e.,$f \leq g$ whenever $f(x) \leq g(x)$ for all $x \in X$. This makes E a Riesz space.

(4) Let μ be Lebesgue measure in $\mathbb{R}^n$. The set $M(\mathbb{R}^n, \mu)$ of all real μ-measurable functions on $\mathbb{R}^n$ is a real vector space (addition and scalar multiplication pointwise). Now write $f \approx g$ for f and g in $M(\mathbb{R}^n, \mu)$ whenever f and g differ only on a null set, i.e., on a set of measure zero. It is easily seen that $\approx$ is an equivalence relation. Let $L_0 = L_0(\mathbb{R}^n, \mu)$ be the set of the corresponding equivalence classes. We denote the equivalence classes of $f, g, \ldots$ by $[f], [g], \ldots$. The set L_0 becomes a real vector space by defining that $[f] + [g] = [f + g]$ and $\alpha[f] = [\alpha f]$ for α real. Observe that these definitions do not depend on the choice of f and g in their equivalence classes. The same is true for the partial order in L_0 if we define $[f] \leq [g]$ to mean that $f(x) \leq g(x)$ holds for all $x \in \mathbb{R}^n$ except on a null set. The vector space L_0 is now a Riesz space. In practice the elements of $L_0 = L_0(\mathbb{R}^n, \mu)$ are usually denoted by $f, g, \ldots$ and treated as if they were functions instead of equivalence classes of functions. Nevertheless, this matter should be treated carefully, as we shall see in the proof that L_0 is Dedekind complete (in Example 12.5(iii)).

(5) Let L_0 be as in the preceding part (4) and let $L_1 = L_1(\mathbb{R}^n, \mu)$ be the linear subspace of L_0 consisting of all μ-summable functions in L_0 (also called μ-integrable functions in L_0 sometimes), i.e.,

$$L_1 = (f : f \in L_0, \int_{\mathbb{R}^n} |f| \, d\mu < \infty).$$

The space L_1, with the ordering inherited from L_0, is a Riesz space on its own. Precisely, for any pair f, g in L_1 the supremum $f \vee g$ and the infimum $f \wedge g$ are the same as in L_0. The space L_1 is called a Riesz subspace of L_0. In part (4) and the present part (5) Lebesgue measure in $\mathbb{R}^n$ may be replaced by any (non- negative and countably additive) measure in an arbitrary point set.

(6) Let $C(X)$ be the vector space of all real continuous functions on the topological space X. The space $C(X)$ is partially ordered by defining that $f \leq g$ holds whenever $f(x) \leq g(x)$ for all $x \in X$. This makes $C(X)$ a Riesz space. For f, g in $C(X)$ the elements $f \vee g$ and $f \wedge g$ are the pointwise maximum and minimum of f and g. The subset $C_b(X)$ of all bounded functions in $C(X)$ is a Riesz subspace. If X is locally compact (such as for example $\mathbb{R}^n$ with the usual topology), the subset $C_0(X)$ of all functions in $C(X)$ having a *compact carrier* is also a Riesz subspace. We recall that the carrier of a continuous function f is the closure of the set on which f differs from zero.

(7) Let Γ be an algebra of subsets of the non-empty point set X and assume that to each $A \in \Gamma$ there is assigned a real number $\mu(A)$ such that

$\mu(A_1 \cup A_2) = \mu(A_1) + \mu(A_2)$ for all disjoint A_1 and A_2 in Γ. The mapping μ from Γ into $\mathbb{R}$ is now called a (real) *finitely additive signed measure* or also a *charge* on Γ. If μ is such a finitely additive measure and if there exists a constant $K > 0$ such that $| \mu(A) | \leq K$ for all $A \in \Gamma$, then μ is said to be *bounded*. The set E_b of all bounded finitely additive signed measures on Γ is a real vector space with respect to the natural definitions of addition and scalar multiplication, i.e.,

$$(\mu_1 + \mu_2)(A) = \mu_1(A) + \mu_2(A), \qquad (\alpha\mu)(A) = \alpha\mu(A)$$

for every $A \in \Gamma$. If we define a partial ordering in E_b by saying that $\mu_1 \leq \mu_2$ whenever $\mu_1(A) \leq \mu_2(A)$ for every $A \in \Gamma$, then E_b becomes a partially ordered vector space. We shall prove that E_b is a Riesz space. For this purpose we shall denote, for $\mu \in E_b$, the number

$$\sup(| \mu(A) |: A \in \Gamma)$$

by $\| \mu \|$. From

$$| (\mu_1 + \mu_2)(A) | \leq | \mu_1(A) | + | \mu_2(A) | \leq \| \mu_1 \| + \| \mu_2 \|,$$

holding for μ_1 and μ_2 in E_b and arbitrary $A \in \Gamma$, it follows that

$$\| \mu_1 + \mu_2 \| \leq \| \mu_1 \| + \| \mu_2 \| .$$

Furthermore, it is evident that $\| \alpha\mu \| = | \alpha | \cdot \| \mu \|$ for $\mu \in E_b$ and α real. Thus, $\| \cdot \|$ is clearly a norm in E_b. We prove now that $\mu_1 \vee \mu_2$ exists in E_b for all μ_1 and μ_2 in E_b. For any $A \in \Gamma$, let the number $\nu(A)$ be defined by

$$\nu(A) = \sup\{\mu_1(B) + \mu_2(A \backslash B) : A \supseteq B \in \Gamma\}.$$

It is not difficult to see that ν is a finitely additive measure on Γ. Since the supremum on the right is less than or equal to $\| \mu_1 \| + \| \mu_2 \|$, we get

$$\sup(| \nu(A) |: A \in \Gamma) \leq \| \mu_1 \| + \| \mu_2 \| .$$

It follows that ν is bounded and $\| \nu \| \leq \| \mu_1 \| + \| \mu_2 \|$. Having thus shown that ν is an element of E_b, we prove now that $\nu = \mu_1 \vee \mu_2$. From the definition of ν it is clear that $\nu(A) \geq \mu_1(A)$ and $\nu(A) \geq \mu_2(A)$ for every $A \in \Gamma$, so ν is an upper bound of μ_1 and μ_2. Let τ be another upper bound. Then, for any $A \in \Gamma$ and $A \supseteq B \in \Gamma$, we have

$$\tau(A) = \tau(B) + \tau(A \backslash B) \geq \mu_1(B) + \mu_2(A \backslash B),$$

so

$$\tau(A) \geq \sup\{\mu_1(B) + \mu_2(A \backslash B) : A \supseteq B \in \Gamma\} = \nu(A).$$

This shows that $\nu = \mu_1 \vee \mu_2$ in E_b. Note now that $\mu_1 \wedge \mu_2$ exists as well, because

$$\mu_1 \wedge \mu_2 = -\{(-\mu_1) \vee (-\mu_2)\}.$$

Hence, E_b is a Riesz space.

Let E be a Riesz space. The subset $E^+ = (f : 0 \leq f \in E)$ is called the *positive cone* of E and the elements of E^+ are called the *positive elements* of E. The positive cone E^+ has the following properties (*cone properties*):

(i) $f \in E^+, g \in E^+$ implies $f + g \in E^+$,

(ii) $f \in E^+$ implies $\alpha f \in E^+$ for any real $\alpha \geq 0$,

(iii) $f \in E^+, -f \in E^+$ implies $f = 0$.

We mention some further properties for which a proof is not necessary.

(a) $f \geq g$ in E is equivalent to $f - g \in E^+$ and to $-g \geq -f$.

(b) $f = f \vee g$ is equivalent to $f \geq g$ and to $g = f \wedge g$.

(c) $f \geq g$ is equivalent to $\alpha f \geq \alpha g$ for $0 < \alpha \in \mathbb{R}$ and also to $\alpha f \leq \alpha g$ for $0 > \alpha \in \mathbb{R}$.

(d) $f \wedge g = -\{(-f) \vee (-g)\}$, and so $f \vee g = -\{(-f) \wedge (-g)\}$.

(e)   Addition formulas, as follows:

$$(f \vee g) + h = (f + h) \vee (g + h), (f \wedge g) + h = (f + h) \wedge (g + h).$$

(f)   Multiplication formulas: For $0 \leq \alpha \in \mathbb{R}$ we have

$$(\alpha f) \vee (\alpha g) = \alpha(f \vee g), (\alpha f) \wedge (\alpha g) = \alpha(f \wedge g).$$

It follows that, for real $\alpha \leq 0$, we have

$$(\alpha f) \vee (\alpha g) = \alpha(f \wedge g), (\alpha f) \wedge (\alpha g) = \alpha(f \vee g).$$

(g) $\{(f \vee g) \vee h\} = \{f \vee (g \vee h)\}$, and similarly for infima.

The extension of these properties to any finite number of terms is evident.

Exercise 4.3. (i) Let E_b be the Riesz space of all bounded finitely additive signed measures on the algebra Γ of subsets of X, as occurring in part (7) of the preceding example. Show that, for μ_1 and μ_2 in E_b and $A \in \Gamma$, we have

$$(\mu_1 \wedge \mu_2)(A) = \inf\{\mu_1(B) + \mu_2(A \backslash B) : A \supseteq B \in \Gamma\}.$$

(ii) Show that if E^+ is a subset of the real vector space E such that E^+ has the above mentioned three cone properties, then E becomes a partially ordered vector space if we define the partial ordering in E by saying that $f \leq g$ holds whenever $g - f \in E^+$.

5. Equalities and Inequalities

At the end of the preceding section we have already become familiar with some simple equalities and inequalities holding in an arbitrary Riesz space. Here and also in the next section we present some more equalities or inequalities, now with proofs. These proofs are elementary, although sometimes a bit tricky. The reader should not believe that this is characteristic for the theory of Riesz spaces as a whole, because such a belief might give the false impression that the theory is not much more than a list of formulas. The formulas are necessary tools, however, for the more interesting subjects that will follow.

Let E be a Riesz space. We introduce some notations, as follows. For any $f \in E$ we shall write

$$f^+ = f \vee 0,\; f^- = (-f) \vee 0 \text{ and } |f| = f \vee (-f).$$

Note already that $-f^- = f \wedge 0$.

Theorem 5.1.
(i) f^+ and f^- are members of E^+; $(-f)^+ = f^-$. and similarly $(-f)^- = f^+$. Furthermore $|-f| = |f|$.
(ii) $f = f^+ - f^-$, $f^+ \wedge f^- = 0$ and $|f| = f^+ + f^-$. Hence $|f| \in E^+$.
(iii) $0 \leq f^+ \leq |f|$ and $0 \leq f^- \leq |f|$. Furthermore $-f^- \leq f \leq f^+$ and $|f| = 0$ if and only if $f = 0$.
(iv) $f \leq g$ if and only if $f^+ \leq g^+$ and $f^- \geq g^-$.

Proof. (i) Evident from the definitions.
(ii) $f^+ - f = (f \vee 0) - f = 0 \vee (-f) = f^-$,so $f^+ - f^- = f$. Also $0 = -f^- + f^- = (f \wedge 0) + f^- = (f + f^-) \wedge f^- = f^+ \wedge f^-$. Finally, $|f| = f \vee (-f) = \{(2f) \vee 0\} - f = 2f^+ - (f^+ - f^-) = f^+ + f^-$.
(iii) $0 \leq f^+ \leq |f|$ and $0 \leq f^- \leq |f|$ follow from $f^+ + f^- = |f|$. The other assertions are now evident.
(iv) $f \leq g$ implies $f^+ = f \vee 0 \leq g \vee 0 = g^+$. Similarly for $f^- \geq g^-$. Conversely, if $f^+ \leq g^+$ and $f^- \geq g^-$,then $f = f^+ - f^- \leq g^+ - g^- = g$.   ∎

The elements f^+ and f^- are called the *positive* and *negative parts* of f, although in view of $f = f^+ - f^-$ it would perhaps have been better to say that $-f^-$ is the negative part of f. The element $|f|$ is called the *absolute value* of f. The names are in accordance with the usual names in case the Riesz space E consists of real functions.

Theorem 5.2. *For f and g in E we have*

$$(f \vee g) + (f \wedge g) = f + g \quad and$$

$$(f \vee g) - (f \wedge g) = |f - g|.$$

Hence

$$f \vee g = \frac{1}{2}(f + g) + \frac{1}{2} \mid f - g \mid \quad and$$

$$f \wedge g = \frac{1}{2}(f + g) - \frac{1}{2} \mid f - g \mid .$$

Proof. Note first that

$$f \vee g = (f - g)^+ + g = (g - f)^+ + f, \tag{1}$$

$$f \wedge g = f - (f - g)^+ = g - (g - f)^+. \tag{2}$$

For the proof of (1) and (2), observe that

$$f \vee g = \{(f - g) \vee 0\} + g = (f - g)^+ + g,$$

$$f \wedge g = f + \{0 \wedge (g - f)\} = f - \{0 \vee (f - g)\} = f - (f - g)^+.$$

The equalities to be proved follow from (1) and (2) by addition and subtraction. Note that for $g = 0$ we find again that $f^+ - f^- = f$ and $f^+ + f^- = \mid f \mid$.∎

If we write $p = \frac{1}{2}(f + g)$ and $q = \frac{1}{2}(f - g)$, i.e., $f = p + q$ and $g = p - q$, we obtain two equalities, equivalent to those in the last theorem, as follows:

$$(p + q) \vee (p - q) = p + \mid q \mid \quad and \quad (p + q) \wedge (p - q) = p - \mid q \mid .$$

There exist also formulas for $\mid f \mid \vee \mid g \mid$ and $\mid f \mid \wedge \mid g \mid$. The proofs are somewhat more difficult.

Theorem 5.3. *For f and g in E we have*

$$\mid f \mid \vee \mid g \mid = \frac{1}{2}\{\mid f + g \mid + \mid f - g \mid\} \quad and$$

$$\mid f \mid \wedge \mid g \mid = \frac{1}{2} \mid\mid f + g \mid - \mid f - g \mid\mid,$$

or equivalently, writing $p = \frac{1}{2}(f + g)$ and $q = \frac{1}{2}(f - g)$,

$$\mid p + q \mid \vee \mid p - q \mid = \mid p \mid + \mid q \mid \quad and$$

$$\mid p + q \mid \wedge \mid p - q \mid = \mid\mid p \mid - \mid q \mid\mid .$$

Proof. In the proof for $\mid f \mid \vee \mid g \mid$ we make repeated use of the formula for $f \vee g$, as follows.

$$\mid f \mid \vee \mid g \mid = \sup(f, -f, g, -g) = \sup\{\sup(f, -g), \sup(g, -f)\} =$$

$$\sup\{\frac{1}{2}(f-g) + \frac{1}{2}\mid f+g \mid, \frac{1}{2}(g-f) + \frac{1}{2}\mid f+g \mid\} =$$

$$\frac{1}{2}\mid f+g \mid + \frac{1}{2}\mid f-g \mid .$$

To prove the formula for $\mid f \mid \wedge \mid g \mid$ we write, as above, $p = \frac{1}{2}(f+g)$ and $q = \frac{1}{2}(f-g)$, so $f = p+q$ and $g = p-q$. Furthermore we use the formula for $f \vee g$ as well as the just established formula for $\mid f \mid \vee \mid g \mid$. We thus get

$$\mid f \mid \wedge \mid g \mid = \mid f \mid + \mid g \mid - \{\mid f \mid \vee \mid g \mid\} =$$

$$\mid f \mid + \mid g \mid - \frac{1}{2}\mid f+g \mid - \frac{1}{2}\mid f-g \mid =$$

$$\mid p+q \mid + \mid p-q \mid - \mid p \mid - \mid q \mid =$$

$$2(\mid p \mid \vee \mid q \mid) - \mid p \mid - \mid q \mid =$$

$$\{\mid p \mid + \mid q \mid + \mid\mid p \mid - \mid q \mid\mid\} - \mid p \mid - \mid q \mid =$$

$$\mid\mid p \mid - \mid q \mid\mid = \mid\mid f+g \mid - \mid f-g \mid\mid .\qquad\blacksquare$$

Corollary 5.4. *The elements f and g in E satisfy $\mid f \mid \wedge \mid g \mid = 0$ if and only if $\mid f+g \mid = \mid f-g \mid$ and also if and only if $\mid f \mid \vee \mid g \mid = \mid f+g \mid$.*

Theorem 5.5. *(Triangle inequality). For f and g in E we have*

$$(f+g)^+ \leq f^+ + g^+, (f+g)^- \leq f^- + g^-,$$

$$\mid\mid f \mid - \mid g \mid\mid \leq \mid f+g \mid \leq \mid f \mid + \mid g \mid .$$

The last inequality is called the triangle inequality.

Proof. It follows from $f^+ + g^+ \geq f+g$ and $f^+ + g^+ \geq 0$ that

$$f^+ + g^+ \geq \sup(f+g, 0) = (f+g)^+.$$

This implies that

$$f^- + g^- = (-f)^+ + (-g)^+ \geq (-f-g)^+ = (f+g)^-,$$

and so

$$\mid f+g \mid = (f+g)^+ + (f+g)^- \leq$$

$$f^+ + g^+ + f^- + g^- = \mid f \mid + \mid g \mid .$$

This was the easy part of the triangle inequality. Now for the remaining part. It follows from $| f | \leq | f - g | + | g |$ that we have $| f | - | g | \leq | f - g |$. Similarly $| g | - | f | \leq | f - g |$. Hence

$$\| f | - | g \| = \sup(| f | - | g |, | g | - | f |) \leq | f - g | .$$

Finally, substituting g for $-g$, we get $\| f | - | g \| \leq | f + g | .$ ∎

The next theorem, in particular the last part, is of frequent use.

Theorem 5.6. *If $f = u - v$ with u and v in E^+ , then $f^+ \leq u$ and $f^- \leq v$. Hence, the decomposition $f = f^+ - f^-$ as a difference of positive elements is a minimal decomposition. In this case, i.e., if $u = f^+$ and $v = f^-$, we have $u \wedge v = 0$ (by (ii) in Theorem 5.1). Conversely, if $f = u - v$ with $u \wedge v = 0$, then this is the minimal decomposition, i.e., $u = f^+$ and $v = f^-$.*

Proof. Let $f = u - v$ with u and v in E^+. Then $f \leq u$ as well as $0 \leq u$, and so $f^+ = f \vee 0 \leq u$. It follows that $f^- = f^+ - f \leq u - f = v$. For the second part, assume that $f = u - v$ with $u \wedge v = 0$. Then, using formula (2) in Theorem 5.2, we have

$$0 = u \wedge v = u - (u - v)^+ = u - f^+,$$

and so $u = f^+$. It follows that $v = u - f = f^+ - f = f^-$.   ∎

Exercise 5.7. There is a curious proof of the formula for $| f | \wedge | g |$, due to D.Z. Djokovic. The proof starts from the observation that

$$2(| f | \vee | g |) = | f + g | + | f - g | . \eqno(3)$$

Now, in (3), replace f and g by $| f + g |$ and $| f - g |$ respectively. After that, return to (3) and replace f and g by $f + g$ and $f - g$ respectively. In both cases the left hand side turns out to be the same. Hence, the right hand sides must become equal as well. This shows that

$$| f + g | + | f - g | + \| f + g | - | f - g \| = 2 | f | + 2 | g | .$$

Use (3) again to derive now that

$$\| f + g | - | f - g \| = 2\{| f | + | g | - | f | \vee | g |\}.$$

6. Distributive Laws, the Birkhoff Inequalities and the Riesz Decomposition Property

Once again we assume that we have a Riesz space E. It will be proved first that, similarly as in a Boolean algebra, the distributive laws hold in E.

Theorem 6.1. *(Infinite distributive laws). Let D be a subset of E possessing a supremum, i.e., $f_0 = \sup D = \sup(f : f \in D)$ exists. Then, for any $g \in E$, we have*

$$f_0 \wedge g = \sup(f \wedge g : f \in D).$$

Similarly, if $f_1 = \inf D$ exists, then

$$f_1 \vee g = \inf(f \vee g : f \in D).$$

Proof. It is evident that $f_0 \wedge g$ is an upper bound of the set $(f \wedge g : f \in D)$. Assume that m is another upper bound of the set. Then, for any $f \in D$,

$$m \geq f \wedge g = f + g - (f \vee g) \geq f + g - (f_0 \vee g).$$

Hence $m - g + (f_0 \vee g) \geq f$ for all $f \in D$. By taking now the supremum on the right, it follows that $m - g + (f_0 \vee g) \geq f_0$, so

$$m \geq f_0 + g - (f_0 \vee g) = f_0 \wedge g.$$

This shows that any arbitrary upper bound of the set $(f \wedge g : f \in D)$ is greater than or equal to $f_0 \wedge g$. In other words, $f_0 \wedge g = \sup(f \wedge g : f \in D)$. The proof for $f_1 \vee g$ is similar. ∎

Corollary 6.2. *(Distributive laws). For f, g and h in E we have*

$$(f \vee g) \wedge h = (f \wedge h) \vee (g \wedge h),$$

$$(f \wedge g) \vee h = (f \vee h) \wedge (g \vee h).$$

As an application we derive an identity due to G.Birkhoff. The *Birkhoff inequalities* in (iii) below are immediate corollaries.

Theorem 6.3. *For f, g and h in E we have*

(i) $| f \vee h - g \vee h | + | f \wedge h - g \wedge h | = | f - g |,$
(ii) $| f \vee h - g \vee h | \leq | f - g |$ *and* $| f \wedge h - g \wedge h | \leq | f - g |,$
(iii) $| f^+ - g^+ | \leq | f - g |$ *and* $| f^- - g^- | \leq | f - g |.$

Proof. In the formula

$$| p - q | = (p \vee q) - (p \wedge q)$$

we first substitute $p = f \vee h, q = g \vee h$ and then $p = f \wedge h, q = g \wedge h$. This gives, after addition,

$$| f \vee h - g \vee h | + | f \wedge h - g \wedge h | =$$

$$f \vee h \vee g \vee h - (f \vee h) \wedge (g \vee h) + (f \wedge h) \vee (g \wedge h) - f \wedge h \wedge g \wedge h =$$

$$f \vee g \vee h - (f \wedge g) \vee h + (f \vee g) \wedge h - f \wedge g \wedge h =$$

$$\{(f \vee g) \vee h + (f \vee g) \wedge h\} - \{(f \wedge g) \vee h + (f \wedge g) \wedge h\} =$$

$$\{(f \vee g) + h\} - \{(f \wedge g) + h\} = (f \vee g) - (f \wedge g) = | f - g | .$$

The inequalities in (ii) follow from (i) and (iii) follows from (ii) by choosing $h = 0$.   ∎

Theorem 6.4. *(Riesz decomposition property). Let the elements u, z_1, z_2 in E^+ satisfy $u \leq z_1 + z_2$. Then there exist elements u_1, u_2 in E^+ such that $u_1 \leq z_1, u_2 \leq z_2$ and $u = u_1 + u_2$.*

Proof. Let $u_1 = u \wedge z_1$ and $u_2 = u - u_1$. Then $u_1 \in E^+$ and $u_1 \leq z_1$. Since $u_1 \leq u$, we have $u_2 = u - u_1 \geq 0$, so $u_2 \in E^+$. Furthermore $u = u_1 + u_2$. It remains to prove that $u_2 \leq z_2$. This follows from

$$u_2 = u - u_1 = u - (u \wedge z_1) =$$

$$u + \{(-u) \vee (-z_1)\} = 0 \vee (u - z_1) \leq 0 \vee z_2 = z_2.$$ ∎

Theorem 6.5. *Let u, v, w be elements in E^+ and let f, g be arbitrary elements in E. Then*

$$(u + v) \wedge w \leq (u \wedge w) + (v \wedge w), \tag{1}$$

$$(f + g) \vee w \leq (f \vee w) + (g \vee w). \tag{2}$$

Furthermore, $v \wedge w = 0$ implies

$$(u + v) \wedge w = u \wedge w, \tag{3}$$

and if $u \wedge v = 0$, then there is equality in (1).

Proof. We use one of the Birkhoff inequalities to show that

$$0 \le \{(u + v) \wedge w\} - u \wedge w \le v.$$

Also $0 \le \{(u + v) \wedge w\} - u \wedge w \le (u + v) \wedge w \le w$, and hence

$$0 \le \{(u + v) \wedge w\} - u \wedge w \le v \wedge w.$$

The proof of (2) is simple; the right hand side is $\ge f + g$ as well as $\ge 2w$, and hence the even stronger inequality

$$(f + g) \vee (2w) \le (f \vee w) + (g \vee w)$$

holds. Note that for arbitrary f, g, h in E we have

$$(f + g) \vee (2h) \le (f \vee h) + (g \vee h),$$

but not necessarily

$$(f + g) \vee h \le (f \vee h) + (g \vee h).$$

Give a simple counterexample. Finally, assume that $u \wedge v = 0$. Then, since $0 \le u \wedge w \le u$ and $0 \le v \wedge w \le v$, we have $(u \wedge w) \wedge (v \wedge w) = 0$. It follows (by Corollary 5.4) that $u + v = u \vee v$ and

$$(u \wedge w) + (v \wedge w) = (u \wedge w) \vee (v \wedge w).$$

Hence

$$(u + v) \wedge w = (u \vee v) \wedge w = (u \wedge w) \vee (v \wedge w) = (u \wedge w) + (v \wedge w). \quad \blacksquare$$

Exercise 6.6. Show that the following chains of inequalities hold for u, v, w in E^+ :

$$(u \wedge w) + (v \wedge w) - w \le (u + v) \wedge w \le (u \wedge w) + (v \wedge w) \le (u \wedge v) + w$$

and

$$(u \vee v) \le (u \vee w) + (v \vee w) - w \le (u + v) \vee w \le (u \vee w) + (v \vee w).$$

Hint: To prove that $(u \vee w) + (v \vee w) - w \le (u + v) \vee w$, subtract (1) from the identity

$$u + v + w = (u + w) + (v + w) - w.$$

Exercise 6.7. Let f, f_0 and g, g_0 be elements of E. It follows easily from the Birkhoff inequalities that

$$| (f \vee f_0) - (g \vee g_0) | \leq | f - g | + | f_0 - g_0 | .$$

There exists a sharper inequality. Show that

$$| (f \vee f_0) - (g \vee g_0) | \leq | f - g | \vee | f_0 - g_0 |,$$

and by induction

$$| \sup(f_1, \ldots, f_n) - \sup(g_1, \ldots, g_n) | \leq \sup(| f_1 - g_1 |, \ldots, | f_n - g_n |).$$

Hint: Write $p = f \vee f_0$ and $q = g \vee g_0$ for convenience. Observe now that $p \wedge q \geq f \wedge g$ and $p \wedge q \geq f_0 \wedge g_0$, so $p \wedge q \geq (f \wedge g) \vee (f_0 \wedge g_0) = r$. Hence

$$| p - q | = (p \vee q) - (p \wedge q) \leq (p \vee q) - r = \{(f \vee g) \vee (f_0 \vee g_0)\} - r =$$

$$\{(f \vee g) - r\} \vee \{(f_0 \vee g_0) - r\} \leq$$

$$\{(f \vee g) - (f \wedge g)\} \vee \{(f_0 \vee g_0) - (f_0 \wedge g_0)\} =$$

$$| f - g | \vee | f_0 - g_0 | .$$

Exercise 6.8. In the Riesz space E, let $u \geq 0, v \geq 0, u + v = z$ and $0 \leq z_1 \leq z_2 \leq \cdots \leq z_n = z$. Show that there exist elements u_k and $v_k (k = 1, \ldots, n)$ such that

$$0 \leq u_1 \leq u_2 \leq \cdots \leq u_n = u, 0 \leq v_1 \leq v_2 \leq \cdots \leq v_n = v$$

and $u_k + v_k = z_k$ for $k = 1, \ldots, n$.

Hint: Let $u_k = u \wedge z_k$ for $k = 1, \ldots, n$. Then $0 \leq u_1 \leq u_2 \leq \cdots \leq u_n = u$, and it is evident now that $v_k = z_k - u_k$ satisfies $v_k \geq 0$ and $u_k + v_k = z_k$ for all k. Since

$$0 \leq u_{k+1} - u_k = (u \wedge z_{k+1}) - (u \wedge z_k) \leq z_{k+1} - z_k$$

by one of the Birkhoff inequalities, we get

$$v_k = z_k - u_k \leq z_{k+1} - u_{k+1} = v_{k+1} \quad \text{for} \quad k = 1, \ldots, n - 1.$$

Hence $0 \leq v_1 \leq v_2 \leq \cdots \leq v_n = z_n - u_n = z - u = v$.

Exercise 6.9. Using the result in the preceding exercise, extend the Riesz decomposition property, as follows. Given that

$$0 \leq u \leq z_1 + \cdots + z_n$$

with $z_k \in E^+$ for $k = 1, \ldots, n$, show that there exist elements $u_1, \ldots, u_n$ in E^+ such that $u = u_1 + \cdots + u_n$ and $u_k \leq z_k$ for $k = 1, \ldots, n$.

Exercise 6.10. Extend the Riesz decomposition property once more, as follows. Given that

$$| f | \leq z_1 + \cdots + z_n$$

with $z_k \in E^+$ for $k = 1, \ldots, n$, show that there exist elements $f_1, \ldots, f_n$ in E such that $f = f_1 + \cdots + f_n$ and $| f_k | \leq z_k$ for $k = 1, \ldots, n$. In particular, if

$$| f | \leq | g_1 + \cdots + g_n |,$$

there exist elements $f_1, \ldots, f_n$ such that $f = f_1 + \cdots + f_n$ and $| f_k | \leq | g_k |$ for $k = 1, \ldots, n$.

Hint: Apply the result in the preceding exercise to f^+ and f^- separately.

Exercise 6.11. Let f and g be elements in the Riesz space E. Show that $| f + g | = | f | + | g |$ if and only if $f^+ \wedge g^- = f^- \wedge g^+ = 0$. Show that this condition is necessary but not sufficient for $| f | \wedge | g | = 0$ to hold.

Hint: Let $f^+ \wedge g^- = f^- \wedge g^+ = 0$. Since $f^+ \wedge f^- = g^+ \wedge g^- = 0$, it follows by means of formula (1) in Theorem 6.5 that

$$(f^+ + g^+) \wedge g^- = 0 \text{ and } (f^+ + g^+) \wedge f^- = 0,$$

and so, once more by the same formula, $(f^+ + g^+) \wedge (f^- + g^-) = 0$. For convenience, write $u = f^+ + g^+$ and $v = f^- + g^-$. Then $f + g = u - v$ with $u \wedge v = 0$, so $u = (f + g)^+$ and $v = (f + g)^-$. It follows that

$$| f + g | = u + v = f^+ + f^- + g^+ + g^- = | f | + | g |.$$

Conversely, if $| f + g | = | f | + | g |$, then

$$(f + g)^+ + (f + g)^- = (f^+ + g^+) + (f^- + g^-).$$

In view of the inequalities in Theorem 5.5 it follows that $(f + g)^+ = f^+ + g^+$ and $(f+g)^- = f^- + g^-$. But $(f+g)^+ \wedge (f+g)^- = 0$, so $(f^+ + g^+) \wedge (f^- + g^-) = 0$. This implies that $f^+ \wedge g^- = 0$ and $g^+ \wedge f^- = 0$.

Exercise 6.12. Let $f_1, \ldots, f_n$ and $g_1, \ldots, g_n$ be elements in the Riesz space E. Show that

$$n\{(f_1 + \cdots + f_n) \vee (g_1 + \cdots + g_n)\} \leq \Sigma_{j,p}(f_j \vee g_p).$$

The numerical factor n can be omitted if all f_j and g_p are positive, but not in general.

Hint: Write $f_1 + \cdots + f_n = s$ and $g_1 + \cdots + g_n = t$. It is evident that

$$s \vee t \leq (f_1 \vee g_1) + \cdots + (f_n \vee g_n).$$

Similarly, by rotating the $g'_p s$,

$$s \vee t \leq (f_1 \vee g_2) + \cdots + (f_{n-1} \vee g_n) + (f_n \vee g_1),$$

and so on.

Exercise 6.13. The result in Exercise 6.11 can be refined. Let f and g be elements in the Riesz space E. Show that

$$|f| + |g| - |f + g| = 2(f^+ \wedge g^- + f^- \wedge g^+),$$

$$|f| + |g| - |f - g| = 2(f^+ \wedge g^+ + f^- \wedge g^-).$$

Hint: From

$$2(f^+ \wedge g^-) = f^+ + g^- - |f^+ - g^-|,$$

$$2(f^- \wedge g^+) = f^- + g^+ - |f^- - g^+|$$

it follows by addition that

$$2(f^+ \wedge g^- + f^- \wedge g^+) = |f| + |g| - (|f^+ - g^-| + |f^- - g^+|).$$

Observe now that

$$f^+ - g^- = f \vee 0 - (-g) \vee 0, \, f^- - g^+ = (-g) \wedge 0 - f \wedge 0.$$

Hence, by the Birkhoff identity,

$$|f^+ - g^-| + |f^- - g^+| = |f - (-g)| = |f + g|$$

This result, due to B. de Pagter and A.R. Schep, is recent.

CHAPTER 3
Ideals, Bands and Disjointness

7. Ideals and Bands

Let E be a Riesz space. Certain linear subspaces of E have special properties and have received special names. Details are contained in the following definition.

Definition 7.1. Let E be a Riesz space. Subsets of E are assumed to inherit the ordering from E.

(i) The linear subspace V of E is called a *Riesz subspace* of E if for all members f and g of V the elements $f \vee g$ and $f \wedge g$ are likewise members of V (where $f \vee g$ and $f \wedge g$ are the supremum and infimum of f and g that exist in E by hypothesis).

(ii) The subset S of E is said to be *solid* if it follows from $f \in S$ and $| g | \leq | f |$ that $g \in S$ (in other words, if it follows from $f \in S$ that the order interval $[- | f |, | f |]$ is a subset of S).

(iii) The subset A of E is called an *ideal* in E if A is a solid linear subspace of E. Sometimes this is called an *order ideal* if it is necessary to distinguish it from an algebraic ideal in a ring.

(iv) The ideal B in E is called a *band* if, whenever a subset of B possesses a supremum in E, this supremum is a member of B, i.e., if it follows from $D \subseteq B$ and $f = \sup D$ that $f \in B$.

We make some remarks concerning these definitions. In the definition of a Riesz subspace it is sufficient to say that V is a Riesz subspace of E if V is a linear subspace of E such that $f \in V, g \in V$ implies $f \vee g \in V$ because this implies already that $f \wedge g \in V$. Indeed, from $f \in V, g \in V$ it follows that $-f \in V, -g \in V$, so

$$f \wedge g = -\{(-f) \vee (-g)\} \in V.$$

The definition of an ideal may be reformulated by saying that an ideal A in E is a linear subspace such that $f \in A, | g | \leq | f |$ implies $g \in A$. This can be reformulated once more by saying that an ideal A in E is a linear subspace in E such that

(i) $f \in A$ if and only if $| f | \in A$,

(ii) $0 \leq g \leq f \in A$ implies $g \in A$.

Note, finally, that condition (ii) is equivalent to

(iii) $0 \leq f \in A$ and $g \in E^+$ implies $f \wedge g \in A$.

The last condition clearly shows the analogy with the definition of a ring ideal in a commutative ring R, where I is an ideal in R whenever I is a subring satisfying the condition that $f \in I, g \in R$ implies $fg \in I$.

Before presenting examples we prove a simple theorem.

Theorem 7.2. *Let E be a Riesz space.*

(i) Every band in E is an ideal and every ideal in E is a Riesz subspace. The set $\{0\}$ consisting only of the null element is a band in E. Similarly, E itself is a band.

(ii) Every Riesz subspace of a Riesz subspace of E is itself a Riesz subspace of E.

(iii) Every ideal in an ideal in E is itself an ideal in E.

(iv) Every band in a band in E is itself a band in E.

Proof. (i) Let A be an ideal in E. We prove that A is a Riesz subspace of E. It is sufficient to show that $f \in A, g \in A$ implies $f \vee g \in A$. Hence, in virtue of $2(f \vee g) = f + g + | f - g |$, it is sufficient to show that $h \in A$ implies $| h | \in A$. This follows from the hypothesis that A is solid.

(ii) Let V_1 be a Riesz subspace of the Riesz subspace V of E and let f and g be members of V_1. We have to prove that $\sup_E(f, g) \in V_1$, where $\sup_E$ denotes that the supremum is taken in E. Since f and g are members of V_1 and V_1 is a subset of V, it is evident that f and g are members of V. But V is a Riesz subspace of E, so $\sup_E(f, g) \in V$. This element is now not only the supremum of f and g in E but also in V (because any upper bound h of f and g in V is an upper bound of f and g in E as well, so $h \geq \sup_E(f, g)$). Hence, $\sup_E(f, g) = \sup_V(f, g)$. But $\sup_V(f, g) \in V_1$ because V_1 is a Riesz subspace of V. Hence $\sup_E(f, g) \in V_1$.

The proofs of parts (iii) and (iv) are simple.   ∎

Example 7.3. (i) Let E be the Riesz space of all real continuous functions polynomials on $[0,1]$ is not a Riesz subspace (see also Exercise 7.5(ii)). The linear subspace of all (real) constants is a Riesz subspace, but not an ideal in E. The linear subspace A of all $f \in E$ satisfying $f(0) = 0$ is an ideal, but not a band because if we define, for $n = 1, 2, \ldots$, the function f_n by $f_n(x) = \inf(nx, e(x))$, then $f_n \in A$ for all n and $\sup f_n = e$, but e is not a member of A. The ideal of all $f \in E$ that vanish on $[0, \frac{1}{2}]$ is a band.

(ii) The Riesz space of all real sequences $x = (x_1, x_2, \ldots)$ with coordinate-wise ordering is usually denoted by (s); the linear subspace of all bounded sequences is denoted by ℓ_∞ ; the linear subspace of all converging sequences by (c) and the linear subspace of all null sequences (i.e.,$x_n \to 0$ as $n \to \infty$) by (c_0). The space ℓ_∞ is an ideal in (s). The space (c) is a Riesz subspace of ℓ_∞, but not an ideal in ℓ_∞. The space (c_0) is an ideal in (c) and also in ℓ_∞. None of these ideals is a band.

(iii) The space $C([0,1])$ of all real continuous functions on [0,1] can be thought of as a linear subspace of the space $L_0 = L_0([0, 1])$ of all real Lebesgue measurable functions on [0,1] provided any function differing from a continuous function f only on a set of measure zero is considered to be the same as f. Then $C([0,1])$ is a Riesz subspace of L_0 but not an ideal in L_0. The linear subspace $L_1([0,1])$ of all Lebesgue summable functions on [0,1] is an ideal in L_0 but not a band in L_0.

We introduce the following conventions. For any non-empty subset D of the Riesz space E we shall denote the set $(f : f \in D, f \geq 0)$ by D^+. Note that D^+ may be empty. Furthermore, if the subset D of E has the property that for any finite number $f_1, \ldots, f_n$ of elements of D the supremum $\sup(f_1, \ldots, f_n)$ is also an element of D, we shall say that D *contains finite suprema* (precisely, D contains all finite suprema of its own elements). Similarly if D contains finite infima. Note that if D is bounded above and D_1 is the set of all finite suprema of elements of D, then D_1 contains D and contains all finite suprema of its own elements. Furthermore, D and D_1 have the same upper bounds. The next theorem is now easy.

Theorem 7.4. *(i) In a Riesz space any intersection of Riesz subspaces (ideals, bands) is a Riesz subspace (ideal, band).*

(ii) By definition, the ideal B in the Riesz space E is a band whenever, for any subset of B possessing a supremum, this supremum is an element of B. It is already sufficient that this condition holds only for subsets of B^+ containing finite suprema. In other words, if for any subset of B^+ containing finite suprema and having a supremum this supremum is an element of B, then B is a band in E.

Proof. (i) Evident from the definitions.

(ii) Assume that the ideal B has the mentioned property for subsets of B^+ and let D be a subset of B such that $f_0 = \sup D$ exists. We have to prove that f_0 is a member of B. Without loss of generality we may assume that D contains finite suprema. Fix an element d of D and let D_1 be the set of all elements $f \vee d, f \in D$. Then D_1 is a (non-empty) subset of D, but this does not affect the supremum, i.e., $\sup D_1 = f_0$. Finally, let $D_2 = (f - d : f \in D_1)$. Then $D_2 \subseteq B^+$ and $\sup D_2 = f_0 - d$. Also, D_2 contains finite suprema. Hence, by hypothesis, $f_0 - d \in B$, which implies immediately that $f_0 \in B$. ∎

Exercise 7.5. (i) In connection with the definition of an order ideal in a Riesz space it was observed that there is a certain analogy with the definition of an algebraic ideal in a commutative ring. In a Riesz space E every ideal in an ideal in E is itself an ideal in E. It may be asked, therefore, if every ideal I_1 in an ideal I in the commutative ring R is itself an ideal in R. Show that this is not always true. As an example, choose for R the ring of all real bounded functions on $[0,1]$, for I the ideal of all $f \in R$ that are continuous at 0 and satisfy $f(0) = 0$, and for I_1 the set of all $f \in I$ that are differentiable at 0.

(ii) Show that the vector space of all real polynomials on $[0,1]$, partially ordered pointwise, is not a Riesz space.

Hint: Let f, g, h be polynomials on $[0,1]$ such that $h \geq f$ as well as $h \geq g$, and f, g are not comparable, i.e., neither $f \geq g$ nor $g \geq f$. Note that $h(x) = f(x)$ holds only for finitely many $x \in [0, 1]$ because otherwise h and f would be identical. Similarly, $h(x) = g(x)$ holds only for finitely many x. Show now that there exists a polynomial k such that $k \leq h$, but k not identically equal to h, and $k \geq f$ as well as $k \geq g$. Precisely, show that k can be chosen as

$$k = h - c(h - f)(h - g),$$

where $c > 0$ is a constant such that $c(h - f) \leq 1$ and $c(h - g) \leq 1$ on $[0,1]$.

For any non-empty subset D of the Riesz space E the intersection of all Riesz subspaces containing D is called the *Riesz subspace generated by D*. The *ideal* and the *band generated by D* are defined similarly. The ideal A_D generated by D can be described explicitly as follows. If $f_1, \ldots, f_n$ are elements of D and $\alpha_1, \ldots, \alpha_n$ are real numbers, then $\mid \alpha_1 f_1 \mid + \cdots + \mid \alpha_n f_n \mid \in A_D$, and so every $g \in E$ satisfying

$$\mid g \mid \leq \mid \alpha_1 f_1 \mid + \cdots + \mid \alpha_n f_n \mid \tag{1}$$

is an element of A_D. On the other hand, the set of all g satisfying an inequality of the form (1), for n variable, $f_1, \ldots, f_n$ variable in D and $\alpha_1, \ldots, \alpha_n$ variable in $\mathbb{R}$, is an ideal in E that contains D. Hence, A_D is exactly the set of all such g. In the particular case that D consists of one element f we write $A_D = A_f$ and A_f is now called the *principal ideal* generated by f. It is evident that

$$A_f = (g \in E : \mid g \mid \leq \mid \alpha f \mid \text{ for some } \alpha \in \mathbb{R}).$$

For some applications it is convenient to note that if D is an infinite subset of E and g is an element of the ideal generated by D, then g is already an element of the ideal generated by some finite subset of D (the finite subset depending on g of course).

We recall that if V_1 and V_2 are subsets of the vector space V, the *algebraic sum $V_1 + V_2$* is defined by

$$V_1 + V_2 = (f_1 + f_2 : f_1 \in V_1, f_2 \in V_2).$$

If V_1 and V_2 are linear subspaces of V, then $V_1 + V_2$ is also a linear subspace. In this case, if $V_1 \cap V_2 = \{0\}$, any $f \in V_1 + V_2$ can be written uniquely as $f = f_1 + f_2$ with $f_1 \in V_1$ and $f_2 \in V_2$, and $V_1 + V_2$ is now called the *direct sum* of V_1 and V_2 (notation $V_1 \oplus V_2$).

Theorem 7.6. *If A_1 and A_2 are ideals in the Riesz space E, then $A_1 + A_2$ is an ideal in E. It is evident that $A_1^+ + A_2^+$ is contained in $(A_1 + A_2)^+$. The converse holds as well, i.e., $(A_1 + A_2)^+ = A_1^+ + A_2^+$. In other words, for any $f \in (A_1 + A_2)^+$ there exist $f_1 \in A_1^+$ and $f_2 \in A_2^+$ (not uniquely determined in general) such that $f = f_1 + f_2$. If $A_1 \cap A_2 = \{0\}$, i.e., if $A_1 + A_2 = A_1 \oplus A_2$, then f_1 and f_2 are uniquely determined. In this case, if f and g are elements of $A_1 \oplus A_2$ having decompositions $f = f_1 + f_2$ and $g = g_1 + g_2$, then $f \leq g$ implies $f_1 \leq g_1$ and $f_2 \leq g_2$.*

Proof. We prove first that $A_1 + A_2$ is an ideal. For this purpose, let $f \in A_1 + A_2$ and let $\mid g \mid \leq \mid f \mid$. We have to prove that $g \in A_1 + A_2$. The element f can be written as $f = f' + f''$ with $f' \in A_1$ and $f'' \in A_2$. Note now that

$$g^+ \leq \mid g \mid \leq \mid f \mid \leq \mid f' \mid + \mid f'' \mid, \tag{2}$$

so in view of the Riesz decomposition theorem there exist elements g' and g'' such that $g^+ = g' + g''$ with $0 \leq g' \leq \mid f' \mid$ and $0 \leq g'' \leq \mid f'' \mid$. Since $f' \in A_1$ and $f'' \in A_2$, it follows that $g' \in A_1$ and $g'' \in A_2$, so $g^+ = g' + g'' \in A_1 + A_2$. Similarly $g^- \in A_1 + A_2$. Hence $g = g^+ - g^- \in A_1 + A_2$.

Assume now that $f \in (A_1 + A_2)^+$. Then $f = f' + f''$ with $f' \in A_1$ and $f'' \in A_2$. From $f = \mid f \mid \leq \mid f' \mid + \mid f'' \mid$ it follows in view of the Riesz decomposition theorem that f is of the form $f = f_1 + f_2$ with $0 \leq f_1 \leq \mid f' \mid$ and $0 \leq f_2 \leq \mid f'' \mid$. Hence $f_1 \in A_1^+$ and $f_2 \in A_2^+$.

The proof of the statement about $f \leq g$ in $A_1 \oplus A_2$ follows by observing that $g - f = (g_1 - f_1) + (g_2 - f_2)$ uniquely, so $g - f \geq 0$ implies that $g_1 - f_1 \geq 0$ and $g_2 - f_2 \geq 0$. $\blacksquare$

The algebraic sum of bands is not always a band. As an example, let $E = C([-1, 1])$, the space of all real continuous functions on the interval $[-1, 1]$, and let the bands B_1 and B_2 be defined by

$$B_1 = (f \in E : f = 0 \text{ on } [0,1]),$$

$$B_2 = (f \in E : f = 0 \text{ on } [-1,0]).$$

Then $B_1 + B_2 = B_1 \oplus B_2 = (f \in E : f(0) = 0)$ is an ideal in E, but not a band. The band generated by $B_1 \oplus B_2$ is the entire space $C([-1, 1])$.

Exercise 7.7. (i) Let u and v be positive elements in the Riesz space E. Prove that for the principal ideals A_u and A_v we have

$$A_{u \wedge v} = A_u \cap A_v \text{ and } A_{\sup(u,v)} = A_{u+v} = A_u + A_v.$$

(ii) Let A and B be ideals in the Riesz space E. Show that $A \cap B = \{0\}$ if and only if $|f| \wedge |g| = 0$ for all $f \in A$ and $g \in B$.

(iii) Let A, B and C be ideals in the Riesz space E. Show that $(A+B) \cap C$ is an ideal satisfying

$$(A+B) \cap C = A \cap C + B \cap C.$$

As we have seen, it is simple to determine the ideal generated by an arbitrary non-empty subset of a Riesz space. An obvious question is now to ask how to determine the band generated by a given ideal. The answer is easy to give but the proof that the answer is correct is a little more difficult. It is convenient to observe first that if D_1 and D_2 are non-empty subsets of the Riesz space E having suprema f_1 and f_2 respectively, then the algebraic sum $D_1 + D_2$ has supremum $f_1 + f_2$. Recall that

$$D_1 + D_2 = (d_1 + d_2 : d_1 \in D_1, d_2 \in D_2).$$

For the proof, let g be an arbitrary upper bound of $D_1 + D_2$, so $d_1 + d_2 \leq g$ for all $d_1 \in D_1$ and $d_2 \in D_2$. Then

$$d_1 + \sup(d_2 : d_2 \in D_2) \leq g,$$

i.e., $d_1 + f_2 \leq g$ for all $d_1 \in D_1$. Hence, by a similar argument, $f_1 + f_2 \leq g$. Since $f_1 + f_2$ is an obvious upper bound, it follows that $f_1 + f_2$ is the supremum of the set $D_1 + D_2$.

Theorem 7.8. *The band $[A]$ generated by the ideal A in the Riesz space E consists of all $f \in E$ satisfying*

$$|f| = \sup D \text{ for some non-empty set } D \in A^+, \tag{3}$$

or, equivalently, $[A]$ consists of all $f \in E$ satisfying

$$|f| = \sup(u : u \in A, 0 \leq u \leq |f|). \tag{4}$$

Proof. It is evident that if f satisfies (4), then f satisfies (3). Conversely, if f satisfies (3), then the set D in (3) is a subset of the set

$$P = (u : u \in A, 0 \leq u \leq |f|).$$

Every upper bound of D is $\geq |f|$ and D is a subset of P, so every upper bound of P is also $\geq |f|$. On the other hand $|f|$ itself is an upper bound of P. Hence $|f| = \sup P$, i.e., f satisfies (4). Having shown thus that (3) and

(4) are equivalent, let us denote by B the set of all f satisfying (3) and (4). We have to prove, therefore, that the set B is the same as the band $[A]$.

From the definition of a band it is evident that the set B is a subset of $[A]$. Hence, it will be sufficient now to prove that B is a band. We prove first that B is a linear subspace of E. Let f_1, f_2 be members of B and let D_1, D_2 be subsets of A^+ such that $\mid f_1 \mid = \sup D_1$ and $\mid f_2 \mid = \sup D_2$. Then the algebraic sum $D_1 + D_2$ is a subset of A^+ having $\mid f_1 \mid + \mid f_2 \mid$ as its supremum, so

$$\sup\{\mid f_1 + f_2 \mid \wedge (d_1 + d_2) : d_1 \in D_1, d_2 \in D_2\} =$$
$$\mid f_1 + f_2 \mid \wedge (\mid f_1 \mid + \mid f_2 \mid) = \mid f_1 + f_2 \mid$$

and

$$\sup(\mid \alpha \mid d_1 : d_1 \in D_1) = \mid \alpha f_1 \mid$$

for every real number α. This shows that B is a linear subspace. Also, if $\mid g \mid \leq \mid f_1 \mid$, it follows from $\mid g \mid \wedge d_1 \in A^+$ for every $d_1 \in D_1$ and

$$\sup(\mid g \mid \wedge d_1 : d_1 \in D_1) = \mid g \mid \wedge \mid f_1 \mid = \mid g \mid$$

that $g \in B$. Hence, B is an ideal.

Finally, to show that B is a band, it is sufficient (by Theorem 7.4) to prove that if D is a non-empty subset of B^+ possessing a supremum u_0 in E, then $u_0 \in B$. For any $u \in D$ we have

$$u = \sup(v : v \in A, 0 \leq v \leq u),$$

and so

$$u_0 = \sup_{u \in D} \{\sup(v : v \in A, 0 \leq v \leq u)\} =$$
$$\sup(v : v \in A, 0 \leq v \leq u \text{ for some } u \in D).$$

On account of (3) it follows that $u_0 \in B$. This concludes the proof. ∎

Exercise 7.9. Show that if $A_1, \ldots, A_n$ are ideals in the Riesz space E and $[A_1], \ldots, [A_n]$ are the corresponding bands, then

$$[\cap_{k=1}^n A_k] = \cap_{k=1}^n [A_k].$$

Hint: The proof in one direction is easy. For the converse direction, it is sufficient to prove that $[A_1] \cap [A_2]$ is a subset of $[A_1 \cap A_2]$. Assume that $0 \leq u \in [A_1] \cap [A_2]$. There exist non-empty sets $D_1 \subseteq A_1^+$ and $D_2 \in A_2^+$ such that

$$u = \sup(v : v \in D_1) = \sup(w : w \in D_2).$$

It is easy to see that $u = \sup D_3$, where $D_3 = (v \wedge w : v \in D_1, w \in D_2)$ is a non-empty subset of $(A_1 \cap A_2)^+$. It follows that $u \in [A_1 \cap A_2]$.

Exercise 7.10. Let $A_1, \ldots, A_n$ be ideals in the Riesz space E. With the notations as in the preceding exercise, show that

$$[[A_1] + \cdots + [A_n]] = [A_1 + \cdots + A_n].$$

Exercise 7.11. Show that if A_1 and A_2 are ideals in the Riesz space E generated by the subsets D_1 and D_2 respectively, then the ideal generated by the union $D_1 \cup D_2$ is $A_1 + A_2$.

Hint: To show that any element in the ideal generated by $D_1 \cup D_2$ is an element of $A_1 + A_2$ use the Riesz decomposition property as in Exercise 6.10.

8. Disjointness

Let E be a Riesz space. The elements f and g in E are said to be *disjoint* if $\mid f \mid \wedge \mid g \mid = 0$. To denote that f and g are disjoint we sometimes write $f \perp g$. Note that $f \perp g$ if and only if $\mid f \mid \perp \mid g \mid$. For any non-empty subset D of E the set

$$D^d = (f \in E : f \perp g \text{ for all } g \in D)$$

is called the *disjoint complement* of D. The disjoint complement $D^{dd} = (D^d)^d$ of D^d is called the *second disjoint complement* of D. If D_1 and D_2 are non-empty subsets of E such that $d_1 \perp d_2$ for every $d_1 \in D_1$ and every $d_2 \in D_2$, then D_1 and D_2 are said to be disjoint (notation $D_1 \perp D_2$).

Theorem 8.1. *Let E be a Riesz space with elements $f, g, \ldots$*

(i)   *If $f \perp g$ and $\mid h \mid \leq \mid f \mid$, then $h \perp g$.*

(ii)   *If $f \perp g$ and $\alpha \in \mathbb{R}$, then $\alpha f \perp g$.*

(iii) *If $f_1 \perp g$ and $f_2 \perp g$, then $(f_1 + f_2) \perp g$.*

(iv)   *$f \perp g$ if and only if $f^+ \perp g$ and $f^- \perp g$.*

(v)   *For every $f \in E$ we have $f^+ \perp f^-$.*

(vi)   *If D_1 and D_2 are non-empty subsets of E such that $D_1 \perp D_2$, then $D_1 \cap D_2$ is either empty or equal to the set $\{0\}$ containing only the null element.*

(vii) *If D is a subset of E such that $f_0 = \sup D$ exists and $g \perp f$ holds for all $f \in D$, then $g \perp f_0$.*

(viii) *If $f_1, \ldots, f_n$ is a set of nonzero and mutually disjoint elements, then this is a linearly independent set.*

Proof. (i) Let $f \perp g$ and $\mid h \mid \leq \mid f \mid$. Then $0 \leq \mid h \mid \wedge \mid g \mid \leq \mid f \mid \wedge \mid g \mid = 0$, so $\mid h \mid \wedge \mid g \mid = 0$, i.e., $h \perp g$.

(ii) Let $f \perp g$ and $\alpha \in \mathbb{R}$. If $\mid \alpha \mid \leq 1$, then $\mid \alpha f \mid \leq \mid f \mid$, so $\alpha f \perp g$. Now, let $\mid \alpha \mid > 1$. From

$$\mid \alpha f \mid \wedge \mid \alpha g \mid = \mid \alpha \mid (\mid f \mid \wedge \mid g \mid) = 0$$

it follows that $\alpha f \perp \alpha g$. Hence, since $\mid g \mid \leq \mid \alpha g \mid$, it is clear that $\alpha f \perp g$.

(iii) Let $f_1 \perp g$ and $f_2 \perp g$. In view of Theorem 6.5 we have

$$\mid f_1 + f_2 \mid \wedge \mid g \mid \leq$$
$$(\mid f_1 \mid + \mid f_2 \mid) \wedge \mid g \mid \leq$$
$$\mid f_1 \mid \wedge \mid g \mid + \mid f_2 \mid \wedge \mid g \mid = 0,$$

so $(f_1 + f_2) \perp g$.

(iv) Let $f \perp g$. Since $0 \leq f^+ \leq \mid f \mid$ and $0 \leq f^- le \mid f \mid$, it follows that $f^+ \perp g$ and $f^- \perp g$. Conversely, if $f^+ \perp g$ and $f^- \perp g$, then $\mid f \mid = f^+ + f^-$ satisfies $\mid f \mid \perp g$, so $f \perp g$.

(v) $f^+ \perp f^-$ holds for every $f \in E$ by Theorem 5.1(ii).

(vi) Let D_1 and D_2 be non-empty subsets of E such that $D_1 \perp D_2$ and assume that $D_1 \cap D_2$ is not empty. Let $f \in D_1 \cap D_2$. Then $f \perp f$, so $\mid f \mid \wedge \mid f \mid = 0$. This implies that $\mid f \mid = 0$, so $f = 0$.

(vii) Let D be a non-empty subset of E such that $f_0 = \sup D$ exists. Observe first that $f_0^+ = \sup(f^+ : f \in D)$ and $f_0^- = \inf(f^- : f \in D)$. The first formula follows from the general observation that $f_0 = \sup(f : f \in D)$ implies $f_0 \vee h = \sup(f \vee h : f \in D)$ for any $h \in E$, in particular for $h = 0$. The proof is simple. The second formula follows from one of the distributive laws stating that $f_0 \wedge h = \sup(f \wedge h : f \in D)$. Hence, for $h = 0$, we get $-f_0^- = \sup(-f^- : f \in D)$, so $f_0^- = \inf(f^- : f \in D)$. Assume now that $g \perp D$, i.e., $g \perp f$ for all $f \in D$, so $g \perp f^+$ and $g \perp f^-$ for every $f \in D$. Then

$$f_0^+ \wedge \mid g \mid = \sup(f^+ \wedge \mid g \mid : f \in D) = \sup 0 = 0,$$
$$f_0^- \wedge \mid g \mid \leq f^- \wedge \mid g \mid = 0 \text{ for every } f \in D,$$

so $f_0^- \wedge \mid g \mid = 0$. Thus $f_0^+ \perp g$ and $f_0^- \perp g$, so $f_0 \perp g$.

(viii) Let $f_1, \ldots, f_n$ be a set of nonzero and mutually disjoint elements. If the set is not linearly independent, then one of the elements, say f_1, is a linear combination of the other elements in the set, say $f_1 = \alpha_2 f_2 + \cdots + \alpha_n f_n$. In view of (ii) and (iii) we have

$$f_1 \perp (\alpha_2 f_2 + \cdots + \alpha_n f_n),$$

i.e. $f_1 \perp f_1$. It follows that $f_1 = 0$, contradicting our hypothesis. Hence, $f_1, \ldots, f_n$ is a linearly independent set. ∎

Theorem 8.2. *As above, let E be a Riesz space with elements $f, g, \ldots$. If $f \perp g$, then*

(1) $\mid f + g \mid = \mid f - g \mid = \mid f \mid + \mid g \mid = \mid\mid f \mid - \mid g \mid\mid = \mid f \mid \vee \mid g \mid,$
(2) $(f + g)^+ = f^+ + g^+ = f^+ \vee g^+ ,$
(3) $(f + g)^- = f^- + g^- = f^- \vee g^- .$

Proof. Let $f \perp g$, so $\mid f \mid \wedge \mid g \mid = 0$. Then

$$0 = 2(\mid f \mid \wedge \mid g \mid) = \mid f \mid + \mid g \mid - \mid\mid f \mid - \mid g \mid\mid .$$

It follows from this and from

$$\mid\mid f \mid - \mid g \mid\mid \leq \mid f + g \mid \leq \mid f \mid + \mid g \mid,$$

$$\mid\mid f \mid - \mid g \mid\mid \leq \mid f - g \mid \leq \mid f \mid + \mid g \mid$$

that

$$\mid f + g \mid = \mid f - g \mid = \mid f \mid + \mid g \mid = \mid\mid f \mid - \mid g \mid\mid .$$

Also, $\mid f \mid + \mid g \mid = (\mid f \mid \vee \mid g \mid) + (\mid f \mid \wedge \mid g \mid) = \mid f \mid \vee \mid g \mid$. This establishes (1). For (2), note first that $\mid f \mid \perp \mid g \mid$ implies $f^+ \perp g^+$, so $f^+ + g^+ = f^+ \vee g^+$. It is sufficient to prove now that $f^+ + g^+ \leq (f + g)^+$, because $(f + g)^+ \leq f^+ + g^+$ was proved in Theorem 5.5. The desired inequality follows from

$$f^+ + g^+ + (f + g)^- \leq f^+ + g^+ + f^- + g^- = \mid f \mid + \mid g \mid =$$

$$\mid f + g \mid = (f + g)^+ + (f + g)^- .$$

The proof for (3) is similar.   ∎

Exercise 8.3. Show that the following conditions for the elements f and g of the Riesz space E are equivalent.

(i) $f \perp g,$
(ii) $\mid f + g \mid = \mid f - g \mid,$
(iii) $\mid f + g \mid = \mid f \mid \vee \mid g \mid,$
(iv) $\mid f - g \mid = \mid f \mid \vee \mid g \mid,$
(v) $\mid f \mid + \mid g \mid = \mid\mid f \mid - \mid g \mid\mid,$
(vi) $(f + g)^+ = f^+ \vee g^+$ and $(f + g)^- = f^- \vee g^- .$

   Hint: Prove that (vi) implies (iii).

Theorem 8.4. *Let D be a non-empty subset of E. Then the disjoint complement D^d is a band in E. Furthermore, D is a subset of D^{dd} and $D^d = D^{ddd}$. Finally $D^d \cap D^{dd} = \{0\}$, so the algebraic sum $D^d + D^{dd}$ is a direct sum. In general this direct sum is an ideal, not necessarily a band.*

Proof. It follows immediately from the disjointness properties proved in Theorem 8.1 that D^d is a linear subspace. Also, if $f \in D^d$ (i.e., $f \perp d$ for all $d \in D$) and $\mid g \mid \leq \mid f \mid$, then $g \perp d$ for all $d \in D$, so $g \in D^d$. This shows that D^d is an ideal. Finally, in view of (vii) in Theorem 8.1, it is clear that the ideal D^d is a band.

From the definition of a disjoint complement it follows immediately that D is a subset of D^{dd}. Hence (replace D by D^d) also $D^d \subseteq D^{ddd}$. Furthermore (as again evident from the definition)$D_1 \subseteq D_2$ implies $D_1^d \supseteq D_2^d$. Hence, $D \subseteq D^{dd}$ implies $D^d \supseteq D^{ddd}$. It follows that $D^d = D^{ddd}$.

In view of Theorem 8.1(vi) it follows from $D^d \perp D^{dd}$ that $D^d \cap D^{dd}$ is either empty or equal to $\{0\}$. In the present case, since $0 \in D^d \cap D^{dd}$, we have $D^d \cap D^{dd} = \{0\}$.   ∎

As we have seen, D^{dd} is a band containing D as a subset. Now, the band $[D]$ generated by D is the smallest band containing D. Thus, $[D]$ is a subset of D^{dd}. The question arises whether D^{dd} can be larger than $[D]$. The answer is that this can indeed happen. By way of example, let E be the lexicographically ordered plane (Example 4.2(ii)), i.e., E is the vector space $\mathbb{R}^2$ of all (x, y) with x and y real numbers, ordered by defining that $(x_1, y_1) \leq (x_2, y_2)$ if either $x_1 < x_2$ or $x_1 = x_2, y_1 \leq y_2$. Let D be the vertical axis,i.e., $D = \{(x, y) : x = 0\}$. Then D is a band in E such that $D^d = \{0\}$. Hence, the band generated by D is D itself, but D^{dd} is the entire space E.

Now, consider the case that D is an ideal A in E. By Theorem 7.8, any $u \in [A]^+$ is the supremum of the set $(v \in A : 0 \leq v \leq u)$. Hence, $[A]$ has the property that for any $f \neq 0$ in $[A]$ there certainly exists an element $g \in A$ such that $0 \neq \mid g \mid \leq \mid f \mid$. Does the possibly larger set A^{dd} still have the same property ? We shall prove that the answer is affirmative. Precisely, A^{dd} is the largest band possessing the stated property.

Theorem 8.5. (i) *For any ideal A in E, the set A^{dd} is the largest among all ideals B in E having the property that for every $0 \neq f \in B$ there exists an element $g \in A$ such that $0 \neq \mid g \mid \leq \mid f \mid$.*

(ii) If A is an ideal in E, then $A^d = \{0\}$, i.e., $A^{dd} = E$, if and only if for every $0 \neq f \in E$ there exists an element $g \in A$ such that $0 \neq \mid g \mid \leq \mid f \mid$.

Proof. (i) It will be proved first that A^{dd} has the mentioned property. Let $0 \neq f \in A^{dd}$ and assume that there does not exist any $g \in A$ satisfying $0 \neq \mid g \mid \leq \mid f \mid$. Then $\mid f \mid \wedge \mid h \mid = 0$ for every $h \in A$ (because if $g_0 = \mid f \mid \wedge \mid h_0 \mid \neq 0$ for some $h_0 \in A$, then $0 \neq g_0 \in A$ and $\mid g_0 \mid \leq \mid f \mid$). It follows that $f \in A^d$. But $f \in A^{dd}$ by hypothesis. Hence $f \in A^d \cap A^{dd} = \{0\}$, which contradicts $f \neq 0$.

Now, let B be an ideal with the property that for any $0 \neq f \in B$ there exists an element $g \in A$ such that $0 \neq \mid g \mid \leq \mid f \mid$. Assume that $B \subseteq A^{dd}$

does not hold. Then there exists an element $f \in B$ not disjoint from A^d, so $k =| f | \wedge | h |\neq 0$ for some $h \in A^d$. It follows that $0 \neq k \in B \cap A^d$. Then, by hypothesis, there exists an element $g \in A$ such that $0 \neq| g |\leq k$, so $0 \neq g \in A \cap A^d$ This, however, is impossible.

(ii) Follows immediately from (i).   ∎

Exercise 8.6. (i) Show that for any non-empty subset D of E we have $(D \cup D^d)^{dd} = E$. Show also that if F is a linear subspace of E, then $(F \oplus F^d)^{dd} = E$.

(ii) Show that if $A_1, \ldots, A_n$ are ideals in E, then $(\cap_{k=1}^n A_k)^{dd} = \cap_{k=1}^n A_k^{dd}$. This is similar to Exercise 7.9.

(iii) Let $\{\sigma\}$ be an arbitrary non-empty set, an arbitrary element of which will be denoted by σ, and to each $\sigma \in \{\sigma\}$ let there be assigned a non-empty subset D_σ of E. Show that $(\cup D_\sigma)^d = \cap D_\sigma^d$ and, assuming that $\cap D_\sigma$ is not empty, show that $(\cap D_\sigma)^d \supseteq \cup D_\sigma^d$, so $(\cap D_\sigma)^{dd} \subseteq \cap D_\sigma^{dd}$. The last inclusion may be proper, even for a countable intersection of ideals. For a finite intersection of ideals there is equality, as seen in part (ii) of this exercise.

Hint: For part (ii), the proof in one direction is easy. For the converse direction, it is sufficient to prove that $A_1^{dd} \cap A_2^{dd}$ is a subset of $(A_1 \cap A_2)^{dd}$. Assume $0 \neq f \in A_1^{dd} \cap A_2^{dd}$. There exists (by the last theorem) an element $h \in A_1$ such that $0 \neq| h |\leq| f |$. Since $0 \neq h \in A_1 \cap A_2^{dd}$, there exists now also an element $g \in A_2$ such that $0 \neq| g |\leq| h |$. Hence $0 \neq g \in A_1 \cap A_2$ and $| g |\leq| f |$. It has been proved thus that the ideal $A_1^{dd} \cap A_2^{dd}$ has the property that for any f in the ideal there exists an element $g \in A_1 \cap A_2$ such that $0 \neq| g |\leq| f |$. Then $A_1^{dd} \cap A_2^{dd}$ is a subset of $(A_1 \cap A_2)^{dd}$, once again in view of the last theorem.

To show in (iii) that the inclusion in $(\cap_1^\infty A_n)^{dd} \subseteq \cap_1^\infty A_n^{dd}$ may be proper (all A_n ideals), let $E = C([0, 1])$, the space of all real continuous functions on $[0, 1]$. Let $r_1, r_2, \ldots$ be the rational numbers in $[0, 1]$ and A_n the ideal of all $f \in E$ satisfying $f(r_n) = 0$. Then $\cap_1^\infty A_n = \{0\}$, so $(\cap A_n)^{dd} = \{0\}$. On the other hand $A_n^d = \{0\}$ for every n, so $\cap A_n^{dd} = E$.

Exercise 8.7. Let $A_1, \ldots, A_n$ be ideals in the Riesz space E. Show that

$$(A_1^{dd} + \cdots + A_n^{dd})^{dd} = (A_1 + \cdots + A_n)^{dd}.$$

This is similar to Exercise 7.10.

Exercise 8.8. Let A be an ideal in the Riesz space E, and let $[A]$ be the band generated by A. Show that A, $[A]$ and A^{dd} have the same disjoint complement, and so $A^{dd} = [A]^{dd}$. More generally, let D be a non-empty subset of E and let A_D and B_D be the ideal and band generated by D. Show that $D^d = (A_D)^d = (B_D)^d$.

Hint: Use Theorem 8.1.

CHAPTER 4
Archimedean Spaces and Convergence

9. Archimedean Riesz spaces

The Riesz space E is said to be *Archimedean* if

$$\inf(n^{-1}u : n = 1, 2, \ldots) = 0$$

holds for every $u \in E^+$. There exist non-Archimedean Riesz spaces. As an example, let $E = \mathbb{R}^2$ with the lexicographical ordering (see Example 4.2(ii)). The element $(0,1)$ in E is a lower bound of the sequence $\{(n^{-1}, n^{-1}) : n = 1, 2, \ldots\}$. Hence, $u = (1, 1)$ does not satisfy the condition that $\inf(n^{-1}u : n = 1, 2, \ldots) = 0$. Actually, the sequence of all $n^{-1}u$ does not have an infimum at all in this case. Before presenting more examples, we prove a simple theorem.

Theorem 9.1. *Let E be a Riesz space.*

(i) The space E is Archimedean if and only if it is true for every $u \in E^+$ that $\inf(\epsilon_n u : n = 1, 2, \ldots) = 0$ holds for any sequence $(\epsilon_n : n = 1, 2, \ldots)$ of non-negative real numbers satisfying $\epsilon_n \to 0$.

(ii) The space E is Archimedean if and only if, given u and v in E^+ such that $0 \leq nv \leq u$ for $n = 1, 2, \ldots$, it follows that $v = 0$. In other words, for any $v \geq 0, v \neq 0$, the sequence $(nv : n = 1, 2, \ldots)$ is not bounded above.

(iii) If E is Archimedean, then every Riesz subspace of E is also Archimedean. In particular, ideals in E and bands in E are Archimedean Riesz spaces on their own.

Proof. Parts (i) and (iii) are evident, so it remains to prove (ii). Let E be Archimedean, and assume that $0 \leq nv \leq u$ for $n = 1, 2, \ldots$. Then $0 \leq v \leq n^{-1}u$ for $n = 1, 2, \ldots$. Since $\inf_n n^{-1}u = 0$, it follows that $v = 0$. Conversely, assume that $0 \leq nv \leq u$ for $n = 1, 2, \ldots$ implies $v = 0$. We have to prove that $\inf_n n^{-1}u_0 = 0$ for any $u_0 \in E^+$. It is sufficient to this end that any lower bound w of the sequence $(n^{-1}u_0)$ satisfies $w \leq 0$. Now, if w is such a lower bound, then $v = w \vee 0$ is still a lower bound of the sequence, so $0 \leq nv \leq u_0$ for $n = 1, 2, \ldots$. It follows by hypothesis that $v = 0$, i.e., $w \vee 0 = 0$, and so $w \leq 0$. ∎

Example 9.2. We return to the spaces in Example 4.2. (i) The space $\mathbb{R}^n$, with the usual coordinatewise ordering, is Archimedean.

(ii) As already observed, the space $\mathbb{R}^2$ with lexicographical ordering is not Archimedean.

(iii) The space E of all real functions on a non-empty point set X (with the algebraic operations and the ordering pointwise) is Archimedean because for any $u \in E^+$ the sequence $(n^{-1}u : n = 1, 2, \ldots)$ converges pointwise to zero as n tends to infinity. The same holds for the space $C(X)$ of all real continuous functions on a topological space X.

(iv) The space $L_0(\mathbb{R}^n, \mu)$ of all real Lebesgue measurable functions on $\mathbb{R}^n$ is Archimedean. More generally, if μ is a non-negative and countably additive measure in the non-empty point set X, then $L_0 = L_0(X, \mu)$ is Archimedean. To see this, let $u \in L_0$. For the proof that $\inf(n^{-1}u : n = 1, 2, \ldots) = 0$ it is permitted to choose a fixed function in the equivalence class u. Denote this function by u again and observe that $n^{-1}u(x) \downarrow 0$ as $n \to \infty$ for every $x \subset X$ (the downwards arrow indicates that the sequence is monotonely decreasing). Since the space $L_1 = L_1(X, \mu)$ of all μ-summable functions is an ideal in $L_0 = L_0(X, \mu)$, the space L_1 is also an Archimedean Riesz space.

(v) Let Γ be an algebra of subsets of the non-empty point set X and let E_b be the Riesz space of all bounded finitely additive signed measures on Γ (see (7) in Example 4.2). It is easy to see that E_b is Archimedean.

(vi) Let E be the vector space of all real functions f on the interval $[0,1]$ such that $f(x) \neq 0$ holds for only finitely many x in $[0,1]$. We introduce a partial ordering in E by defining that $f \leq g$ holds when either $f(0) < g(0)$ or $f(0) = g(0), f(x) \leq g(x)$ for all $x \in [0, 1]$. Then E is a non-Archimedean Riesz space.

As we shall prove, Archimedean Riesz spaces are characterized by the property that bands and disjoint complements in the space are the same. It was proved in Theorem 8.4 that, in any arbitrary Riesz space, every disjoint complement is a band. To say, in the converse direction, that in an Archimedean space E every band is a disjoint complement is equivalent to saying that $B = B^{dd}$ holds for every band B in E. Indeed, if $B = B^{dd}$ holds for the band B, then B is obviously a disjoint complement. Conversely, if B can be written as $B = D^d$ for some non-empty subset D of E, then (since $D^d = D^{ddd}$ by Theorem 8.4) we have $B = D^d = D^{ddd} = (D^d)^{dd} = B^{dd}$. We shall now first prove the easier half of the theorem that E is Archimedean if and only if $B = B^{dd}$ holds for every band B in E. This will be followed by a discussion of further properties of spaces $C(X)$ and $L_0(X, \mu)$. The discussion of $C(X)$ spaces will touch upon some subjects which are not immediately necessary for our present purposes. These spaces received considerable attention, however, some thirty years ago and, although the attention is not so large any more now, it is still worth while to mention a few points of interest.

Theorem 9.3. *If the Riesz space E is Archimedean, then $B = B^{dd}$ holds for every band B in E.*

Proof. As we have seen, B is a subset of B^{dd} for any band B in E. Assume now that E is Archimedean and we have a band B which is a proper subset of B^{dd}, i.e., there exists some $f \in B^{dd}$ such that f is not contained in B. The same is true then for $u = \mid f \mid$, so $0 < u \in B^{dd}$ and $u \notin B$ (where $0 < u$ stands for $0 \leq u, 0 \neq u$). Let

$$M_u = (v \in B : 0 < v < u).$$

Note that in view of Theorem 8.5(i) the set M_u is not empty. Obviously, u is an upper bound of M_u but u is not the supremum of M_u, because then it would follow from $u = \sup M_u$ and $M_u \subseteq B$, together with the fact that B is a band, that $u \in B$, which is not true. Hence, since u is not the supremum of M_u, not every upper bound w of M_u satisfies $w \geq u$. In other words, there exists an upper bound w^* of M_u such that $w^* \geq u$ does not hold. Let $w = u \wedge w^*$. Then w is an upper bound of M_u such that $w < u$ (and so $0 < w < u$). It follows that $w \in B^{dd}$ (since $u \in B^{dd}$) and $u - w > 0$. Since $0 < u - w \in B^{dd}$, there exists (in view of Theorem 8.5(i) again) an element $z \in B$ such that $0 < z < u - w$. For every v in M_u we have now that $z + v \in B$ and

$$0 < z + v \leq z + w < u,$$

so $z + v \in M_u$. In particular, since $z \in M_u$, we have $2z \in M_u$. Repeating the operation of adding z to each element of M_u, we obtain $3z \in M_u$. Generally, $nz \in M_u$ for $n = 1, 2, \ldots$, i.e., $0 < nz < u$ for $n = 1, 2, \ldots$. This is impossible since E is Archimedean. Hence, it is impossible for any band B to be a proper subset of B^{dd}. It follows that $B = B^{dd}$ holds for every band B in E.   ∎

Example 9.4. Let X be a compact Hausdorff space and E the Riesz space $C(X)$ of all real continuous functions on X (see (6) in Example 4.2). Recall that now X is completely regular, i.e., if F is a closed subset of X and x_0 is a point of X not contained in F, then there exists a real continuous function f on X such that $f(x_0) = 1$ and $f(x) = 0$ for all $x \in F$. If desired, the reader may restrict himself (herself) to the case that X is a bounded interval $[a, b]$ in $\mathbb{R}$ with the usual definitions for the distance between points and continuity for functions. Let A be an ideal in E and let N be the closed subset of X, defined by

$$N = \{x \in X : f(x) = 0 \text{ for all } f \in A\}.$$

For the present purposes, the complement $O = X \backslash N$ of N will be called the *open carrier* of the ideal A. Note already that the open carrier of A^d is the interior of N (which we shall denote by "int N"). Precisely,

$$A^d = \{f \in E = C(X) : f(x) = 0 \text{ for all } x \in X\backslash\text{int}N\}.$$

Note that we use here that X is completely regular.

(i) Different ideals in $E = C(X)$ may have the same open carrier. As an example, let $X = [-1, 1]$ and let $u_1(x) = 0$ for $-1 \leq x \leq 0$ and $u_1(x) = x$ for $0 \leq x \leq 1$, whereas $u_2(x) = 0$ for $-1 \leq x \leq 0$ and $u_2(x) = x^2$ for $0 \leq x \leq 1$. The principal ideals generated by u_1 and u_2 respectively are different and have the same open carrier.

(ii) The open carrier O of the ideal A in $C(X)$ is exactly the set of all $x \in X$ such that $f(x) > 0$ for at least one $f \in A$. It is not necessarily true, however, that there exists a function $f \in A^+$ such that $f(x) > 0$ for all points $x \in O$. As an example, let $X = [0, 1]$ and, for $n = 1, 2, \ldots$, let

$$X_n = (x \in X : (n + 1)^{-1} < x < n^{-1}).$$

Now let A consist of all $f \in C(X)$ such that the set of all x at which $f(x) \neq 0$ is contained in the union of at most finitely many X_n.

(iii) For any open subset O of X there exists a largest ideal A having O as its open carrier. This ideal is given by

$$A = \{f \in C(X) : f(x) = 0 \text{ for all } x \in N = X\backslash O\}.$$

From what was observed above about A^d, we may conclude that A^d is the largest ideal having int N as its open carrier. In the case that $X = [a, b]$ and A is the largest ideal having the given open subset O of X as its open carrier, there exists a function $u \in A^+$ such that $u(x) > 0$ for all $x \in O$. We mention a function satisfying this condition. For each $x \in X$, let $u(x)$ be the distance from the point x to the set $N = X\backslash O$, i.e.,

$$u(x) = \inf(|\,x - y\,| : y \in N).$$

(iv) If O_1 and O_2 are open subsets of X with A_1 and A_2 the largest ideals having O_1 and O_2 as their open carriers, then $A_1 = A_2$ if and only if $O_1 = O_2$. Not every largest ideal of the kind introduced here is a band. Every band, however, is such a largest ideal. For an example of a largest ideal which fails to be a band, let $X = [0, 1]$ and let O be the union of the open intervals $(^1/_4, ^1/_2)$ and $(^1/_2, ^3/_4)$. If B is a band in $C(X)$, then $B = B^{dd}$ since $C(X)$ is Archimedean, and so B can be written as $B = A^d$ for $A = B^d$. As observed about disjoint complements already, $B = A^d$ is therefore a largest ideal.

(v) For any subset S of X we denote the interior of S by int S and the closure of S by S^-. Recall that int S is the largest open set contained in S and S^- is the smallest closed set containing S. It follows immediately that $(X\backslash S)^- = X\backslash$ int S and $\text{int}(X\backslash S) = X\backslash S^-$. Also, $O \subseteq \text{int}O^-$ for any open set O and $F \supseteq (\text{int}F)^-$ for any closed set F. Easy examples show that these inclusions may be proper. Any open set O satisfying $O = \text{int } O^-$ is said to be *regularly open* and any closed set F satisfying $F = (\text{int}F)^-$ is said to be *regularly closed.* The set O is regularly open if and only if its complement $F = X\backslash O$ is regularly closed. As an aside we prove that the interior of an arbitrary closed set is always regularly open, i.e., if F is closed, then $\mathrm{int}(\mathrm{int}F)^- = \mathrm{int}F$. From $F \supseteq (\mathrm{int}F)^-$ it follows immediately that $\mathrm{int}F \supseteq \mathrm{int}(\mathrm{int}F)^-$. Conversely, from $\mathrm{int}F \subseteq (\mathrm{int}F)^-$ it follows that $\mathrm{int}F = \mathrm{int}(\mathrm{int}F) \subseteq \mathrm{int}(\mathrm{int}F)^-$. For convenience, let us denote now the open carrier of the ideal A in $C(X)$ by $c(A)$ and the complement of $c(A)$ by $n(A)$. As shown above,

$$c(A^d) = \mathrm{int}\,n(A) = X\backslash\{c(A)\}^-, \text{ so } n(A^d) = X\backslash c(A^d) = \{c(A)\}^-.$$

It follows that

$$c(A^{dd}) = \mathrm{int}\,n(A^d) = \mathrm{int}\{c(A)\}^-.$$

Hence, A and A^{dd} have the same open carrier, i.e., $c(A) = c(A^{dd})$, if and only if $c(A) = \mathrm{int}\{c(A)\}^-$, that is to say, if and only if $c(A)$ is regularly open. Assume now that A is the largest ideal having a given open set O as its open carrier, so $c(A) = O$. Recall that A^{dd}, being a disjoint complement, is the largest ideal having $c(A^{dd})$ as open carrier. Largest ideals are identical if and only if they have the same open carriers. Hence, $A = A^{dd}$ if and only if $c(A) = c(A^{dd})$, i.e., if and only if $O = c(A)$ is regularly open. But $A = A^{dd}$ if and only if A is a band (since $C(X)$ is Archimedean). Thus, we may conclude that if A is the largest ideal having the open set O as its carrier, then A is a band if and only if O is regularly open.

Example 9.5. Let μ be a non-negative countably additive measure in the non- empty point set X. As before, we denote the Archimedean Riesz space of all real μ-measurable functions on X by $L_0(X,\mu)$, briefly L_0. The space $L_0(\mathbb{R}^n,\mu)$, with μ Lebesgue measure, is a particular case. Recall that the members of $L_0(X,\mu)$ are equivalence classes of functions, although we usually speak about these members as if they were functions. The functions f_1 and f_2 are in the same equivalence class if and only if they differ only on a subset of X of measure zero. This is often expressed by saying that f_1 and f_2 are *equal almost everywhere* or, even simpler, saying that f_1 and f_2 are *almost equal.* Equivalently, $f_1 - f_2$ and the function identically zero are in the same equivalence class. It will be clear now what it means to say that a function $f \in L_0$ vanishes almost everywhere on a certain subset X_1 of X. Precisely, it means that, no matter which member f_1 of the equivalence class f we choose, we have $f_1(x) = 0$ for all $x \in X_1$, except on a subset of X_1 of measure zero. Observe however that if f_2 is another member of the same equivalence class f, then the exceptional subset of X_1 of measure zero may be different from the one for f_1. In the case that the empty set is the only set of measure zero in X all these complications disappear because then each equivalence class contains only one function, so that now functions are almost equal if and only if they are identical.

Let A be an ideal in L_0. The measurable subset X_1 of X is sometimes called a *null set for* A if it is true for every $f \in A$ that f vanishes almost everywhere on X_1. In general, and in particular if μ is Lebesgue measure, it makes no sense to say that there exists a point $x_0 \in X_1$ such that $f(x_0) = 0$ holds for all $f \in A$ simultaneously. In other words, the definition of a null set is not a pointwise definition. For future use, note already that a finite union of null sets is a null set. What we should like is to define a maximal null set $n(A)$ for A, because this set would be analogous to the set $n(A)$ in the preceding example, where A is an ideal in $C(X)$. It is evident from the remarks above that a pointwise definition is not possible, i.e., we cannot simply define $n(A)$ as the set of all $x \in X$ such that $f(x) = 0$ holds for all $f \in A$ simultaneously. Also, we cannot define $n(A)$ as the union of all null sets for A, because in general the union of an uncountable number of measurable sets need not be measurable. Moreover, even if the union turns out to be measurable, it may be much too large (in the case of Lebesgue measure in $\mathbb{R}^n$ each point is a null set, so the union of these points is already the whole of $\mathbb{R}^n$). In the case that μ is a *finite measure*, i.e., $\mu(X)$ is a finite number, all difficulties can be overcome, as follows. Assume that $\mu(X)$ is finite, and let Γ be the collection of all null sets for A. Then the number

$$\alpha = \sup\{\mu(S) : S \in \Gamma\}$$

is a finite non-negative number, so there exists a sequence $(X_n : n = 1, 2, \ldots)$ of sets in Γ such that $\mu(X_n)$ converges to α. Defining $S_n = \cup_{k=1}^n X_k$ for $n = 1, 2, \ldots$, we obtain an increasing sequence $(S_n : n = 1, 2, \ldots)$ of sets in Γ such that $\mu(S_n) \uparrow \alpha$. The set $n(A) = \lim S_n = \cup_{n=1}^\infty S_n$ is now a member of Γ satisfying $\mu\{n(A)\} = \alpha$. It follows easily that the complement $c(A) = X \backslash n(A)$ does not have a subset of positive measure that is a member of Γ. The set $n(A)$ and its complement $c(A)$ are uniquely determined by the condition that $n(A) \in \Gamma$ and $c(A)$ does not have a subset of positive measure which is a member of Γ (unique modulo sets of measure zero of course). The set $c(A)$ is called the *carrier* of the ideal A. Exactly as in the preceding example, different ideals may have the same carrier and there may be an ideal A such that there does not exist a function $u \in A^+$ satisfying $u(x) > 0$ for almost every $x \in c(A)$. Obviously, the set $n(A)$ is the carrier of A^d. Precisely, A^d is the set of all $f \in L_0$ such that f vanishes almost everywhere on $c(A)$. For any given measurable subset X_1 of X there exists a largest ideal A having X_1 as its carrier, given by

$$A = (f \in L_0 : f(x) = 0 \text{ for almost every } x \in X \backslash X_1).$$

The ideal A is such a largest ideal if and only if A is a band, i.e., if and only if $A = A^{dd}$. All these results can be extended to the case that, although $\mu(X)$ may be equal to infinity, the set X is a countable union $\cup_1^\infty X_n$ of sets X_n of finite measure. It may be assumed that the sets X_n are mutually disjoint. The measure μ is now called a σ-*finite measure*. Given the ideal A in L_0, we define the ideals $A_n (n = 1, 2, \ldots)$ by $A_n = (f \chi_{X_n} : f \in A)$, where χ_{X_n} is the characteristic function of X_n (see Example 3.4 for the definition of a characteristic function). The sets $n(A)$ and $c(A)$ are now the unions $\cup_1^\infty n(A_n)$ and $\cup_1^\infty c(A_n)$ respectively.

After these rather elaborately worked out examples we shall now complete the proof that a Riesz space is Archimedean if and only if $B = B^{dd}$ holds for every band B in the space. The first part of the proof is contained in Theorem 9.3, where it was shown that if the Riesz space E is Archimedean, then $B = B^{dd}$ holds for every band B in E. This result will be of frequent use (as already seen in the last two examples), more so than its converse, because in most of what follows the Riesz spaces we shall discuss will be Archimedean by hypothesis. Nevertheless, to be complete, we include the proof of the converse result.

Theorem 9.6. *Let E be a Riesz space. Then E is Archimedean if and only if $B = B^{dd}$ holds for every band B in E.*

Proof. It is sufficient to prove that if $B = B^{dd}$ holds for every band B in E, then E is Archimedean. To this end, assume that $B = B^{dd}$ for every band B, but E s not Archimedean. Then there exist elements u and v in E such that $0 < nv \leq u$ for $n = 1, 2, \ldots$. Denote by A_v the principal ideal generated by v, and let $A = A_v \oplus A_v^d$. It is easy to see that $A^d = \{0\}$, because any $f \perp A$ satisfies $f \perp A_v$ and $f \perp A_v^d$, so $f \in A_v^d \cap A_v^{dd} = \{0\}$. Hence $A^{dd} = E$, i.e., $[A]^{dd} = E$, where $[A]$ is the band generated by A. We use here that A and $[A]$ have the same disjoint complement (see Exercise 8.8). Since $[A] = [A]^{dd}$ by hypothesis, it follows that $[A] = E$. Hence $0 < u \in [A]$, which implies (by Theorem 7.8) that $u = \sup M_u$, where

$$M_u = (w : w \in A, 0 \leq w \leq u).$$

We shall prove now that $w + v \in M_u$ for every $w \in M_u$, i.e., $w + v \leq u$. Indeed, if $w \in M_u$, then $w = w_1 + w_2$, where $w_1 \in A_v, w_2 \in A_v^d$ and where $0 \leq w_1 \leq w$ and $0 \leq w_2 \leq w$ (see Theorem 7.6). Since A_v consists of all $f \in E$ satisfying $| f | \leq nv$ for some natural number n, we have $w_1 \leq nv$ for some n, so

$$w_1 + v \leq (n + 1)v \leq u$$

with $w_1 + v \in A_v$. Furthermore $0 \leq w_2 \leq w \leq u$ and $w_2 \perp (w_1 + v)$ on account of $w_1 + v \in A_v$ and $w_2 \in A_v^d$, so

$$w + v = (w_1 + v) + w_2 = \sup(w_1 + v, w_2) \leq u.$$

Then $w \leq u - v$ for every $w \in M_u$, i.e., $u - v$ is an upper bound of the set M_u. Since $v > 0$, this contradicts the result that $u = \sup M_u$. Hence, E must be Archimedean. ∎

Exercise 9.7. Let $\mathcal{B}$ be the set of all bands in the Riesz space E. The set $\mathcal{B}$ is partially ordered by inclusion, i.e., $B_1 \leq B_2$ whenever $B_1 \subseteq B_2$. The bands $\{0\}$ and E are the smallest and largest elements in $\mathcal{B}$. Evidently, $\mathcal{B}$ is a lattice. For any pair B_1 and B_2 in $\mathcal{B}$ the supremum and infimum of B_1 and B_2 exist, given by

$$B_1 \vee B_2 = [B_1 + B_2] \text{ and } B_1 \wedge B_2 = B_1 \cap B_2,$$

where (as before) $[B_1 + B_2]$ denotes the band generated by the ideal $B_1 + B_2$.

(i) Show that the lattice $\mathcal{B}$ is distributive, i.e., show that if A, B_1 and B_2 are bands, then

$$A \wedge (B_1 \vee B_2) = (A \wedge B_1) \vee (A \wedge B_2).$$

It is evident that the right hand side is a subset of the left hand side. For the converse, note first that

$$A \wedge (B_1 \vee B_2) = A \cap [B_1 + B_2] = [A] \cap [B_1 + B_2] = [A \cap (B_1 + B_2)],$$

where Exercise 7.9 is used to establish the last equality. Let now $0 \leq u \in A \cap (B_1 + B_2)$. The element u can be written as

$$u = u_1 + u_2 \text{ with } u_1 \in B_1^+, u_2 \in B_2^+ \text{ and } u_1 \in A^+, u_2 \in A^+.$$

Show that it follows now that $u \in A \cap B_1 + A \cap B_2$, and so $A \cap (B_1 + B_2)$ is a subset of $A \cap B_1 + A \cap B_2$. The desired result follows.

(ii) Show that the lattice $\mathcal{B}$ is a Boolean algebra if E is Archimedean. Precisely, show that for any band B the band B^d is the complement of B in $\mathcal{B}$. Observe at which point in the proof it is used that E is Archimedean.

Exercise 9.8. Let E be a Riesz space. Show that E is Archimedean if and only if, for every non-empty subset D of E, the band generated by D is equal to D^{dd}.

Hint: Observe that if $[D]$ is the band generated by D, then $D^d = [D]^d$ and so we have $D^{dd} = [D]^{dd}$.

10. Order Convergence and Uniform Convergence

Let E be a Riesz space. The sequence $(f_n : n = 1, 2, \ldots)$ in E is said to be *increasing* if $f_1 \leq f_2 \leq \cdots$ and *decreasing* if $f_1 \geq f_2 \geq \ldots$. This will be denoted by $f_n \uparrow$ or $f_n \downarrow$ respectively. If $f_n \uparrow$ and $f = \sup f_n$ exists in E, we shall write $f_n \uparrow f$. Similarly, if $f_n \downarrow$ and $f = \inf f_n$ exists, we write $f_n \downarrow f$. If $f_n \uparrow f$ or $f_n \downarrow f$, we say sometimes that f_n *converges monotonely* to f as n tends to infinity. Note already that if $f_n \uparrow f$, then $f_n + g \uparrow f + g$ for every

$g \in E$. Similarly, $f_n \downarrow f$ implies $f_n + g \downarrow f + g$. Also, $f_n \uparrow f$ if and only if $(-f_n) \downarrow (-f)$ and $f_n \downarrow f$ if and only if $(-f_n) \uparrow (-f)$. It follows that $f_n \uparrow f$ if and only if $(f - f_n) \downarrow 0$ and $f_n \downarrow f$ if and only if $(f_n - f) \downarrow 0$.

We prove a simple lemma.

Lemma 10.1. (i) *If* $f_n \uparrow f$, *then* $f_{n_k} \uparrow f$ *for every subsequence* $(f_{n_k} : k = 1, 2, \ldots)$ *such that* $n_1 < n_2 < \ldots$. *This holds in particular for any subsequence* $(f_n : n \geq n_0)$, *where* n_0 *is an arbitrary but fixed natural number. Similarly for a decreasing sequence.*

(ii) *If* $f_n \uparrow f$ *and* $\alpha \geq 0$ *is real, then* $\alpha f_n \uparrow \alpha f$. *Similarly for a decreasing sequence.*

(iii) *If* $p_n \downarrow 0$ *and* $q_n \downarrow 0$, *then* $p_n + q_n \downarrow 0$.

(iv) *If* $0 \leq r_n \leq p_n \downarrow 0$, *then* $\inf r_n = 0$. *Hence* $r_n \downarrow 0$ *if* r_n *is decreasing. Also,* $r_n = 0$ *for all* n *if* r_n *is increasing.*

Proof. (i) Let $f_n \uparrow f$ and let $(f_{n_k} : k = 1, 2, \ldots)$ be a subsequence such that $n_1 < n_2 < \ldots$. Evidently f is an upper bound of the subsequence. If g is another upper bound of the subsequence and f_m is an arbitrary member of the original sequence, there exists a member $n_{k,0}$ of the index set $(n_k : k = 1, 2, \ldots)$ such that $m \leq n_{k,0}$, so $f_m \leq f_{n_{k,0}} \leq g$. This shows that $f_m \leq g$ holds for all m, so $f = \sup f_m \leq g$. It follows that $f = \sup f_{n_k}$.

(ii) Evident.

(iii) Let $p_n \downarrow 0$ and $q_n \downarrow 0$. The sequence of all $p_n + q_n$ is decreasing and 0 is a lower bound of the sequence. If h is another lower bound and m, n are natural numbers, say $m \leq n$, then $h \leq p_n + q_n \leq p_m + q_n$, and so $h \leq \inf_n (p_m + q_n) = p_m$. This holds for every m, so $h \leq \inf p_m = 0$. Hence $\inf(p_n + q_n) = 0$, i.e., $p_n + q_n \downarrow 0$.

(iv) Let $0 \leq r_n \leq p_n \downarrow 0$ with r_n increasing. Note that, for any n_0, the sequence of all r_n and the sequence $(r_n : n \geq n_0)$ have the same upper bounds. For $n \geq n_0$ we have $r_n \leq p_n \leq p_{n_0}$, so p_{n_0} is an upper bound of the sequence of all r_n. This holds for every n_0, so $0 = \inf p_{n_0}$ is likewise an upper bound. But then $r_n = 0$ for all n. ∎

Monotone convergence is a particular case of a more general notion of convergence which we now introduce. The sequence $(f_n : n = 1, 2, \ldots)$ in E is said to *converge in order* to f if there exists a sequence $p_n \downarrow 0$ such that $|f - f_n| \leq p_n$ holds for all n, i.e., $(f - p_n) \leq f_n \leq (f + p_n)$ for all n. We shall denote this by $f_n \to f$ or, if necessary, by $f_n \to f$ (ord). The main properties of order convergence are listed in the following theorem.

Theorem 10.2. (i) *If $f_n \to f$ and $f_n \to g$, then $f = g$. In other words, order limits are uniquely determined.*

(ii) *If $f_n \uparrow f$ or $f_n \downarrow f$, then $f_n \to f$.*

(iii) *If $f_n \to f, g_n \to g$ and α, β are real numbers, then $\alpha f_n + \beta g_n \to \alpha f + \beta g$. Furthermore $f_n \vee g_n \to f \vee g$ and $f_n \wedge g_n \to f \wedge g$. In particular, if $f_n \to f$, then $f_n^+ \to f^+, f_n^- \to f^-$ and $\mid f_n \mid \to \mid f \mid$.*

(iv) *If $f_n \to f$ and $f_n \geq g$ holds for all n, then $f \geq g$.*

(v) *If $f_n \uparrow$ or $f_n \downarrow$ and $f_n \to f$, then $f_n \uparrow f$ or $f_n \downarrow f$ respectively.*

(vi) *If $f_n \to f$, then $f_{n_k} \to f$ for every subsequence $(f_{n_k} : k = 1, 2, \ldots)$ such that $n_1 < n_2 < \ldots$.*

Proof. (i) Let $\mid f - f_n \mid \leq p_n \downarrow 0$ and $\mid g - f_n \mid \leq q_n \downarrow 0$. Then, using part (iii) of the lemma, we have

$$\mid f - g \mid \leq \mid f - f_n \mid + \mid g - f_n \mid \leq p_n + q_n \downarrow 0,$$

so $\mid f - g \mid = 0$, i.e., $f = g$.

(ii) Evident.

(iii) Let $\mid f - f_n \mid \leq p_n \downarrow 0$ and $\mid g - g_n \mid \leq q_n \downarrow 0$, and let α, β be real numbers. Then

$$\mid (\alpha f + \beta g) - (\alpha f_n + \beta g_n) \mid \leq$$

$$\mid \alpha \mid \cdot \mid f - f_n \mid + \mid \beta \mid \cdot \mid g - g_n \mid \leq \mid \alpha \mid \cdot p_n + \mid \beta \mid \cdot q_n \downarrow 0,$$

so $\alpha f_n + \beta g_n \to \alpha f + \beta g$. Furthermore, using now one of the Birkhoff inequalities in Theorem 6.3, we have

$$\mid f \vee g - f_n \vee g_n \mid \leq$$

$$\mid f \vee g - f_n \vee g \mid + \mid f_n \vee g - f_n \vee g_n \mid \leq$$

$$\mid f - f_n \mid + \mid g - g_n \mid \leq p_n + q_n \downarrow 0,$$

so $f_n \vee g_n \to f \vee g$. The proof for $f \wedge g$ is similar.

(iv) Let $f_n \to f$ and $f_n \geq 0$ for all n, so $f_n^- = 0$ for all n. Since $f_n^- \to f^-$, it follows that $f^- = 0$, i.e., $f \geq 0$. If $f_n \to f$ and $f_n \geq g$ for all n, then $f_n - g \to f - g$ and $f_n - g \geq 0$ for all n, so $f - g \geq 0$, i.e., $f \geq g$.

(v) Let $f_n \uparrow$ and $\mid f - f_n \mid \leq p_n \downarrow 0$. Then $(f - f_n)^-$ is increasing and, on the other hand, $0 \leq (f - f_n)^- \leq \mid f - f_n \mid \leq p_n \downarrow 0$. Hence, by part (iv) of the lemma, $(f - f_n)^- = 0$ for all n, so $f - f_n = \mid f - f_n \mid$. It follows that $f - f_n \downarrow 0$, so $f_n \uparrow f$.

(vi) Follows from part (i) of the lemma. ∎

There is still another type of convergence, analogous to uniform convergence for sequences of functions on an interval. Let $0 < u \in E$. The sequence $(f_n : n = 1, 2, \ldots)$ in E is said to *converge u-uniformly* to f if for any number $\epsilon > 0$ there exists an index $n(\epsilon)$ such that $| f - f_n | \leq \epsilon u$ for all $n \geq n(\epsilon)$. Equivalently, there exists a sequence of numbers $\epsilon_n \downarrow 0$ such that $| f - f_n | \leq \epsilon_n u$ for all n. Similarly, the sequence is called an *u-uniform Cauchy sequence* if for any $\epsilon > 0$ there exists an index $n(\epsilon)$ such that $| f_m - f_n | < \epsilon u$ for all $m, n \geq n(\epsilon)$. Evidently, every u-uniformly converging sequence is u-uniformly Cauchy. If f_n converges u-uniformly to f, we shall write $f_n \to f(u\text{-un})$. If $0 < u \leq v$ and $f_n \to f(u\text{-un})$, then $f_n \to f(v\text{-un})$. If $u > 0$ and $v > 0$, and $f_n \to f(u\text{-un})$ as well as $f_n \to f(v\text{-un})$, then $f_n \to f(u \wedge v\text{-un})$. The sequence $(f_n : n = 1, 2, \ldots)$ in E is said to *converge relatively uniformly* to f if there exists an element $u > 0$ in E such that f_n converges u-uniformly to f. This will be denoted by $f_n \to f$ (un). The element u is then called the *regulator* of the relatively uniform convergence. Note that if $f_n \to f$ (un) and $g_n \to g$ (un), these sequences may have different regulators, say u and v. Of course, both sequences are then $(u \vee v)$-uniformly convergent.

The notion of relatively uniform convergence does not make much sense in a non-Archimedean Riesz space, because the same sequence may then have many different limits. In the next theorem, therefore, let us assume that E is an Archimedean Riesz space.

Theorem 10.3. *Let E be an Archimedean Riesz space.*

(i) If $f_n \to f$ (un), then $f_n \to f$ in order. Hence, if f_n converges relatively uniformly to f as well as to g, then $f = g$. In other words, relatively uniform limits are uniquely determined. In general, order convergence of f_n to f does not imply relatively uniform convergence.

(ii) If $f_n \to f$ (un), $g_n \to g$ (un) and α, β are real numbers, then $\alpha f_n + \beta g_n \to \alpha f + \beta g$ (un). Furthermore $f_n \vee g_n \to f \vee g$ (un) and $f_n \wedge g_n \to f \wedge g$ (un). In particular, if $f_n \to f$ (un), then $f_n^+ \to f^+$ (un), $f_n^- \to f^-$ (un) and $| f_n | \to | f |$ (un).

(iii) If $f_n \to f$ in order and $(f_n : n = 1, 2, \ldots)$ is an u-uniform Cauchy sequence, then f_n converges u-uniformly to f. Hence, if for $u > 0$ and $v > 0$ we have $f_n \to f(v\text{-un})$ and $(f_n : n = 1, 2, \ldots)$ is an u-uniform Cauchy sequence, then $f_n \to f(u\text{-un})$.

Proof. (i) Let $f_n \to f$ (un), i.e., there exists an element $u > 0$ in E and a sequence of numbers $(\epsilon_n : n = 1, 2, \ldots)$ such that we have $\epsilon_n \downarrow 0$ and $| f - f_n | \leq \epsilon_n u$ for all n. Since E is Archimedean, we see that $\epsilon_n u \downarrow 0$. This shows that f_n converges to f in order. For an example of a sequence in an Archimedean space which converges in order but not relatively uniformly, let E be the space ℓ_∞ of all bounded real sequences and, for $n = 1, 2, \ldots$, let f_n be the element in ℓ_∞ having the first n coordinates equal to zero and all other coordinates equal to one. Then $f_n \downarrow 0$ in order, but there does not exist any $u \in (\ell_\infty)^+$ for which f_n converges u-uniformly to zero.

(ii) Let $f_n \to f$ (un) and $g_n \to g$ (un). To prove the desired result we may assume without loss of generality that the convergence in the two sequences is with respect to the same regulator. The proof is then similar to the corresponding proof for order convergence.

(iii) Let $(f_n : n = 1, 2, \ldots)$ be an u-uniform Cauchy sequence and let $f_n \to f$ (ord). For any given $\epsilon > 0$ there exists an index $n(\epsilon)$ such that $\mid f_m - f_n \mid \leq \epsilon u$ for all $m \geq n \geq n(\epsilon)$, i.e.,

$$f_n - \epsilon u \leq f_m \leq f_n + \epsilon u \text{ for } m \geq n \geq n(\epsilon).$$

Now, $f_m \to f$ in order as m tends to infinity, so (by part (iv) of the preceding theorem) the limit f satisfies

$$f_n - \epsilon u \leq f \leq f_n + \epsilon u \text{ for } n \geq n(\epsilon).$$

In other words, $\mid f - f_n \mid \leq \epsilon u$ for all $n \geq n(\epsilon)$. This shows that f_n converges u-uniformly to f.   ∎

Exercise 10.4. Let $u > 0$ be given in the Archimedean Riesz space E and assume that every monotone u-uniform Cauchy sequence has an u-uniform limit. Show that every (not necessarily monotone)u-uniform Cauchy sequence has an u-uniform limit.

Hint: Let $(f_n : n = 1, 2, \ldots)$ be an arbitrary u-uniform Cauchy sequence. There exists a subsequence $(f_{n_k} : k = 1, 2, \ldots)$ such that $n_1 < n_2 < \ldots$ and

$$\mid f_{n_{k+1}} - f_{n_k} \mid \leq 2^{-k} u \text{ for } k = 1, 2, \ldots.$$

The partial sums of the series

$$f_{n_1}^+ + (f_{n_2} - f_{n_1})^+ + (f_{n_3} - f_{n_2})^+ + \cdots,$$

$$f_{n_1}^- + (f_{n_2} - f_{n_1})^- + (f_{n_3} - f_{n_2})^- + \cdots$$

form increasing u-uniform Cauchy sequences, having therefore limits g_1 and g_2 respectively. Then the partial sums of

$$f_{n_1} + (f_{n_2} - f_{n_1}) + (f_{n_3} - f_{n_2}) + \cdots$$

converge u-uniformly to $g = g_1 - g_2$. Thus, $f_{n_k} \to g(u\text{-un})$ as k tends to infinity. It follows easily that $f_n \to g(u\text{-un})$ as n tends to infinity.

The Archimedean Riesz space E is said to be *uniformly complete* if, for every $u > 0$ in E, every u-uniform Cauchy sequence has a limit. Sometimes there exists a particular element $u_0 > 0$ in E such that if we know only that every u_0-uniform Cauchy sequence has a limit we may already conclude that E is uniformly complete. To mention an example where this is the case, note that if for some $u > 0$ every u-uniform Cauchy sequence has a limit, and v is either a positive multiple of u or $v \leq u$, then every v-uniform Cauchy sequence has a limit. Combining these cases, we may conclude that if $v > 0$ and v is an element in the principal ideal A_u generated by u, then every v-uniform Cauchy sequence has a limit. Hence, if $A_u = E$ and every u-uniform Cauchy sequence has a limit, then E is uniformly complete. An element $u > 0$ for which $A_u = E$ is called a *strong unit* for E (precisely, a *strong order unit*).

Besides order convergent sequences there are order Cauchy sequences. The sequence $(f_n : n = 1, 2, \ldots)$ in the arbitrary Riesz space E is said to be an *order Cauchy sequence* if there exists a sequence $p_n \downarrow 0$ such that $\mid f_m - f_n \mid \leq p_n$ for all $m \geq n \geq 1$. Note that any subsequence of an order Cauchy sequence is again order Cauchy. It is evident that any order convergent sequence is order Cauchy. The space E is *order complete* if every order Cauchy sequence has an order limit. Some details are contained in the next exercises.

Exercise 10.5. Show that for the Riesz space E the following conditions are equivalent.

(i) E is order complete.

(ii) If $p_n \uparrow, q_n \downarrow$ and $q_n - p_n \downarrow 0$ in E, then there exists an element $s \in E$ such that $p_n \leq s \leq q_n$ for all n. In this case $s = \sup(p_n) = \inf(q_n)$.

(iii) Every monotone order Cauchy sequence has an order limit.

Hint: For the proof that (i) implies (ii), assume that $p_n \uparrow, q_n \downarrow$ and $q_n - p_n \downarrow 0$. Then $(p_n : n = 1, 2, \ldots)$ is order Cauchy, and so p_n order converges to some $s \in E$. But then $p_n \uparrow, p_n \rightarrow s$, which implies $p_n \uparrow s$ by Theorem 10.2(v). Since $p_m \leq q_n$ for all m and fixed n, it follows that $s \leq q_n$.

For the proof in the converse direction, assume that (ii) holds and let $\mid f_m - f_n \mid \leq u_n \downarrow 0$ for $m \geq n$. We have to prove that f_n converges in order. Set $q_1 = f_1 + u_1$. Then $f_n \leq q_1$ for $n = 1, 2, \ldots$. Set $q_2 = q_1 \wedge (f_2 + u_2)$. Then $q_2 \leq q_1$ and $f_n \leq q_2$ for $n = 2, 3, \ldots$. Set $q_3 = q_2 \wedge (f_3 + u_3)$. Then $q_3 \leq q_2$ and $f_n \leq q_3$ for $n = 3, 4, \ldots$. Proceed by induction. Thus we obtain a sequence $(q_n : n = 1, 2, \ldots)$ such that $q_n \downarrow$ and $f_m \leq q_n$ for $m \geq n$. Similarly we obtain a sequence $p_n \uparrow$ such that $f_m \geq p_n$ for $m \geq n$. From $q_n - p_n \leq 2u_n$ it follows that $q_n - p_n \downarrow 0$. In view of hypothesis (ii) there exists now an element $s \in E$ such that $p_n \leq s \leq q_n$ for all n. Finally, derive from $p_n \leq f_n \leq q_n$ that $\mid s - f_n \mid \leq q_n - p_n \downarrow 0$. This shows that f_n converges in order to s.

For (iii), observe that (i) implies (iii) trivially and (iii) implies (ii).

Exercise 10.6. This is a sequel. Show that the following conditions for the Riesz space E are equivalent.

(i) E is order complete.

(iv) Every increasing order Cauchy sequence has a supremum.

(v) Every order Cauchy sequence has a supremum.

As an illustration, assume that E is an order complete Riesz space and (f_n) is a sequence of mutually disjoint elements in E^+ such that $s = \sup f_n$ exists. Show that the sequence (f_n), and hence every subsequence, converges in order to the null element. Derive from this that every subsequence of (f_n) has a supremum.

Hint: It is easy to see that (iv) is equivalent to condition (ii) in the preceding exercise. It remains to prove that (iv) implies (v). Assume for this that (f_n) is an arbitrary order Cauchy sequence; $\mid f_n - f_{n+m} \mid \leq p_n \downarrow 0$ for $n, m \geq 1$. Let $s_n = f_1 \vee \ldots \vee f_n$ for $n = 1, 2, \ldots$. The sequence (s_n) is increasing and has the same upper bounds as (f_n). It is sufficient, therefore, to prove that (s_n) is order Cauchy. This follows from

$$\mid s_n - s_{n+m} \mid = \mid (s_{n-1} \vee f_n) - (s_{n-1} \vee f_n \vee \ldots \vee f_{n+m}) \mid \leq$$

$$\mid f_n - (f_n \vee \ldots \vee f_{n+m}) \mid \leq \mid f_n - f_{n+1} \mid \vee \ldots \vee \mid f_n - f_{n+m} \mid \leq p_n,$$

where we have used one of the Birkhoff inequalities and Exercise 6.7.

For the example mentioned as an illustration, observe that if $s_n = f_1 + \cdots + f_n = f_1 \vee \ldots \vee f_n$ for all n, then (s_n) has the same upper bounds as (f_n). Hence $\sup s_n = s$, i.e., $s_n \uparrow s$. Note now that $f_n \leq s - s_n \downarrow 0$.

A monotone sequence is a special example of the more general notion of a directed set. We present the definition of a directed set.

Definition 10.7. Let $\{\tau\}$ be a non-empty set, the elements of which will be denoted simply by τ (in general $\{\tau\}$ will be an infinite set, countable or uncountable), and let E be a Riesz space. Assume furthermore that to each $\tau \in \{\tau\}$ there is assigned an element $f_\tau \in E$, i.e., there is a mapping $\tau \Rightarrow f_\tau$ from $\{\tau\}$ into E. The set $\{\tau\}$ serves, therefore, as the index set for the set (f_τ). The set $(f_\tau : \tau \in \{\tau\})$ is said to be *upwards directed* if for any τ_1 and τ_2 in $\{\tau\}$ there exists $\tau_3 \in \{\tau\}$ such that $f_{\tau_3} \geq f_{\tau_1} \vee f_{\tau_2}$ and the set is called *downwards directed* if there exists $\tau_3 \in \{\tau\}$ such that $f_{\tau_3} \leq f_{\tau_1} \wedge f_{\tau_2}$. This is denoted by $f_\tau \uparrow$ or $f_\tau \downarrow$ respectively. If $f_\tau \uparrow$ and $f = \sup f_\tau$ exists, we write $f_\tau \uparrow f$. If $f_\tau \downarrow$ and $f = \inf f_\tau$ exists, we write $f_\tau \downarrow f$.

The upwards directed sets (f_τ) and (g_τ), indexed by means of the same index set $\{\tau\}$, are said to be *equidirected* if for any τ_1 and τ_2 in $\{\tau\}$ there exists an index $\tau_3 \in \{\tau\}$ such that $f_{\tau_3} \geq f_{\tau_1} \vee f_{\tau_2}$ and $g_{\tau_3} \geq g_{\tau_1} \vee g_{\tau_2}$ hold simultaneously. The sets are equidirected in particular if and only if $f_{\tau_1} \leq f_{\tau_2}$ holds if and only if $g_{\tau_1} \leq g_{\tau_2}$ holds. The definition for equidirected downwards directed sets is similar.

Note that a downwards directed sequence is not the same as a decreasing sequence. As a simple example in the space $\mathbb{R}$ of real numbers, let $a_n = n^{-1}$ for $n = 1, 3, 5, \ldots$ and $a_n = n^{-1} + 1$ for $n = 2, 4, 6, \ldots$. Then the sequence

$(a_n : n = 1, 2, \ldots)$ is downwards directed, but not monotone. Furthermore, $a_n \downarrow 0$ holds as a directed sequence, but not in the familiar sequential sense. Also, if $(b_n : n = 1, 2, \ldots)$ is an arbitrary decreasing sequence of real numbers, then (a_n) and (b_n) are equidirected, although it is not so that $a_m \leq a_n$ holds if and only $b_m \leq b_n$ holds.

Theorem 10.8. *The following results (and the proofs) are analogous to those for sequences in Lemma 10.1 and Theorem 10.2.*

(i) If $f_\tau \uparrow f$, then $(f_\tau : f_\tau \geq f_{\tau_0}) \uparrow f$ for any τ_0.

(ii) If $f_\tau \uparrow f$ and $\alpha \geq 0$ is real, then $\alpha f_\tau \uparrow \alpha f$.

(iii) If $p_\tau \downarrow 0, q_\tau \downarrow 0$ and $(p_\tau), (q_\tau)$ are equidirected, then $p_\tau + q_\tau \downarrow 0$.

(iv) If $0 \leq r_\tau \leq p_\tau \downarrow 0$, then $\inf r_\tau = 0$. Hence, if (r_τ) is downwards directed, then $r_\tau \downarrow 0$.

(v) If $f_\tau \uparrow f, g_\tau \uparrow g$ with $(f_\tau), (g_\tau)$ equidirected and if $\alpha \geq 0, \beta \geq 0$ are real, then $\alpha f_\tau + \beta g_\tau \uparrow \alpha f + \beta g$. Also $f_\tau \vee g_\tau \uparrow f \vee g$ and $f_\tau \wedge g_\tau \uparrow f \wedge g$. In particular $f_\tau^+ \uparrow f^+$ and $f_\tau^- \uparrow f^-$.

Besides indexed subsets there are also some non-empty and non-indexed subsets D of E for which it makes sense to say that D is upwards or downwards directed.

Definition 10.9. The non-empty subset D of the Riesz space E is said to be *upwards directed* if for any two elements f and g in D there exists an element $h \in D$ such that $h \geq f \vee g$. The definition of a *downwards directed set* is similar.

Note that if D is an arbitrary non-empty subset of E and we adjoin to D all finite suprema of elements of D, then the thus enlarged set is upwards directed.

CHAPTER 5
Projections and Dedekind Completeness

11. Projection Bands

As we have seen in section 7, the algebraic sum of two ideals in the Riesz space E is again an ideal, but the algebraic sum of two bands (even if these are disjoint) is not necessarily a band. In the converse direction it may be asked what can be said concerning disjoint ideals A and B in the Riesz space E if it is given that $A \oplus B$ is a band in E. In this case we may just as well assume that $A \oplus B = E$, and an answer to the above question is contained in the following theorem.

Theorem 11.1. *If A and B are ideals in the Riesz space E such that $A \oplus B = E$, then $B = A^d$ and $A = B^d$, so $A = A^{dd}$ and $B = B^{dd}$. Hence, A and B are bands, each the disjoint complement of the other one.*

Proof. It follows from $A \oplus B = E$ that A and B are disjoint, so $B \subseteq A^d$. It will be sufficient to prove that, conversely, A^d is a subset of B. For this purpose, assume that $0 \leq u \in A^d$. By hypothesis, u can be written as $u = u_1 + u_2$ with $0 \leq u_1 \in A$ and $0 \leq u_2 \in B$. Since $0 \leq u_1 \leq u \in A^d$ holds now, we have $u_1 \in A^d$. On the other hand we have $u_1 \in A$. It follows that $u_1 = 0$, and so $u = u_2 \in B$. This shows that $A^d \subseteq B$. ∎

There is a variant of the theorem in which we assume A to be merely a linear subspace of E, but now B is immediately assumed to be the disjoint complement of A.

Exercise 11.2. Show that if A is a linear subspace of the Riesz space E such that $A \oplus A^d = E$, then $A = A^{dd}$.

Hint: To show that A^{dd} is a subset of A, let $f \in A^{dd}$ be given. By hypothesis $f = f_1 + f_2$ with $f_1 \in A \subseteq A^{dd}$ and $f_2 \subseteq A^d$. This is also the unique decomposition of f in $A^{dd} \oplus A^d$, so $f_2 = 0$. It follows that $f = f_1 \in A$.

There is still another variant for the formulation of which it is convenient to introduce the following definition. The ideal C in the Riesz space E is called the *order direct sum* of the linear subspaces A and B if $C = A \oplus B$ and, for any $u \geq 0$ in C, the (unique) decomposition $u = v + w$ with $v \in A$ and $w \in B$

is such that $v \geq 0$ and $w \geq 0$. If A and B are disjoint ideals and $C = A \oplus B$, then this is an order direct sum by Theorem 7.6. In the converse direction we have the following theorem.

Theorem 11.3. *If the ideal C in the Riesz space E is the order direct sum of the linear subspaces A and B, then A and B are ideals. Hence, if E is the order direct sum of the linear subspaces A and B, then A and B are bands such that $A = B^d$ and $B = A^d$.*

Proof. Assume that the ideal C is the order direct sum of the linear subspaces A and B. We prove first that $0 \leq v \leq u \in A^+$ implies $v \in A^+$. To this end, set $w = u - v$, so $u = v + w$ with $v \geq 0$ and $w \geq 0$. Since $u \in A^+ \subseteq C^+$ and the ideal C is the order direct sum of A and B, there exist decompositions $v = v_1 + v_2$ and $w = w_1 + w_2$ with v_1, w_1 in A^+ and v_2, w_2 in B^+. It follows that

$$v_2 + w_2 = u \quad (v_1 + w_1) \in A \cap B = \{0\},$$

so $v_2 + w_2 = 0$, and hence $v_2 = w_2 = 0$. This shows that $v = v_1 \in A^+$.

In the next step of the proof we show that every $f \in A$ can be written as $f = p_1 - q_1$ with p_1 and q_1 in A^+. For this, observe that $f \in A \subseteq C$, so f^+ and f^- are members of C (since C is an ideal). Now, let $f^+ = p_1 + p_2$ and $f^- = q_1 + q_2$ with p_1, q_1 in A^+ and p_2, q_2 in B^+. Then

$$f = f^+ - f^- = (p_1 - q_1) + (p_2 - q_2),$$

so

$$p_2 - q_2 = f - (p_1 - q_1) \in A \cap B = \{0\}.$$

It follows that $f = p_1 - q_1$ with p_1 and q_1 in A^+.

For the final step, observe now that if $f \in A$ is given, then $f = p_1 - q_1$ with p_1 and q_1 in A^+ as well as $f = f^+ - f^-$ with $0 \leq f^+ \leq p_1$ and $0 \leq f^- \leq q_1$ (see Theorem 5.6). Hence, f^+ and f^- are members of A^+ in view of the first part of the proof, so $|f| \in A^+$. It has been proved thus that $0 \leq v \leq u \in A^+$ implies $v \in A^+$ and also that $f \in A$ if and only if $|f| \in A$. Therefore, A is an ideal. Similarly, B is an ideal. $\blacksquare$

For a further investigation of the situation we may assume now that B is a band in the Riesz space E and we may ask for conditions under which $B \oplus B^d = E$. Any band B satisfying this condition is called a *projection band*. In this case, if $f = f_1 + f_2$ is the decomposition of an arbitrary $f \in E$ as the sum of $f_1 \in B$ and $f_2 \in B^d$, the elements f_1 and f_2 are called the *components* of f in B and in B^d respectively. If B is a projection band, then so is B^d. The converse does not necessarily hold (if E is the lexicographically ordered space $\mathbb{R}^2$ and B is the "vertical axis", i.e., B is the set of all points $(0, y)$, then

$B^d = \{0\}$ is a projection band, but B is not because $B = B^{dd}$ does not hold). If, however, E is Archimedean, then B is a projection band if and only if B^d is so (if B^d is a projection band, then B^{dd} is a projection band and $B = B^{dd}$ holds).

Before stating the next theorem we recall that the mapping T from the (real) vector space V into the (real) vector space W is called a *linear mapping* or *linear transformation* if

$$T(\alpha f + \beta g) = \alpha T f + \beta T g$$

for all f, g in V and all real α, β. Usually such a linear mapping is called a *linear operator* or, briefly, an *operator*. If T_1 and T_2 are operators from V into V itself, the operator $T_1 T_2$ is defined by $(T_1 T_2)f = T_1(T_2 f)$ for all $f \in V$. Instead of TT we write T^2. If the operator T (from V into V) satisfies $T^2 = T$, then T is said to be *idempotent* and T is called a *projection*. These definitions and notations are of course well-known from linear algebra.

Theorem 11.4. (i) *The band B in the Riesz space E is a projection band if and only if, for any $u \in E^+$, the element*

$$u_1 = \sup(v : v \in B, 0 \le v \le u)$$

exists, and u_1 is then the component of u in B. Similarly,

$$u_2 = \sup(w : w \in B^d, 0 \le w \le u)$$

is then the component of u in B^d. Hence, these suprema satisfy $u = u_1 + u_2$ with $u_1 \in B$ and $u_2 \in B^d$.

(ii) For any $f \in E$, denote the component of f in the projection band B by $P_B f$. The mapping P_B from E into E itself has the following properties:

(a) P_B is linear and idempotent (i.e., $P_B^2 = P_B$), so P_B is a projection.
(b) $0 \le P_B u \le u$ holds for every $u \in E^+$.

On account of these properties P_B is called the band projection or order projection on the projection band B.

(iii) In the converse direction, if P is a projection mapping E into E itself and such that $0 \le Pu \le u$ holds for every $u \in E^+$, there exists a projection band B such that P is the band projection on B.

Proof. (i) Assume first that B is a projection band, so $E = B \oplus B^d$, and let $u \in E^+$ have the decomposition $u = u_1 + u_2$. Define D by

$$D = (v : v \in B, 0 \le v \le u).$$

For any $v \in D$, the element $u - v \ge 0$ has the decomposition

$$u - v = (u_1 - v) + u_2,$$

and so (since positive elements have positive components) we have $u_1 - v \geq 0$, i.e., $v \leq u_1$. This shows that u_1 is an upper bound of D. On the other hand $u_1 \in D$. Hence $u_1 = \sup D$.

Assume now, conversely, that B is a band such that, for any $u \in E^+$,

$$u_1 = \sup(v : v \in B, 0 \leq v \leq u)$$

exists. Then $u_1 \in B$(since B is a band), so it will be sufficient for the proof of $B \oplus B^d = E$ to show that $u_2 = u - u_1$ is a member of B^d. If not, we have $p = u_2 \wedge z > 0$ for some $z \in B$(recall that $p > 0$ denotes that $p \geq 0$ and $p \neq 0$). Then $0 < p \in B$, and also $p \leq u_2$. It follows that $u_1 + p \in B$ and $u_1 + p \leq u_1 + u_2 = u$, so

$$u_1 + p \in (v : v \in B, 0 \leq v \leq u).$$

But then

$$u_1 + p \leq \sup(v : v \in B, 0 \leq v \leq u) = u_1,$$

so $p \leq 0$, which contradicts $p > 0$. As desired, it follows that $u_2 \in B^d$.

(ii) This is evident.

(iii) This is a rather immediate consequence of the result in Theorem 11.3. Let B be the range of the operator P, i.e., $B = (Pf : f \in E)$. Similarly, let A be the range of the operator $I - P$, where I denotes the identity mapping, so $A = (f - Pf : f \in E)$. Evidently, B and A are linear subspaces of E. Furthermore we have $B \cap A = \{0\}$. To prove this, observe that any $f \in B \cap A$ satisfies $f = Pg$ for some $g \in E$ as well as $f = h - Ph$ for some $h \in E$, and so

$$f = Pg = P^2 g = P(Pg) = P(h - Ph) = Ph - P^2 h = 0.$$

The algebraic sum $B + A$ is therefore a direct sum $B \oplus A$. Since $f = Pf + (f - Pf)$ holds for every $f \in E$, we see that $B \oplus A = E$. Finally, since by hypothesis $u \geq 0$ implies $Pu \geq 0$ and $u - Pu \geq 0$, the space E is the order direct sum of B and A. It follows now from Theorem 11.3 above that B is a band such that $B^d = A$, and so $B \oplus B^d = E$. Thus we see that B is a projection band, and evidently P is the band projection on B.   ∎

If B is a principal band we can say more. Recall that the ideal A_v generated by the element $v \in E^+$ consists of all $f \in E$ satisfying $\mid f \mid \leq nv$ for some natural number $n = n_f$. The band $[A_v]$ generated by A_v is called the *principal band* generated by v. Let now $u \in E^+$ be given and let D and D^* be the subsets of $[A_v]^+$, defined by

$$D = (w : w \in [A_v], 0 \leq w \leq u),$$

$$D^* = (u \wedge nv : n = 1, 2, \ldots).$$

We shall prove that D and D^* have the same upper bounds. Since D^* is a subset of D it is evident that any upper bound of D is an upper bound of D^*. To prove the converse, note that any $w \in D$ satisfies

$$w = \sup(z : z \in A_v, 0 \leq z \leq w)$$

by Theorem 7.8. Denote the set $(z : z \in A_v, 0 \leq z \leq w)$ by S_w, so $w = \sup S_w$. Now, any of these $z \in S_w$ satisfies $z \leq n_z v$ for some natural number n_z, so it follows from $0 \leq z \leq u$ and $0 \leq z \leq n_z v$ that $0 \leq z \leq u \wedge (n_z v)$. This shows that any upper bound s of D^* is an upper bound of all $z \in S_w$, and so s is an upper bound for $w = \sup S_w$, i.e., $s \geq w$. This holds for every $w \in D$, so s is an upper bound of D. Thus, every upper bound of D^* is an upper bound of D. It follows that D and D^* have the same upper bounds. In particular, $\sup D^*$ exists if and only if $\sup D$ exists and these suprema are then the same. Now, in view of our last theorem, the band $[A_v]$ is a projection band if and only if, for any $u \in E^+$, the element $u_1 = \sup D$ exists and in this case u_1 is then the component of u in $[A_v]$. Since $\sup D = \sup D^*$, the following result has been proved therefore.

Theorem 11.5. *The principal band $[A_v]$, generated by $v \in E^+$, is a projection band if and only if for any $u \in E^+$ the element*

$$u_1 = \sup(u \wedge (nv) : n = 1, 2, \ldots)$$

exists, and in this case u_1 is the component of u in $[A_v]$.

Corollary 11.6. *Let $u \in [A_v]^+$, where $[A_v]$ is the principal band generated by $v \in E^+$. Then*

$$u = \sup(u \wedge (nv) : n = 1, 2, \ldots).$$

Proof. Follows from the last theorem by observing that $[A_v]$ is a Riesz space on its own and $[A_v]$ itself is a projection band in this space. Note that $[A_v]$ is not necessarily a projection band in E. ∎

At this point we introduce a notion intermediate between an ideal and a band. Recall that a band in the Riesz space E is an ideal A with the special property that if D is any subset of A such that the element $\sup D$ exists in E, then $\sup D \in A$. If this does not necessarily hold for every subset D of A possessing a supremum but only if D is at most countable, then A is called a σ–*ideal*. In other words, the ideal A in E is a σ–ideal whenever it follows from $f_n \in A$ for $n = 1, 2, \ldots$ and $f = \sup f_n$ that $f \in A$. Every band is a σ-ideal but, in general, a σ-ideal is not a band. Example: Let E be the Riesz space of all real functions on the interval $[0,1]$, with the algebraic operations and the ordering pointwise, and let A be the ideal of all $f \in E$ such that $f(x) \neq 0$

holds for only finitely or countably many $x \in [0,1]$. The ideal A is a σ-ideal, but A is not a band.

The σ-ideal A_σ generated by the ideal A is, by definition, the smallest σ-ideal having A as a subset, i.e., A_σ is the intersection of all σ-ideals containing A as a subset. For the positive elements in A_σ there is a characterization similar to the characterization of positive elements in the band $[A]$ (see Theorem 7.8).

Theorem 11.7. *Let A_σ be the σ-ideal generated by the ideal A in the Riesz space E. Then A_σ^+ consists of all $u \in E^+$ for which there exists a sequence $(u_n : n = 1, 2, \ldots)$ in A^+ such that $0 \leq u_n \uparrow u$.*

Proof. Let B be the set of all $u \in E^+$ such that $u = \sup u_n^*$ for some sequence $(u_n^* : n = 1, 2, \ldots)$ in A^+. Then, if $u_n = u_1^* \vee \ldots \vee u_n^*$ for all n, we have $u_n \in A^+$ and $0 \leq u_n \uparrow u$ holds for the increasing sequence (u_n) in A^+. The proof that $A_\sigma^+ = B$ is now exactly the same as the corresponding proof in Theorem 7.8.∎

Corollary 11.8. *Every principal σ-ideal is already a band, i.e., the set of all principal σ-ideals is the same as the set of all principal bands.*

Proof. Let $(A_v)_\sigma$ and $[A_v]$ be the σ-ideal and the band generated by $v \in E^+$. We have to prove that any $u \geq 0$ in $[A_v]$ is already contained in $(A_v)_\sigma$. As proved in Corollary 11.6, any $u \geq 0$ in $[A_v]$ satisfies

$$u = \sup(u \wedge (nv) : n = 1, 2, \ldots).$$

Hence, writing $u_n = u \wedge (nv)$ for $n = 1, 2, \ldots$, we have $u_n \in A_v$ for all n and $0 \leq u_n \uparrow u$. In view of the last theorem this shows that $u \in (A_v)_\sigma$. ∎

If every band in the Riesz space E is a projection band, then E is said to have the *projection property*. Similarly, if every principal band is a projection band, then E is said to have the *principal projection property*.

As a preparation for a more general result in the next section (Theorem 12.3), we shall prove already that if E has the principal projection property, then E is Archimedean. Also, as an example, we show that the space $C([0,1])$ of all real continuous functions on the interval $[0,1]$ does not have any projection bands at all, except (0) and $C([0,1])$ itself.

Theorem 11.9. *If the Riesz space E has the principal projection property, then E is Archimedean.*

Proof. Let $0 \leq nv \leq u$ for $n = 1, 2, \ldots$. We have to prove that $v = 0$. It follows from $0 \leq nv \leq u$ that $u \wedge (nv) = nv$ for $n = 1, 2, \ldots$ and it follows from the principal projection property that the component u_0 of u in the principal band $[A_v]$ exists. The component u_0 is given by

$$u_0 = \sup_n \{u \wedge (nv)\}.$$

In other words, $u_0 = \sup_n nv$ in this case. But then

$$2u_0 = \sup_n 2nv = \sup_n nv = u_0,$$

so $u_0 = 0$. It follows that $v = 0$. ∎

Example 11.10. Let $E = C([0,1])$ be the Riesz space of all real continuous functions on [0,1] and assume that B is a projection band in E different from (0) and E itself. Let $e \in E$ be the function satisfying $e(x) = 1$ for all $x \in [0,1]$. Note that the band generated by e is E itself (the ideal generated by e is already equal to E). Since B is a projection band, we have $e = e_1 + e_2$ with $0 \le e_1 \in B$ and $0 \le e_2 \in B^d$. If e_1 would be identically zero on [0,1], we should have $e = e_2 \in B^d$, so $B^d = E$, which would imply $B = (0)$, contrary to hypothesis. Hence, e_1 is not identically zero. Similarly, e_2 is not identically zero, so e_1 is not identically one. Since $e_1 \wedge e_2 = 0$ and $e_1(x) + e_2(x) = 1$ for every $x \in [0,1]$, it is also impossible that e_1 assumes on $[0,1]$ a value strictly between zero and one. Hence, e_1 is a continuous function on [0,1] assuming the values zero and one and no intermediate value. This is impossible. It follows that (0) and $C([0,1])$ are the only projection bands in $C([0,1])$. Note that $C([0,1])$ is an example of an Archimedean space not possessing the principal projection property.

12. Dedekind Completeness

We recall the definitions of Dedekind completeness and Dedekind σ–completeness, as given in Definition 1.3 for a partially ordered set. For the case that the partially ordered set is a Riesz space E, the space E is Dedekind complete whenever every non-empty subset D of E that is bounded above has a supremum. As follows from Theorem 1.4 or alternatively, as follows from the observation that

$$\inf(f : f \in D) = -\sup(-f : f \in D),$$

it is evident that every non-empty subset of E that is bounded below has then an infimum. The space E is Dedekind σ-complete whenever every non-empty countable subset of E that is bounded above has a supremum. Here it follows from the formula mentioned above that every non-empty subset that is bounded below has then an infimum (Theorem 1.4 cannot be applied now). We prove first of all that in these defining conditions we may restrict ourselves to upwards directed subsets of positive elements. The definition of an upwards directed set was given in Definition 10.9.

Theorem 12.1. (i) *The Riesz space E is Dedekind complete if and only if every non-empty subset of E^+ that is upwards directed and bounded above has a supremum.*

(ii) The Riesz space E is Dedekind σ-complete if and only if every increasing sequence in E^+ that is bounded above has a supremum (in other words, if and only if it follows from $0 \leq u_n \leq v_0$ that there exists an element $u_0 \in E$ such that $0 \leq u_n \uparrow u_0$).

Proof. (i) If E is Dedekind complete, the condition in the statement of the theorem is evidently satisfied. Conversely, assume this condition satisfied and assume that D is a non-empty subset of E such that D is bounded above. We have to prove that $\sup D$ exists. Without loss of generality we may also assume that D contains all finite suprema of its elements (because adjoining to D all finite suprema of its elements does not have any effect on the set of upper bounds of D). Now choose an element $f_0 \in D$ and denote the set $(f \vee f_0 : f \in D)$ by D_1. The sets D and D_1 have the same upper bounds, so it is sufficient to show that $\sup D_1$ exists. Note that D_1 is upwards directed and every $f \in D_1$ satisfies $f \geq f_0$. Let D_2 be the subset of E^+ defined by $D_2 = (f - f_0 : f \in D_1)$. Then D_2 is upwards directed and bounded above. Hence, by hypothesis, $\sup D_2$ exists. This implies that $\sup D = \sup D_1 = \sup D_2 + f_0$ exists as well.

(ii) The proof is similar, based on the observation that if $(f_n : n = 1, 2, \ldots)$ is a sequence that is bounded above and $g_n = f_1 \vee \ldots \vee f_n$ for all n, then $(g_n : n = 1, 2, \ldots)$ is an increasing sequence having the same upper bounds as $(f_n : n = 1, 2, \ldots)$. $\blacksquare$

Before proving now that every Dedekind complete Riesz space has the projection property, we observe that Dedekind completeness trivially implies the property to be Archimedean. Indeed, the Riesz space E is Archimedean whenever $0 \leq nv \leq u(n = 1, 2, \ldots)$ implies $v = 0$. Assume now that E is Dedekind complete and let $0 \leq nv \leq u$ for $n = 1, 2, \ldots$. Then, in view of the Dedekind completeness, $u_0 = \sup_n nv$ exists. Hence $2u_0 = \sup_n 2nv = \sup_n nv = u_0$, so $u_0 = 0$, which implies $v = 0$.

In general, as shown already in section 7, the direct sum of disjoint bands in a Riesz space E is not a band. If E is Dedekind complete, however, a direct sum of disjoint bands is again a band. Moreover, if the bands are each other's disjoint complements, their direct sum is E, i.e., every band is a projection band.

Theorem 12.2. *Let E be a Dedekind complete Riesz space. Then the following holds.*

(i) If B_1 and B_2 are disjoint bands in E, then $B_1 \oplus B_2$ is a band.

(ii) Every band B in E satisfies $B \oplus B^d = E$, i.e., every band B in E is a projection band.

Proof. (i) Let B_1 and B_2 be disjoint bands in E and let D be an upwards directed (non-empty) set in $(B_1 \oplus B_2)^+$ such that $\sup D = \sup(u : u \in D) = u_0$ exists in E. It is sufficient to show that $u_0 \in B_1 \oplus B_2$ (see Theorem 7.4(ii)). For any $u \in D$ there exists a unique decomposition $u = v_u + w_u$ with $0 \le v_u \in B_1$ and $0 \le w_u \in B_2$. Note here that the sets $D = (u : u \in D), D_1 = (v_u : u \in D)$ and $D_2 = (w_u : u \in D)$ are indexed by the set $D = (u : u \in D)$ itself. Since $u_1 \le u_2$ (with u_1 and u_2 in D) implies $v_{u_1} \le v_{u_2}$ and $w_{u_1} \le w_{u_2}$ by Theorem 7.6, it is clear that the sets $D_1 = (v_u : u \in D)$ and $D_2 = (w_u : u \in D)$ are upwards directed. More precisely, the sets D, D_1 and D_2 are equidirected. Note here that if $u_3 \ge u_1$ and $u_3 \ge u_2$, then

$$v_{u_3} \ge v_{u_1} \vee v_{u_2} \text{ as well as } w_{u_3} \ge w_{u_1} \vee w_{u_2}.$$

With the notations of Definition 10.7 and Theorem 10.8 we have $0 \le u \uparrow u_0, 0 \le v_u \le u_0$ and $0 \le w_u \le u_0$ and hence, in view of the Dedekind completeness of E, there exist elements u_1 and u_2 in E such that $v_u \uparrow u_1$ and $w_u \uparrow u_2$. All v_u are members of the band B_1 and $u_1 = \sup v_u$, so $u_1 \in B_1$. Similarly, $u_2 \in B_2$. It follows now by Theorem 10.8(v) from the equidirectedness of the sets (v_u) and (w_u) that $u = v_u + w_u \uparrow u_1 + u_2$. On the other hand we have $u \uparrow u_0$, and so $u_0 = u_1 + u_2$. This shows that $u_0 \in B_1 \oplus B_2$, which is the desired result.

(ii) Let B be a band in E. Then $C = B \oplus B^d$ is a direct sum of disjoint bands, so C is a band (in view of the just established result in part (i)). Any f disjoint to C is disjoint to B as well as to B^d, so $f \in B \cap B^d$, i.e., $f = 0$. This shows that $C^d = (0)$, and so $C = C^{dd} = E$ (the equality $C = C^{dd}$ holds since E is Archimedean). ∎

The next theorem is sometimes referred to as the "main inclusion theorem".

Theorem 12.3. *The following holds in any Riesz space E. Dedekind completeness implies Dedekind σ–completeness and it also implies the projection property. Each of these last two mentioned properties implies the principal projection property and the principal projection property implies that E is Archimedean.*

Proof. It is evident that Dedekind completeness implies Dedekind σ-completeness and that the projection property implies the principal projection property. Furthermore, it was proved in the last theorem that Dedekind completeness implies the projection property and it was proved in Theorem 11.9 that if E has the principal projection property, then E is Archimedean. It remains only to prove that Dedekind σ–completeness implies the principal projection property. Hence, assume E Dedekind σ-complete and let B_v be the principal band in E generated by $v \in E^+$. To show that B_v is a projection band we have to show that for any $u \in E^+$ the element

$$\sup(u \wedge nv : n = 1, 2, \ldots)$$

exists (see Theorem 11.5). Writing $w_n = u \wedge nv$ for $n = 1, 2, \ldots$, we have $0 \le w_n \le u$, and so $\sup w_n$ exists since E is Dedekind σ-complete.   ∎

It can be proved that in the main inclusion theorem no implication in the converse direction holds and also that Dedekind σ–completeness and the projection property are independent properties. In Example 11.10 it was shown already, for example, that $C([0,1])$ is an Archimedean Riesz space not possessing the principal projection property. Several other examples can be found in the exercises. First, however, it will be proved now that if E has any of the properties mentioned in the main inclusion theorem and A is an ideal in E, then A has the same property.

Theorem 12.4. *If the Riesz space E has one of the properties in the main inclusion theorem and A is an ideal in E, then A has the same property.*

Proof. Assume E Dedekind complete. To show that the ideal A is Dedekind complete, assume that D is a (non-empty) subset of A^+, bounded above by $v \in A^+$, say. Then $u_0 = \sup D$ exists in E, and it follows from $0 \le u_0 \le v$ that $u_0 \in A$. This shows that A is Dedekind complete. The proof for Dedekind σ–completeness is similar.

The proof for the projection property is a little more difficult. Assume that E has the projection property and let B be a band in A. We have to show that B is a projection band in A, i.e., we have to show that for any $u \in A^+$ the set

$$V = (v : 0 \le v \le u, v \in B)$$

has a supremum in A. It will be sufficient to show that V has a supremum u_0 in E such that u_0 is a member of A, because then u_0 is obviously the supremum of V in A. Let $[B]$ be the band in E generated by B. Since E has the projection property, $[B]$ is a projection band, and so the set

$$W = (w : 0 \le w \le u, w \in [B])$$

has a supremum u_0 in E. It is evident that $u_0 \in A$(on account of $0 \le u_0 \le u \in A$). Hence, it will be sufficient to show that the sets V and W are identical. For this, since V is a subset of W, we have to show that any $w \in W$ is a member of V. Let $w \in W$. Since $0 \le w \le u \in A$ and A is an ideal, we have $w \in A$, so $w \in A \cap [B]$. But w is the supremum of a set of positive elements in $A \cap B$ and B is a band in A, so w itself is a member of $A \cap B$, i.e., $w \in V$.

The proof for the principal projection property is similar; note that if B is the principal band in A generated by the element $v \in A^+$, then $[B]$ is the principal band in E generated by v.

The proof for the Archimedean property is trivial.   ∎

Example 12.5. (i) The n-dimensional real number space $\mathbb{R}^n$, with the usual coordinatewise ordering, is a Dedekind complete Riesz space. The space $\mathbb{R}^2$, with the lexicographical ordering, is not even Archimedean, and therefore does not have any of the properties in the main inclusion theorem.

(ii) The Riesz space of all real functions on the arbitrary (non-empty) point set X, with pointwise ordering, is Dedekind complete. The supremum of any (non-empty) set D of functions such that D is bounded above is the pointwise supremum. In the case that $X = (1, 2, \ldots)$, the corresponding Riesz space is the space (s) of all real sequences. Hence, (s) is Dedekind complete and so is any ideal in (s), such as the space ℓ_p of all bounded sequences or the space (c_0) of all sequences converging to zero. Another class of ideals is the class of all sequence spaces ℓ_p, where $0 < p < \infty$. The space ℓ_p is the linear subspace of (s) consisting of all sequences $f = (f_1, f_2, \ldots)$ such that $\sum_n | f_n |^p$ converges. The proof that ℓ_p is linear follows immediately from the simple inequality

$$(a + b)^p \leq 2^p \{\max(a, b)\}^p = 2^p \max(a^p, b^p) \leq 2^p(a^p + b^p),$$

holding for $a \geq 0, b \geq 0$. It is evident that for $0 < p \leq q < \infty$ the space ℓ_p is a linear subspace of ℓ_q.

(iii) As in Example 9.5, let μ be a non-negative countably additive measure in the non-empty point set X and let $L_0 = L_0(X, \mu)$ be the corresponding Riesz space of all real μ-measurable functions on X (with identification of μ-almost equal functions). If the measure is σ-finite, i.e., if X is a finite or countable union of sets of finite measure, then L_0 is Dedekind complete. For the proof, assume first that μ is a finite measure, i.e., $0 < \mu(X) < \infty$.. Let D be a non-empty subset of L_0^+, bounded above by the function $v \in L_0^+$, where we assume first that v is a bounded function (so $\int_X v d\mu$ exists as a finite number). For the proof that $\sup D$ exists in L_0 it may be assumed without loss of generality that D contains all finite suprema of its elements. The set P of non-negative numbers

$$\left(\int_X u d\mu : u \in D \right)$$

is bounded above by $\int_X v d\mu$, so $\alpha = \sup P$ exists and there is a sequence $(u_n : n = 1, 2, \ldots)$ in D such that $\int_X u_n d\mu$ converges to α. Again without loss of generality we may assume that $u_n \uparrow$ (replace u_n by $u_1 \vee \cdots \vee u_n$ if necessary). On account of $0 \leq u_n \leq v$ the supremum $u_0 = \sup u_n$ exists in L_0 (as an ordinary pointwise supremum) and $\int_X u_0 d\mu = \alpha$. We assert that $u_0 = \sup D$. For the proof, choose an arbitrary $u^* \in D$. Then $u^* \vee u_n \in D$ for all n and $u^* \vee u_n \uparrow u^* \vee u_0$, which implies

$$\int_X (u^* \vee u_0) d\mu = \lim_n \int_X (u^* \vee u_n) d\mu \geq \lim_n \int_X u_n d\mu = \alpha.$$

On the other hand, the integral on the left is a member of the set P of all numbers $(\int_X u d\mu : u \in D)$, so the integral cannot be larger that α, and therefore is equal to α. This implies that

$$\int_X \{(u^* \vee u_0) - u_0\} d\mu = 0,$$

and so $(u^* \vee u_0) - u_0 = 0$ almost everywhere, i.e., $u^* \leq u_0$. This holds for every $u^* \in D$, so $u_0 = \sup D$.

Assume now that the upper bound $v \in L_0^+$ of the set D is not necessarily a bounded function. Let e be the function identically one on X and, for $n = 1, 2, \ldots$, let $v_n = v \wedge ne$ and $D_n = (u \wedge ne : u \in D)$. Then v_n is a bounded function and v_n is an upper bound of D_n, so $u_{n0} = \sup D_n$ exists in L_0. From $u_{n0} \leq v$ it follows that $u_0 = \sup_n u_{n0}$ exists and it is easily seen that

$$u_0 = \sup(u \wedge ne : u \in D, n = 1, 2, \ldots) = \sup(u : u \in D).$$

This concludes the proof for the case that $\mu(X)$ is finite.

Now let $\mu(X) = \infty$, where X is a countable union of sets $X_n (n = 1, 2, \ldots)$ of finite measure. Then, if $S_n = \cup_{k=1}^n X_k$ for $n = 1, 2, \ldots$, the sequence (S_n) is increasing and $X = \cup_n S_n$. Hence, if $Y_1 = S_1 = X_1$ and $Y_n = S_n \backslash S_{n-1}$ for $n = 2, 3, \ldots$, the sets Y_n are of finite measure and mutually disjoint, and $X = \cup_n Y_n$. For brevity, denote the characteristic function of Y_n by χ_n. As before, let D be a non-empty subset of L_0^+, bounded above by the function $v \in L_0^+$. Now, let $D_n = (u\chi_n : u \in D)$ for $n = 1, 2, \ldots$. Then the set D_n is non-empty and bounded above by the function $v\chi_n$. Hence, in view of what was proved already, $u_{n0} = \sup D_n$ exists in L_0. Evidently, the function

$$u_0^* = \sup_n u_{n0} = \sum_n u_{n0}$$

is the supremum of D. This concludes the proof.

It follows that every ideal in L_0 is Dedekind complete, such as for example the space $L_\infty = L_\infty(X, \mu)$ of all bounded μ-measurable functions. To say that $f \in L_0$ is bounded means that there exists a constant $M > 0$ such that $| f(x) | \leq M$ holds for almost every $x \in X$ (the exceptional set depending on the choice of f in its equivalence class). Another important class of ideals in L_0 is the class of all spaces of L_p-type. If $p(0 < p < \infty)$ is given, $L_p = L_p(X, \mu)$ is the space of all μ-measurable functions on X such that $\int_X | f(x) |^p d\mu$ is finite. The linearity of L_p follows (as in the preceding example) from the inequality $(a + b)^p \leq 2^p(a^p + b^p)$, holding for all $a \geq 0, b \geq 0$.

Note that if $X = (1, 2, \ldots)$ and μ is the *counting measure* in X (i.e., for any arbitrary subset Y of X the measure $\mu(Y)$ is the number of points in Y), then $L_0(X, \mu)$ is the space (s) of all real sequences, the space $L_\infty(X, \mu)$ is ℓ_∞ and, for $0 < p < \infty$, the space $L_p(X, \mu)$ is ℓ_p.

(iv) Let E_b be the Riesz space of all bounded finitely additive signed measures on the algebra Γ of subsets of the non-empty set X, as introduced in (7) of Example 4.2. The space E_b is Dedekind complete. For the proof, assume that D is an upwards directed subset of E^+, bounded above by the element $\lambda \in E_b^+$ (i.e., $\mu \le \lambda$ for all $\mu \in D$). It is easy to see that if $\mu_0(A)$ is defined now for any $A \in \Gamma$ as the supremum of the set $(\mu(A) : \mu \in D)$, then $\mu_0 \in E_b^+$ and $\mu_0 = \sup(\mu : \mu \in D)$ holds in E_b.

(v) Let $[a, b]$ be a closed interval in $\mathbb{R}$ and let E be the vector space of all (real) linear functions on $[a, b]$, i.e., E consists of all functions of the form $f(x) = \alpha x + \beta (\alpha, \beta$ real) for $x \in [a, b]$. Note that every $f \in E$ is uniquely determined by its values at the endpoints of $[a, b]$. The space E is ordered pointwise, i.e., if f and g are elements of E, we define $f \le g$ whenever $f(x) \le g(x)$ for all $x \in [a, b]$. This makes E into a Riesz space; the function $h = f \vee g$ is the linear function satisfying

$$h(a) = \max\{f(a), g(a)\} \text{ and } h(b) = \max\{f(b), g(b)\}.$$

It is now easy to see that E is Dedekind complete.

As a simple exercise we determine the set of all bands in E. Let $e \in E^+$ be defined by $e(x) = 1$ for all $x \in [a, b]$, let $e_1 \in E^+$ be defined by $e_1(a) = 0, e_1(b) = 1$ and let $e_2 \in E^+$ be defined by $e_2(a) = 1, e_2(b) = 0$. Then $e_1 + e_2 = e$ and $e_1 \wedge e_2 = 0$ (so e_1 and e_2 are disjoint in E). Let E_1 and E_2 be the bands generated by e_1 and e_2 respectively. Then $E_1 \oplus E_2 = E$. Furthermore any element in E_1 must be a multiple of e_1, so the only bands contained in E_1 are (0) and E_1 itself. Similarly for E_2. It follows easily that $((0), E_1, E_2, E)$ is the set of all bands in E.

In the exercise which follows now it is indicated how to prove that in the main inclusion theorem there are no inclusions in the converse direction.

Exercise 12.6. (i) We define a Dedekind σ-complete Riesz space not possessing the projection property, and therefore the space is not Dedekind complete. For this, let E be the Riesz space of all (real) bounded functions on $[0,1]$ such that $f(x) \ne f(0)$ holds for at most countably many x, with pointwise ordering. Show that if $0 \le u_n \le v$ holds for the sequence $(u_n : n = 1, 2, \ldots)$ in E, the function $\sup u_n$ exists in E (it is simply the pointwise limit of the sequence). Hence, E is Dedekind σ-complete. To show that E does not have the projection property, let (for example) A be the band consisting of all $f \in E$ vanishing on $[0, {}^1/{}_2]$, so that, therefore, A^d is the band of all $f \in E$ vanishing on $({}^1/{}_2, 1]$. Show that every f in $A \oplus A^d$ satisfies $f(0) = 0$, which implies that $A \oplus A^d \ne E$. Note that E has the principal projection property (because E is Dedekind σ-complete), so this example shows also that the principal projection property does not imply the projection property.

(ii) We now define a Riesz space possessing the projection property, but the space is not Dedekind σ-complete, and therefore not Dedekind complete.

Let E be the Riesz space of all (real) bounded functions on $[0,1]$ assuming only a finite number of different values, with pointwise ordering. For the proof that E has the projection property, let A be an arbitrary band in E. Set

$$X_1 = (x : f(x) \neq 0 \text{ for at least one } f \in A)$$

and $X_2 = [0,1]\backslash X_1$, and denote the characteristic functions of X_1 and X_2 by χ_1 and χ_2 respectively. Show that A^d consists of all $f \in E$ vanishing on X_1, and so $A = A^{dd}$ consists of all $f \in E$ vanishing on X_2. Show then that $f = f\chi_1 + f\chi_2$ with $f\chi_1 \in A$ and $f\chi_2 \in A^d$ holds for every $f \in E$, so $E = A \oplus A^d$. To show that E is not Dedekind σ-complete, take (for example) the sequence $(u_n : n = 1, 2, \ldots)$ in E, where $u_n(x) = x$ for $x = 1, {}^1/_2, \ldots, {}^1/_n$ and $u_n(x) = 0$ for all other x. Then $0 \leq u_n \leq e$ holds in E, where $e(x) = 1$ for all x. Evidently, $\sup u_n$ does not exist in E. Note that E has the principal projection property, so this example shows that the principal projection property does not imply Dedekind σ–completeness.

(iii) We finally define a Riesz space possessing the principal projection property, but neither the projection property nor the property of being Dedekind σ-complete. Let $X = (0, 1, 2, \ldots)$ and let E be the Riesz space of all (real) bounded functions on X such that $f(x) \neq f(0)$ holds for at most finitely many $x \in X$, with pointwise ordering. To prove that E has the principal projection property, let $v \in E^+$ be given and let A be the principal band generated by v. Set

$$X_1 = (x : f(x) \neq 0 \text{ for at least one } f \in A) = (x : v(x) > 0)$$

and $X_2 = X\backslash X_1$, and denote the characteristic functions of these sets by χ_1 and χ_2 respectively. If $v(0) = 0$, then X_1 consists of a finite number of points and the point $0 \in X$ is contained in X_2. If $v(0) > 0$, then X_1 contains all but a finite number of points and the point $0 \in X$ is contained in X_1. Show that in both cases $f = f\chi_1 + f\chi_2$ with $f\chi_1 \in A$ and $f\chi_2 \in A^d$ holds for every $f \in E$, so $E = A \oplus A^d$. Show, similarly as in part (ii), that E is not Dedekind σ-complete. To show that E does not have the projection property, let (for example) A be the band defined by

$$A = (f \in E : f(x) = 0 \text{ for } x = 1, 3, 5, \ldots).$$

Then

$$A^d = (f \in E : f(x) = 0 \text{ for } x = 2, 4, 6, \ldots),$$

so every $f \in A \oplus A^d$ satisfies $f(0) = 0$, and hence $A \oplus A^d$ is a proper subset of E. Precisely, show that $A \oplus A^d$ is the set of all $f \in E$ for which $f(0) = 0$.

We insert a brief interlude about the notions of the lim sup (limes superior) and the lim inf (limes inferior) of a sequence in a Riesz space. Let $(f_n : n = 1, 2, \ldots)$ be a sequence in the Riesz space E with the property that

$g_n = \sup(f_k : k \geq n)$ exists in E for $n = 1, 2, \ldots$. Evidently, $(g_n : n = 1, 2, \ldots)$ is a decreasing sequence. If there exists an element $g \in E$ such that $g_n \downarrow g$, we write $g = \limsup f_n$. Similarly, we write $h = \liminf f_n$ if $h_n = \inf(f_k : k \geq n)$ exists for $n = 1, 2, \ldots$ and $h_n \uparrow h$. If both $\limsup f_n$ and $\liminf f_n$ exist, then $\liminf f_n \leq \limsup f_n$, and if they coincide, say $\liminf f_n = \limsup f_n = f$, then f_n converges in order to f. For the proof, note that

$$(f - f_n)^+ \leq (f - h_n)^+ = f - h_n \downarrow 0,$$

$$(f - f_n)^- = (f_n - f)^+ \leq (g_n - f)^+ = g_n - f \downarrow 0,$$

so $\mid f - f_n \mid \leq (f - h_n) + (g_n - f) \downarrow 0$, i.e., f_n converges in order to f. If E is Dedekind σ-complete, there is a more complete result.

Theorem 12.7. *Let the sequence $(f_n : n = 1, 2, \ldots)$ in the Dedekind σ-complete Riesz space E be bounded (that is to say, bounded above as well as bounded below). Then $\limsup f_n$ and $\liminf f_n$ exist and f_n converges in order to f if and only if*

$$f = \limsup f_n = \liminf f_n.$$

Proof. We have only to prove that $\mid f_n - f \mid \leq u_n \downarrow 0$ implies $f = \limsup f_n = \liminf f_n$. As above, let $g_n = \sup(f_k : k \geq n)$ and $h_n = \inf(f_k : k \geq n)$ for $n = 1, 2, \ldots$. The existence of g_n and h_n is clear in view of the Dedekind σ–completeness of E. It follows from $\mid f - f_n \mid \leq u_n \downarrow 0$ that

$$f - u_n \leq h_n \leq g_n \leq f + u_n$$

for all n. Then

$$g_n \leq f + u_n \downarrow f \text{ and } g_n \downarrow \limsup f_n,$$

so $\limsup f_n \leq f$. Also

$$h_n \geq f - u_n \uparrow f \text{ and } h_n \uparrow \liminf f_n,$$

so $\liminf f_n \geq f$. Combining these results, we see that $f = \limsup f_n = \liminf f_n$. $\blacksquare$

As defined, the Archimedean Riesz space E is uniformly complete if, for every $u > 0$ in E, every u-uniform Cauchy sequence has a u-uniform limit (see section 10). In view of the result in Exercise 10.4 it is already sufficient for uniform completeness that, for every $u > 0$ in E, every monotone u-uniform Cauchy sequence in E^+ has a u-uniform limit. This enables us to prove readily that every Dedekind σ-complete Riesz space is uniformly complete.

Theorem 12.8. *If E is a Dedekind σ-complete Riesz space, then E is uniformly complete.*

Proof. Let $u > 0$ be given in the Dedekind σ-complete space E and let $0 \leq v_n$ be a u-uniform Cauchy sequence in E^+. We have to prove that the sequence has a u- uniform limit. For any $\epsilon > 0$ there exists a natural number n_ϵ such that $| v_m - v_n | \leq \epsilon u$ for $m, n \geq n_\epsilon$, so

$$v_n - \epsilon u \leq v_m \leq v_n + \epsilon u \tag{1}$$

for $m \geq n \geq n_\epsilon$. It follows that the sequence $(v_m : m = 1, 2, \ldots)$ is bounded above, and so (by the Dedekind σ–completeness) $v = \sup v_m$ exists in E^+. But then, in view of (1), we have

$$v_n - \epsilon u \leq v \leq v_n + \epsilon u$$

for all $n \geq n_\epsilon$, i.e., $v - v_n \leq \epsilon u$ as well as $v_n - v \leq \epsilon u$, and so $| v - \dot{v}_n | \leq \epsilon u$ for all $n \geq n_\epsilon$. This shows that v is the u-uniform limit of the sequence $(v_n : n = 1, 2, \ldots)$. $\blacksquare$

Dedekind σ–completeness, and even the principal projection property, is by no means necessary for uniform completeness, as shown by the space $C([0, 1])$ of all real continuous functions on the interval $[0,1]$.

Exercise 12.9. Show that there exists a Riesz space possessing the projection property but such that the space is not uniformly complete.

CHAPTER 6
Complex Riesz Spaces

13. Complex Riesz spaces

All Riesz spaces in the preceding sections are real Riesz spaces. We shall now define complex Riesz spaces and then extend a considerable part of the theory to these complex spaces. Recall first that the Cartesian product $X \times Y$ of the non-empty sets X and Y is the set of all ordered pairs (x, y) such that $x \in X$ and $y \in Y$. In the case that $X = Y = V$, where V is a real vector space, we can equip the Cartesian product $V \times V$ with a vector space structure by defining

$$(f_1, g_1) + (f_2, g_2) = (f_1 + f_2, g_1 + g_2),$$

$$(\alpha + i\beta)(f, g) = (\alpha f - \beta g, \beta f + \alpha g) \text{ for } \alpha, \beta \text{ real numbers,}$$

and the so defined complex vector space is denoted by $V + iV$. Note that $(f, 0) + (g, 0) = (f + g, 0)$ and $\alpha(f, 0) = (\alpha f, 0)$ for α real. Hence, identifying $f \in V$ and $(f, 0) \in V + iV$, the space V is embedded in $V + iV$ as a real-linear subspace. Note also that $i(g, 0) = (0, g)$ by the above definition, so

$$(f, g) = (f, 0) + (0, g) = (f, 0) + i(g, 0).$$

Hence, in view of the mentioned identification, we may now write $f + ig$ instead of (f, g). If $h = f + ig$ with f and g in V, we write $f = Reh$ and $g = Imh$.

If V is the space $\mathbb{R}$ of real numbers, then $V + iV = \mathbb{R} + i\mathbb{R}$ is the space $\mathbb{C}$ of complex numbers with the usual addition and multiplication (of course, in an expression such as αf we have to see α as a complex multiplier and f as an element of the space $\mathbb{C}$, but the outcome αf is the familiar product of the complex numbers α and f).

Let now E be a (real) Riesz space and let $E + iE$ be its complexification. The space $E + iE$ can be partially ordered coordinatewise, i.e., $f_1 + ig_1 \leq f_2 + ig_2$ whenever $f_1 \leq f_2$ and $g_1 \leq g_2$. Then $E + iE$ is a Riesz space and, for $h = f + ig$ (f and g in E), the element $\mid h \mid$ is given by $\mid h \mid = \mid f \mid + i \mid g \mid$. The role of multiplication by complex numbers, however, is completely obscured in this manner. In fact, $E + iE$ is now nothing else but the space $\mathbb{R}^2$ in disguise.

Technically speaking, $E + iE$ and $\mathbb{R}^2$ are isomorphic Riesz spaces. Moreover, in the case of complex numbers (i.e., $E + iE$ is $\mathbb{R} + i\mathbb{R} = \mathbb{C}$), the absolute value of $z = \alpha + i\beta(\alpha, \beta \in \mathbb{R})$ would be $\mid \alpha \mid + i \mid \beta \mid$ instead of $(\alpha^2 + \beta^2)^{1/2}$. The familiar absolute value $\mid z \mid = (\alpha^2 + \beta^2)^{1/2}$ is a non- negative real number, also if z is not real, and such that if z itself is real, then $\mid z \mid = z \vee (-z)$ in accordance with the definition in a real Riesz space. Analogously, if E is an arbitrary Riesz space and $h = f + ig$ is any element of $E + iE$, we wish to define an absolute value $\mid h \mid$ of h such that $\mid h \mid \in E^+$ and such that if h itself is an element of E, then $\mid h \mid = h \vee (-h)$. As we shall see, this is not always possible. It is possible, however, if E is Archimedean and uniformly complete. In general, it will not be appropriate to define $\mid h \mid$ for $h = f + ig$ by $\mid h \mid = (f^2 + g^2)^{1/2}$, simply because in most Riesz spaces squares and square roots are not defined. There is an alternative definition, however. In the case of complex numbers we can define $\mid z \mid$, for z a complex number, by

$$\mid z \mid = \sup\{Re(ze^{-i\theta}) : 0 \leq \theta \leq 2\pi\}.$$

Similarly, if E is a Riesz space and $h = f + ig \in E + iE$ with f and g in E, we define

$$\mid h \mid = \sup\{Re(he^{-i\theta}) : 0 \leq \theta \leq 2\pi\},$$

whenever this supremum exists. Note that $\mid h \mid \geq 0$ in this case, since the elements $Re(he^{-i\theta})$ occur in pairs, for θ and $\theta + \pi$, and such that

$$Re(he^{-i(\theta+\pi)}) = -Re(he^{-i\theta}).$$

Hence, if the supremum exists, then

$$\mid h \mid = \sup\{\mid Re(he^{-i\theta}) \mid : 0 \leq \theta \leq 2\pi\},$$

and so $\mid h \mid \geq 0$. For the case that $h \in E$, we have $Re(he^{-i\theta}) = h \cos \theta$, so the supremum exists and is equal to the already existing supremum $\mid h \mid = h \vee (-h)$. Before proceeding, observe that for $h = f + ig$ with f and g in E, we have

$$Re(he^{-i\theta}) = f \cos \theta + g \sin \theta.$$

Example 13.1. We consider the complexification of the Riesz space $C([0, 1])$ of all real continuous functions on the interval [0,1]. If $h = f + ig$ is a complex continuous function on [0,1] with f and g real, then the familiar absolute value $\mid h \mid$ satisfies

$$\mid h \mid = (f^2 + g^2)^{1/2} \in C([0, 1]),$$

but also

$$\mid h \mid = \sup_{\theta}[Re(he^{-i\theta})] = \sup_{\theta}(f \cos \theta + g \sin \theta),$$

where the supremum must be taken with respect to the (pointwise) partial ordering in the Riesz space $C([0,1])$. That this supremum exists and is equal to the function $(f^2 + g^2)^{1/2}$ is seen by observing that for any fixed $x \in [0,1]$ the number $Re[h(x)e^{-i\theta}]$ is less than or equal to the number

$$| h(x) |= \{f^2(x) + g^2(x)\}^{1/2},$$

with equality holding for exactly one value of θ (depending on x) if $h(x) \neq 0$, and equality for all θ if $h(x) = 0$. It follows, therefore, that the function $| h |$, defined by $| h | (x) =| h(x) |$ for every $x \in [0,1]$, is a member of $C([0,1])$ and $| h |$ is on $[0,1]$ the pointwise supremum of the functions $Re(he^{-i\theta}) = f\cos\theta + g\sin\theta$, and so $| h |$ is then also the supremum of this set of functions with respect to the ordering in $C([0,1])$.

Example 13.2. The argument in the preceding example shows that if $h = f + ig$ (f and g real) is any complex function on a point set X, then $p = (f^2 + g^2)^{1/2}$ is the pointwise supremum of the set of functions

$$(f\cos\theta + g\sin\theta : 0 \leq \theta \leq 2\pi).$$

Hence, if E is a Riesz space of real functions on X (with pointwise ordering) and $h = f + ig \in E + iE$, then the function $p = (f^2 + g^2)^{1/2}$ satisfies

$$p = \sup_\theta(f\cos\theta + g\sin\theta) = \sup_\theta[Re(he^{-i\theta})]$$

with respect to the ordering in E if and only if $p \in E$. The function p is not always a member of E. By way of example, let E be the space of all real continuous functions f on $[0,1]$ such that $f(x) = \alpha + \beta x$ (α and β real) for all x in a neighbourhood of zero (the neighbourhood depending on f). Then E is a Riesz space with respect to the familiar pointwise ordering (i.e., the value of $f \vee g$ at the point x is the maximum of $f(x)$ and $g(x)$). The function $h(x) = (1 + x) + i(1 - x)$ is a member of $E + iE$, but

$$p(x) = \{(1 + x)^2 + (1 - x)^2\}^{1/2} = (2 + 2x^2)^{1/2}$$

is not a member of E. On the other hand, if E is the Riesz space of all real continuous functions on a topological space X or if E is the Riesz space of all real functions on an arbitrary point set X (pointwise ordering) and $h = f + ig \in E + iE$, then $| h |= \sup_\theta[Re(he^{-i\theta})]$ exists in E^+ and $| h |$ satisfies $| h | (x) = \{f^2(x) + g^2(x)\}^{1/2}$ for all $x \in X$.

Finally, let (X, μ) be a σ-finite measure space and let E be an ideal in the space $L_0(X, \mu)$ of all real μ-measurable functions on X (with the usual identification of μ-almost equal functions). If $h = f + ig \in E + iE$ and $| h | (x) = \{f^2(x) + g^2(x)\}^{1/2}$ for all $x \in X$, then $| h |\in E$. Note here that although f^2 and g^2 may not be members of E, we have $| h |\in E$ on account of $| h |\leq| f | + | g |$. Hence

$$\mid h \mid = (f^2 + g^2)^{1/2} = \sup_{\theta}[Re(he^{-i\theta})]$$

holds with respect to the ordering in E.

We prove now that if the Riesz space E is Archimedean and uniformly complete, then

$$\sup(f\cos\theta + g\sin\theta : 0 \leq \theta \leq 2\pi)$$

exists for all f and g in E. The proof is divided into several steps. Recall first that the sequence $(x_n : n = 1, 2, \ldots)$ of points in the interval $[a, b]$ is said to be dense in $[a, b]$ if every subinterval of $[a, b]$ contains at least one point (and hence infinitely many points) of the sequence. Alternatively, for any $x_0 \in [a, b]$ there is a subsequence of (x_n) converging to x_0.

Lemma 13.3. *Let E be an Archimedean and uniformly complete Riesz space. For $h = f + ig \in E + iE$ and $0 \leq \theta \leq 2\pi$, let $h(\theta) = \mid f\cos\theta + g\sin\theta \mid$ and $e = \mid f \mid + \mid g \mid$. Then, if $\theta_n(n = 1, 2, \ldots)$ converges to θ, the sequence $h(\theta_n)$ converges e-uniformly to $h(\theta)$. Also, if $(\theta_n : n = 1, 2, \ldots)$ is dense in $[0, 2\pi]$ and $s = \sup_n h(\theta_n)$ exists in E, then $s = \sup\{h(\theta) : 0 \leq \theta \leq 2\pi\}$.*

Proof. Let θ_n converge to θ. Then, if $\epsilon > 0$ is given, we have

$$\mid h(\theta) - h(\theta_n) \mid \leq \mid f(\cos\theta - \cos\theta_n) + g(\sin\theta - \sin\theta_n) \mid$$

$$\leq \epsilon(\mid f \mid + \mid g \mid) = \epsilon e$$

for all sufficiently large n. This shows that $h(\theta_n)$ converges e-uniformly to $h(\theta)$.

Assume now that $(\theta_n : n = 1, 2, \ldots)$ is dense in $[0, 2\pi]$ and let $s = \sup_n h(\theta_n)$. Note first now that the set of all $h(\theta)$, for $0 \leq \theta \leq 2\pi$, is bounded above by $e = \mid f \mid + \mid g \mid$, so e is an upper bound of the set. We have to show that s is the least upper bound of the set of all $h(\theta)$. Obviously, if s_0 is an arbitrary upper bound, we have $s_0 \geq h(\theta_n)$ for all n, so $s_0 \geq \sup_n h(\theta_n) = s$. We prove now that $s \geq h(\theta)$ for all θ, so s is an upper bound of the set of all $h(\theta)$, and s has the property (as proved above) that $s \leq s_0$ for any other upper bound s_0. It will follow then that s is the least upper bound, i.e., $s = \sup_\theta h(\theta)$. For the proof that $s \geq h(\theta)$, let $(\tau_k : k = 1, 2, \ldots)$ be a subsequence of $(\theta_n : n = 1, 2, \ldots)$ such that τ_k converges to θ, so $h(\tau_k)$ converges e-uniformly to $h(\theta)$. Then $h(\tau_k) \to h(\theta)$ in order (see Theorem 10.3(i)). Every $h(\tau_k)$ satisfies $h(\tau_k) \leq s$ and $h(\tau_k) \to h(\theta)$. It follows that $h(\theta) \leq s$ (see Theorem 10.2(iv)). This concludes the proof. ∎

Theorem 13.4. *If the Riesz space E is Archimedean and uniformly complete, then*

$$| h | = \sup\{Re(he^{-i\theta}) : 0 \le \theta \le 2\pi\} = \sup(f \cos\theta + g \sin\theta : 0 \le \theta \le 2\pi)$$

exists in E for every $h = f + ig \in E + iE$.

Proof. As in the lemma, let $h(\theta) = | f \cos\theta + g \sin\theta |$ and $e = | f | + | g |$. We have to show that $\sup(h(\theta) : 0 \le \theta \le 2\pi)$ exists. For $n = 1, 2, \ldots$, let P_n be the partition of $[0, 2\pi]$, defined by

$$P_n = (\theta_1, \ldots, \theta_{2^n}) \text{ with } \theta_k = 2\pi k/2^n \text{ for } k = 1, \ldots, 2^n,$$

and let $h(P_n) = \sup\{h(\theta_1), \ldots, h(\theta_{2^n})\}$. For every n, the partition P_{n+1} is a refinement of the partition P_n, and hence the sequence $(h(P_n) : n = 1, 2, \ldots)$ in E is increasing. We prove that the sequence is an e-uniform Cauchy sequence. For this purpose, let m and n be natural numbers such that $m < n$, and let P_m and P_n be the corresponding partitions. Hence

$$P_m = (\theta_1, \ldots, \theta_{2^m}) \text{ with } \theta_k = 2\pi k/2^m \text{ for } k = 1, \ldots, 2^m,$$

$$P_n = (\theta'_1, \ldots, \theta'_{2^n}) \text{ with } \theta = 2\pi j/2^n \text{ for } j = 1, \ldots, 2^n.$$

To compare $h(P_m)$ and $h(P_n)$ more easily, we write P_m somewhat differently, without changing $h(P_m)$. The modified P_m (call it P_m^*) consists of the same points as P_m, but each point repeated 2^{n-m} times. Hence

$$P_m^* = (\theta_1^*, \ldots, \theta_{2^n}^*)$$

with $\theta_1^* = \cdots = \theta_{2^{n-m}}^* = \theta_1, \theta_{2^{n-m}+1}^* = \cdots = \theta_{2^{n-m+1}}^* = \theta_2$, and so on. Note that $h(P_m^*) = h(P_m)$, as asserted. Now recall Exercise 6.7, according to which

$$| \sup(f_1, \ldots, f_n) - \sup(g_1, \ldots, g_n) | \le \sup(| f_1 - g_1 |, \ldots, | f_n - g_n |)$$

for arbitrary elements $f_1, \ldots, f_n$ and $g_1, \ldots, g_n$ in a Riesz space. Applying this to $h(P_n) - h(P_m^*)$, we get

$$| h(P_n) - h(P_m^*) | \le \sup(| h(\theta'_j) - h(\theta_j^*) | : j = 1, \ldots, 2^n).$$

Observe now that $| \theta'_j - \theta_j^* | \le 2\pi/2^m$ for all j. Hence, as shown in the first lines of the proof of the preceding lemma, for any $\epsilon > 0$ we have $| h(\theta'_j) - h(\theta_j^*) | \le \epsilon e$ for all j if m is sufficiently large, and therefore also $| h(P_n) - h(P_m) | \le \epsilon e$ for $n \ge m$ and m sufficiently large. It follows that the sequence $(h(P_n) : n = 1, 2, \ldots)$ is e-uniformly Cauchy. Hence, since E is uniformly complete, the uniform limit s of the sequence exists in E. Since this is also the order limit, and since the sequence is increasing, we have $s = \sup_n h(P_n)$. Observe finally that the union of all partition points of all P_n is a countable set (say $\tau_1, \tau_2, \ldots$) lying dense in $[0, 2\pi]$. Since $s = \sup_n h(P_n) = \sup_k h(\tau_k)$, it follows now from the preceding lemma that

$$s = \sup(h(\theta) : 0 \leq \theta \leq 2\pi).$$

This is the desired result.   ∎

In the remaining part of the present section we shall assume that E is an Archimedean Riesz space having the property that

$$\mid h \mid = \sup\{Re(he^{-i\theta}) : 0 \leq \theta \leq 2\pi\} = \sup(f \cos\theta + g \sin\theta : 0 \leq \theta \leq 2\pi)$$

exists for every $h = f + ig \in E + iE$. As seen above, a sufficient condition for E to possess these properties is that E is Archimedean and uniformly complete. Since any Dedekind σ-complete space is Archimedean and uniformly complete (see Theorem 12.8), Dedekind σ-completeness is certainly a sufficient condition.

We shall prove now that the absolute value in $E + iE$ has similar properties as the familiar absolute value for complex numbers.

Theorem 13.5. *If $h = f + ig \in E + iE$, then*

$$\mid f \mid \leq \mid h \mid, \mid g \mid \leq \mid h \mid \quad and \quad \mid h \mid \leq \mid f \mid + \mid g \mid .$$

Furthermore, $\mid h \mid = 0$ if and only $h = 0$, $\mid \lambda h \mid = \mid \lambda \mid \cdot \mid h \mid$ for any complex number λ, and the triangle inequality

$$\mid h_1 + h_2 \mid \leq \mid h_1 \mid + \mid h_2 \mid \quad and \quad \mid \mid h_1 \mid - \mid h_2 \mid \mid \leq \mid h_1 - h_2 \mid$$

holds for h_1 and h_2 in $E + iE$.

Proof. Choosing $\theta = 0$ and $\theta = \pi/2$ in

$$\mid h \mid = \sup(\mid f \cos\theta + g \sin\theta \mid : 0 \leq \theta \leq 2\pi),$$

we get $\mid f \mid \leq \mid h \mid$ and $\mid g \mid \leq \mid h \mid$. Since $\mid f \cos\theta + g \sin\theta \mid \leq \mid f \mid + \mid g \mid$ for every θ, we get $\mid h \mid \leq \mid f \mid + \mid g \mid$. The proof for $\mid h_1 + h_2 \mid \leq \mid h_1 \mid + \mid h_2 \mid$ is immediate and the second form for the triangle inequality follows by observing that

$$\mid h_1 \mid - \mid h_2 \mid \leq \mid h_1 - h_2 \mid \quad and \quad \mid h_2 \mid - \mid h_1 \mid \leq \mid h_2 - h_1 \mid = \mid h_1 - h_2 \mid . ∎$$

We proceed to the proof of some further properties, but first we present a definition. The elements h_1 and h_2 in $E + iE$ are said to be *disjoint* if $\mid h_1 \mid \perp \mid h_2 \mid$, i.e., $\mid h_1 \mid \wedge \mid h_2 \mid = 0$.

Theorem 13.6. (i) *If h_1 and h_2 in $E + iE$ are disjoint, then*

$$\mid h_1 - h_2 \mid = \mid h_1 + h_2 \mid = \mid h_1 \mid + \mid h_2 \mid = \mid\mid h_1 \mid - \mid h_2 \mid\mid = \mid h_1 \mid \vee \mid h_2 \mid .$$

(ii) If $h = f + ig \in E + iE$ with f and g positive, then

$$\mid h \mid = \sup(f \cos\theta + g \sin\theta : 0 \leq \theta \leq \pi/2).$$

(iii) If $h = f + ig$ and $h^* = \mid f \mid + i \mid g \mid$ with f and g in E, then $\mid h \mid = \mid h^* \mid$.

(iv) If $h_1 = f_1 + ig_1$ and $h_2 = f_2 + ig_2$ with $\mid f_1 \mid \leq \mid f_2 \mid$ and $\mid g_1 \mid \leq \mid g_2 \mid$, then $\mid h_1 \mid \leq \mid h_2 \mid$.

Proof. (i) It was proved in Corollary 5.4 that for f and g in E we have $\mid f \mid \wedge \mid g \mid = 0$ if and only if $\mid f + g \mid = \mid f \mid \vee \mid g \mid$. Hence, in the present case where $\mid h_1 \mid \wedge \mid h_2 \mid = 0$, we have

$$\mid h_1 \mid + \mid h_2 \mid = \mid\mid h_1 \mid - \mid h_2 \mid\mid = \mid h_1 \mid \vee \mid h_2 \mid .$$

Furthermore

$$\mid h_1 - h_2 \mid \leq \mid h_1 \mid + \mid h_2 \mid = \mid\mid h_1 \mid - \mid h_2 \mid\mid \leq \mid h_1 - h_2 \mid,$$

so $\mid h_1 - h_2 \mid = \mid h_1 \mid + \mid h_2 \mid$. Then also $\mid h_1 + h_2 \mid = \mid h_1 \mid + \mid h_2 \mid$.

(ii) Since f and g are positive, we may restrict ourselves to $0 \leq \theta \leq \pi/2$.

(iii) Let θ be fixed temporarily in $[0, \pi/2]$. Besides the element $f \cos\theta + g \sin\theta$ we consider the corresponding elements for $-\theta, \pi - \theta$ and $-(\pi - \theta)$, and we compute the supremum p_θ of these four elements. Taking first the supremum for θ and $-\theta$, we get

$$\sup(f \cos\theta + g \sin\theta, f \cos\theta - g \sin\theta) =$$

$$f \cos\theta + (g \vee -g)(\sin\theta) =$$

$$f \cos\theta + \mid g \mid \sin\theta.$$

Similarly, the supremum of the two remaining elements is

$$f \cos(\pi - \theta) + \mid g \mid \sin(\pi - \theta) = -f \cos\theta + \mid g \mid \sin\theta.$$

Hence

$$p_\theta = (f \cos\theta) \vee (-f \cos\theta) + \mid g \mid \sin\theta = \mid f \mid \cos\theta + \mid g \mid \sin\theta.$$

Since it is evident that $\mid h \mid = \mid f + ig \mid = \sup(p_\theta : 0 \leq \theta \leq \pi/2)$ and since also

$$\mid h^* \mid = \mid\mid f \mid + i \mid g \mid\mid = \sup(p_\theta : 0 \leq \theta \leq \pi/2)$$

by part (ii), it follows that $\mid h \mid = \mid h^* \mid$.

(iv) Let $h_1 = f_1 + ig_1$ and $h_2 = f_2 + ig_2$ with $\mid f_1 \mid \leq \mid f_2 \mid$ and $\mid g_1 \mid \leq \mid g_2 \mid$. Then

$$\mid h_1 \mid = \sup(\mid f_1 \mid \cos\theta + \mid g_1 \mid \sin\theta : 0 \leq \theta \leq \pi/2) \leq$$

$$\sup(\mid f_2 \mid \cos\theta + \mid g_2 \mid \sin\theta : 0 \leq \theta \leq \pi/2) = \mid h_2 \mid . \qquad \blacksquare$$

Corollary 13.7. *The elements $h_1 = f_1 + ig_1$ and $h_2 = f_2 + ig_2$ are disjoint if and only if f_1 and g_1 are disjoint from both f_2 and g_2. Also, if $h = \Sigma_{k=1}^n \lambda_k h_k$ with all λ_k complex and all h_k mutually disjoint, then*

$$\mid h \mid = \mid \Sigma_1^n \lambda_k h_k \mid = \Sigma_1^n \mid \lambda_k \mid \cdot \mid h_k \mid .$$

The subset A of $E + iE$ is called an *ideal* if A is a complex linear subspace of $E + iE$ and if any element of $E + iE$ majorized in absolute value by an element of A is itself an element of A. This is the same definition, therefore, as in the real case. The set of all real elements in the ideal A will be denoted by A_r, i.e., $A_r = A \cap E$. Evidently, if $h = f + ig$ (with f and g in E) is a member of A, then $\mid h \mid$ is a member of A_r (by definition), and so f and g are members of A_r as well (since $\mid f \mid$ and $\mid g \mid$ are majorized by $\mid h \mid$).

Theorem 13.8. *If A is an ideal in $E + iE$, then A_r is an ideal in E, and $A = A_r + iA_r$. Conversely, if B is an ideal in E, then $B + iB$ is an ideal in $E + iE$, and $(B + iB)_r = B$.*

Proof. Let A be an ideal in $E + iE$. It is easy to see that $A_r = A \cap E$ is an ideal in E. Also, for f and g in E, we have $f + ig \in A$ if and only if f and g are members of A_r. Hence $A = A_r + iA_r$.

Conversely, let B be an ideal in E and $A = B + iB$. Then A is a complex linear subspace of $E + iE$. For the proof that A is an ideal, assume that $h = f + ig$ and $h_1 = f_1 + ig_1$ with $h \in A$ and $\mid h_1 \mid \leq \mid h \mid$. We have to prove that $h_1 \in A$. The elements f and g are members of B, so $\mid f \mid + \mid g \mid \in B$. Furthermore, $\mid f_1 \mid \leq \mid h_1 \mid \leq \mid h \mid \leq \mid f \mid + \mid g \mid$. It follows that $f_1 \in B$. Similarly, $g_1 \in B$. Hence $h_1 = f_1 + ig_1 \in B + iB = A$. $\qquad \blacksquare$

If A and B are ideals in $E + iE$, the algebraic sum $A + B$ satisfies

$$A + B = A_r + iA_r + B_r + iB_r = (A_r + B_r) + i(A_r + B_r).$$

This shows that $A + B$ is an ideal in $E + iE$ and $(A + B)_r = A_r + B_r$. Furthermore, we have $A \perp B$ (i.e., $\mid f \mid \perp \mid g \mid$ for all $f \in A$ and all $g \in B$) if and only if $A \cap B = (0)$. In this case, let f and g be elements of $A \oplus B$ having the decompositions $f = f_1 + f_2$ and $g = g_1 + g_2$. Then $\mid f \mid$ and $\mid g \mid$ have the decompositions $\mid f \mid = \mid f_1 \mid + \mid f_2 \mid$ and $\mid g \mid = \mid g_1 \mid + \mid g_2 \mid$ (see Corollary 13.7). Hence, if $\mid f \mid \leq \mid g \mid$, then $\mid f_1 \mid \leq \mid g_1 \mid$ and $\mid f_2 \mid \leq \mid g_2 \mid$ (see Theorem 7.6).

The ideal A in $E + iE$ is called a *band* if the real part of A_r is a band in E. For any non-empty subset D of $E + iE$ the *disjoint complement* D^d of D is defined, similarly as in the real case, by

$$D^d = (h \in E + iE : | h | \perp | f | \text{ for all } f \in D).$$

It is evident (from the disjointness properties in Theorem 8.1) that D^d is an ideal in $E + iE$. We prove that D^d is a band. For this purpose, let P be a set of positive elements in $(D^d)_r$ such that $h_0 = \sup P$ exists in E. We have to prove that $h_0 \in (D^d)_r$. Every $h \in P$ satisfies $0 \le h \perp | f |$ for all $f \in D$, so $h_0 = \sup P$ satisfies $h_0 \perp | f |$ for all $f \in D$ (see Theorem 8.1(vii)). This shows that $h_0 \in D^d$. More precisely, since h_0 is real, $h_0 \in (D^d)_r$.

For any non-empty subset D of E we have to distinguish between the disjoint complement D^d of D in $E + iE$ and the disjoint complement $D^\perp$ of D in E. It is evident that $D^\perp = (D^d)_r$. Several properties of $A_1^{dd}, \ldots, A_n^{dd}$ for ideals $A_1, \ldots, A_n$ will be discussed in the next exercise. It will be used explicitly that E is Archimedean.

Exercise 13.9. (i) If A is an ideal in $E + iE$, then $A^d = (A_r)^d$, and hence

$$(A^d)_r = ((A_r)^d)_r = (A_r)^\perp.$$

Furthermore $(A^{dd})_r = (A_r)^{\perp\perp}$. Prove this.

(ii) Show that every band A in $E + iE$ satisfies $A = A^{dd}$.

(iii) Show that if $A_1, \ldots, A_n$ are ideals in $E + iE$, then

$$(\cap_{j=1}^n A_j)^{dd} = \cap_{j=1}^n A_j^{dd}.$$

Hint: (i) Let A be an ideal in $E + iE$. It follows from $A_r \subseteq A$ that $A^d \subseteq (A_r)^d$. Conversely, if $f \in (A_r)^d$, then $| f | \perp | g |$ for all $g \in A_r$. It follows that $| f | \perp | h |$ for all $h \in A$, so $f \in A^d$. This shows that $A^d = (A_r)^d$. But then

$$(A^d)_r = ((A_r)^d)_r = (A_r)^\perp.$$

Replacing A by A_d in this formula, we obtain

$$(A^{dd})_r = ((A^d)_r)^\perp = ((A_r)^\perp)^\perp = (A_r)^{\perp\perp}.$$

(ii) Let A be a band in $E + iE$. The real part A_r is a band in E, so $(A_r)^{\perp\perp} = A_r$ since E is Archimedean. It follows (in view of part (i)) that $(A^{dd})_r = A_r$, i.e., the bands A^{dd} and A have the same real parts. This implies that $A^{dd} = A$.

(iii) In Exercise 8.6(ii) it was indicated how to prove that if $A_1, \ldots, A_n$ are ideals in E, then $(\cap_1^n A_j)^{\perp\perp} = \cap_1^n A_j^{\perp\perp}$. The proof for the complex case follows from this result. Let $A_1, \ldots, A_n$ be ideals in $E + iE$. Then $\cap_1^n A_j$ is also an ideal in $E + iE$, so

$$((\cap_1^n A_j)^{dd})_r = ((\cap_1^n A_j)_r)^{\perp\perp}$$

by part (i). Using the mentioned result for the real case and part (i) once more, the right hand side becomes

$$((\cap_1^n A_j)_r)^{\perp\perp} = \cap_1^n((A_j)_r)^{\perp\perp} = \cap_1^n(A_j^{dd})_r = (\cap_1^n A_j)_r.$$

Hence $((\cap_1^n A_j)^{dd})_r = (\cap_1^n A_j^{dd})_r$, and so $(\cap_1^n A_j)^{dd} = \cap_1^n A_j^{dd}$.

The ideal A in $E + iE$ is a subset of the band B if and only if the ideal A_r in E is a subset of the band B_r. It follows that B is the band generated by A (i.e., B is the smallest band containing A) if and only if B_r is the band generated by A_r. Now, since E is Archimedean, the band generated by A_r is $(A_r)^{\perp\perp}$ (see Exercise 9.8). Hence, B is the band generated by A if and only if $B_r = (A_r)^{\perp\perp}$. But $(A_r)^{\perp\perp} = (A^{dd})_r$ by part (i) of the last theorem. It follows, finally, that B is the band generated by A if and only if $B_r = (A^{dd})_r$, i.e., if and only if $B = A^{dd}$. In particular, the band generated by A is the space $E + iE$ itself (i.e., $A^{dd} = E + iE$) if and only if $(A^{dd})_r = E$, i.e., $(A_r)^{\perp\perp} = E$, that is to say, if and only if the band generated by A_r in E is the space E itself.

The band A in $E + iE$ is called a *projection band* if A_r is a projection band in E. In this case it follows from $A_r \oplus (A_r)^d = E$ that $A \oplus A^d = E + iE$ (where we have used that $(A_r)^\perp = (A^d)_r$). If $h = h_1 + h_2$ is the decomposition of an arbitrary $h \in E + iE$ into $h_1 \in A$ and $h_2 \in A^d$, then $| h |=| h_1 | + | h_2 |$ is the decomposition of $| h |$ (see Corollary 13.7), and so (by Theorem 11.4(i))

$$| h_1 |= \sup(| w |: w \in A_r, | w |\leq| h |).$$

Observe now that the supremum on the right is equal to

$$\sup(| v |: v \in A, | v |\leq| h |).$$

Indeed, for any $v \in A$ such that $| v |\leq| h |$, the element $w =| v |$ satisfies $w \in A_r$ and $| w |\leq| h |$. Hence

$$| h_1 |= \sup(| v |: v \in A, | v |\leq| h |), \tag{1}$$

$$| h_2 |= \sup(| w |: w \in A^d, | w |\leq| h |).$$

Conversely, if A is a band in $E + iE$ with the property that every $h \in E + iE$ admits a decomposition $h = h_1 + h_2$ such that $| h_1 |$ and $| h_2 |$ satisfy (1), then $| h_1 |\in A_r$ and $| h_2 |\in (A_r)^d$, so $h_1 \in A$ and $h_2 \in A^d$. It follows that in this case A is a projection band.

Finally, note that if E has the projection property, then $E + iE$ has the projection property, i.e., $A \oplus A^d = E + iE$ holds for every band A in $E + iE$.

Exercise 13.10. (i) Let X be a point set consisting of two points and let E be the Riesz space of all real functions on X, with pointwise ordering. Given $h = f + ig$ in $E + iE$, determine $|\, h\,|$.

(ii) In Example 12.5(v) we have defined the Riesz space E of all real linear functions f on the interval $[a, b]$, so $f(x) = \alpha x + \beta$ (with α, β real) for $x \in [a, b]$. Given $h = f + ig$ in $E + iE$, determine $|\, h\,|$.

CHAPTER 7
Normed Riesz Spaces and Banach Lattices

14. Normed Spaces and Banach Spaces

Let V be a real or complex vector space and assume that to each element $f \in V$ there is assigned a real number $\| f \|$ such that

(i) $\| f \| \geq 0$ for every $f \in V$ and $\| f \| = 0$ if and only if $f = 0$,

(ii) $\| \lambda f \| = | \lambda | \cdot \| f \|$ for every $f \in V$ and every (real or complex) number λ,

(iii) $\| f + g \| \leq \| f \| + \| g \|$ for all f and g in V (triangle inequality).

The number $\| f \|$ is now called the *norm* of f and V is said to be a *normed vector space* (or briefly a *normed space*). Writing now $d(f,g) = \| f - g \|$, it follows immediately that $d(f,g) = d(g,f) \geq 0$ for all f,g and $d(f,g) = 0$ if and only if $f = g$. Furthermore

$$d(f,g) \leq d(f,h) + d(h,g)$$

for all f, g and h. This shows that $d(f,g)$, as a function of f and g, has all the familiar properties of a distance. More formally stated, V is a *metric space* with respect to $d(f,g)$ as the *distance* between f and g. Several definitions, familiar for ordinary twodimensional or threedimensional space, carry over to the present situation. For any $f_0 \in V$ and $r > 0$, the set $(f : \| f - f_0 \| < r)$ is the *open ball* with centre f_0 and radius r. The subset D of V is called an *open set* if for each point f in D there exists an open ball with centre f such that the ball is completely contained in D. It is easy to see (using the triangle inequality) that every open ball is an open set. Any subset F of V such that the complement $V \backslash F$ is open is said to be a *closed set*. The empty subset of V is considered to be open, so V itself and the empty set are both open and closed. The sequence $(f_n : n = 1, 2, \ldots)$ in V is said to *converge in norm* to f if $\| f - f_n \| \to 0$ and the sequence is called a *norm Cauchy sequence* if for every $\epsilon > 0$ there exists a natural number $n(\epsilon)$ such that $\| f_m - f_n \| < \epsilon$ for all $m, n \geq n(\epsilon)$. If, conversely, every norm Cauchy sequence is converging in norm, i.e., if V is *norm complete*, the space V is called a *Banach space* (after the Polish mathematician S. Banach who, in the years around 1930, extensively investigated the properties of these spaces).

We prove some simple lemmas.

Lemma 14.1. *The subset F of V is norm closed if and only if it follows from $f_n \in F$ for $n = 1, 2, \ldots$ and $\| f - f_n \| \to 0$ that $f \in F$.*

Proof. Assume F is closed, i.e., $D = V \backslash F$ is open. Let $f_n \in F$ for $n = 1, 2, \ldots$ and $\| f - f_n \| \to 0$. We have to show that $f \in F$. If not, then $f \in D = V \backslash F$, so there exists an open ball with centre f completely contained in D, and therefore no f_n is an element of this ball (because all f_n are in F). This contradicts the hypothesis that $\| f - f_n \| \to 0$. Hence $f \in F$.

Conversely, assume that if $f_n \in F$ for $n = 1, 2, \ldots$ and $\| f - f_n \| \to 0$, then $f \in F$. We have to show that $D = V \backslash F$ is open. If not, there exists a point $f_0 \in D$ such that, for $n = 1, 2, \ldots$, the open ball with centre f_0 and radius n^{-1} contains a point of F, say $f_n \in F$. But then $\| f_0 - f_n \| \to 0$ with $f_n \in F$ for $n = 1, 2, \ldots$, so $f_0 \in F$ by hypothesis. Contradiction. Hence $D = V \backslash F$ is open, i.e., F is closed. ∎

Another lemma, which is frequently useful, is as follows.

Lemma 14.2. *If $(f_n : n = 1, 2, \ldots)$ is a norm Cauchy sequence and there exists a subsequence $(f_{n_k} : n_1 < n_2 < \cdots)$ converging to f, then f_n converges to f.*

Proof. Let $\epsilon > 0$ be given. There exists n_0 such that $\| f_m - f_n \| < \epsilon/2$ for $m, n \geq n_0$ and there exists k_0 such that $\| f - f_{n_k} \| < \epsilon/2$ for all $k \geq k_0$. Hence, for all $n \geq n_0$ and $n_k \geq \max(n_0, n_{k_0})$, we have

$$\| f - f_n \| \leq \| f - f_{n_k} \| + \| f_{n_k} - f_n \| < (\epsilon/2) + (\epsilon/2) = \epsilon,$$

so $\| f - f_n \| \to 0$. ∎

The series $\Sigma_1^\infty f_n$, with all f_n elements of the normed space V, is said to converge in norm to $f \in V$ if the sequence of partial sums $s_n = \Sigma_{k=1}^n f_k$ converges to f in norm. The series is said to *converge absolutely* if $\Sigma_1^\infty \| f_n \|$ converges. In this case, the sequence of partial sums $s_n = \Sigma_1^n f_k$ is a norm Cauchy sequence since

$$\| s_m - s_n \| \leq \Sigma_{n+1}^\infty \| f_k \| \to 0 \text{ for } m > n \to \infty.$$

Therefore, if V is a Banach space and $\Sigma \| f_n \|$ converges, then Σf_n converges (in norm). Conversely, if V is a normed space having the property that convergence of $\Sigma \| f_n \|$ implies convergence of Σf_n, then V is a Banach space. To prove this, assume that $(f_n : n = 1, 2, \ldots)$ is a Cauchy sequence. There exists a subsequence $(f_{n_k} : k = 1, 2, \ldots)$ such that $n_1 < n_2 < \ldots$ and $\| f_{n_{k+1}} - f_{n_k} \| \leq 2^{-k}$ for all k. Then $\Sigma(f_{n_{k+1}} - f_{n_k})$ converges by hypothesis, i.e., the sequence $(f_{n_k} : k = 1, 2, \ldots)$ converges to some element $f_0 \in V$. Hence, by the last lemma, f_n converges to f_0.

Exercise 14.3. Show that if f_n converges in norm to f, then the limit f is uniquely determined.

Hint: If f_n converges to f as well as to g, then

$$\| f - g \| \leq \| f - f_n \| + \| f_n - g \| \to 0 \text{ as } n \to \infty.$$

15. Normed Riesz Spaces and Banach Lattices

Let E be a (real) Riesz space, equipped with a norm. The norm in E is called a *Riesz norm* if $| f | \leq | g |$ in E implies $\| f \| \leq \| g \|$. Note that this implies that for any $f \in E$ the elements f and $| f |$ have the same norm. Any Riesz space, equipped with a Riesz norm, is called a *normed Riesz space*. If the normed Riesz space E is norm complete (i.e., if every norm Cauchy sequence has a norm limit), E is called a *Banach lattice*. As a first result, observe immediately that any normed Riesz space is Archimedean. Indeed, if $0 \leq nu \leq v$ for $n = 1, 2, \ldots$ in E, then $\| u \| \leq n^{-1} \| v \|$ for $n = 1, 2, \ldots$, so $\| u \| = 0$, i.e., $u = 0$. Also, note already that if f_n converges in norm to f, then $| f_n |$ converges in norm to $| f |$ (since $\| | f | - | f_n | \| \leq \| f - f_n \|$).

We shall now compare order convergence, relatively uniform convergence and norm convergence. For this purpose, let E be an Archimedean Riesz space, not yet immediately a normed Riesz space. Recall that $f_n \to f, f_n \to f(u\text{-un})$ and $f_n \to f$ (un) denote that f_n converges to f in order, u-uniformly or relatively uniformly respectively (see section 10). In a normed Riesz space norm convergence of f_n to f is denoted by $f_n \to f$ (norm). In all these cases of convergence the limit f is uniquely determined (for u-uniform convergence and relatively uniform convergence because E is Archimedean). Furthermore $f_n \to f$ (un) implies $f_n \to f$ (see Theorem 10.3(i)).

The subset D of E is said to be *order closed* or *un-closed* respectively if it follows from $f_n \in D(n = 1, 2, \ldots)$ and $f_n \to f$ (in order or un) that $f \in D$. A similar statement holds for a norm closed set (as shown in Lemma 14.1). The three methods of convergence have some properties in common, as will be proved in the following theorem.

Theorem 15.1. *(i) If $f_n \to f$ and $g_n \to g$ (in order, un or in norm), then $f_n \vee g_n \to f \vee g$ and $f_n \wedge g_n \to f \wedge g$. In particular, $f_n^+ \to f^+, f_n^- \to f^-$ and $| f_n | \to | f |$. The mappings $f \to f^+, f \to f^-$ and $f \to | f |$ are, therefore, sequentially continuous mappings from E into itself.*

(ii) If $f_n \to f$ (in order, un or in norm) and $f_n \geq g$ for all n, then $f \geq g$. Hence, if $f_n \to f$ and $f_n \geq 0$ for all n, then $f \geq 0$. This shows that the positive cone E^+ is closed.

(iii) If D is a subset of E and $f_n \to f$ (in order, un or in norm) such that $f_n \perp g$ for all $g \in D$ and all n, then $f \perp g$ for all $g \in D$. Hence, every disjoint complement is closed or, equivalently (since E is Archimedean), every band in E is closed.

Proof. (i) The proof for $f_n \vee g_n$ follows immediately from

$$| f \vee g - f_n \vee g_n | \leq$$

$$| f \vee g - f \vee g_n | + | f \vee g_n - f_n \vee g_n | \leq | g - g_n | + | f - f_n |,$$

and similarly for $f_n \wedge g_n$.

(ii) It may be assumed that $g = 0$. Since $| f^- - f_n^- | \leq | f - f_n |$, the sequence $(f_n^- : n = 1, 2, \ldots)$ converges to f^-. But $f_n^- = 0$ for all n, so $f^- = 0$. In other words, $f \geq 0$.

(iii) Since $| f_n | \wedge | g | = 0$ for all n and every $g \in D$, we have $| f | \wedge | g | = 0$ for every $g \in D$. ∎

It was already observed above that $f_n \to f$ (un) implies $f_n \to f$. It follows that every order closed set is relatively uniformly closed. If E is a normed Riesz space, then $f_n \to f(u\text{-un})$ implies $f_n \to f$ (norm). To prove this, observe that for any $\epsilon > 0$ there exists $n(\epsilon)$ such that $| f - f_n | \leq \epsilon u$ for all $n \geq n(\epsilon)$, so $\| f - f_n \| \leq \epsilon \| u \|$ for all $n \geq n(\epsilon)$. This shows that $f_n \to f$ (norm). It follows that every norm closed subset of E is relatively uniformly closed.

Having seen thus that relatively uniform convergence implies order convergence as well as norm convergence, it may be asked if there is any relation between order convergence and norm convergence. We shall prove, by means of examples, that order convergence does not imply norm convergence and norm convergence does not imply order convergence.

Example 15.2. (i) Let E be the Riesz space ℓ_∞ of all bounded sequences $f = (f(1), f(2), \ldots)$ with norm $\| f \| = \sup_n | f(n) |$, the *supremum norm*. For $n = 1, 2, \ldots$, let u_n be the element in ℓ_∞ with the first n coordinates equal to zero and all other coordinates one. Then $u_n \downarrow 0$, but the sequence does not converge in norm, because $\| u_m - u_n \| = 1$ for $m \neq n$, so the sequence is not even a norm Cauchy sequence. This shows that, even for monotone sequences, order convergence does not imply convergence in norm.

(ii) Let μ be Lebesgue measure in the closed interval $[0,1]$ and let E be the Riesz space $L_1([0, 1], \mu)$ of all Lebesgue integrable functions on $[0,1]$ with norm $\| f \| = \int_0^1 | f(x) | \, d\mu$ (usually written as $\int_0^1 | f(x) | \, dx$). Note that if f_n converges in order to zero in E, then $f_n(x) \to 0$ for almost every $x \in [0,1]$, i.e., f_n converges pointwise to zero almost everywhere on $[0,1]$. Now let $(X_n : n = 1, 2, \ldots)$ be the sequence of intervals

$[0, 1/2], [1/2, 1], [0, 1/3], [1/3, 2/3], [2/3, 1], [0, 1/4], \ldots$, and let f_n be the characteristic function of X_n for all n. Then f_n converges to zero in norm, but f_n does not converge in order. To see this, assume that $f_n \to f_0$ for some f_0, i.e., there exists a sequence $p_n \downarrow 0$ such that $\mid f - f_n \mid \leq p_n \downarrow 0$. Then, for $m \geq n$, we have

$$\mid f_m - f_n \mid \leq p_m + p_n \leq 2p_n.$$

Now, choose $m > n$ in such a manner that X_m and X_n have no points in common (which is possible in many ways). Then $\mid f_m - f_n \mid (x) = 1$ for all $x \in X_n$, so $p_n(x) \geq 1/2$ for (almost) all $x \in X_n$. This holds for every n, so if $m \geq n$, then $p_n(x) \geq p_m(x) \geq 1/2$ for (almost) every $x \in X_m$. It follows immediately (varying m) that $p_n(x) \geq 1/2$ for (almost) every $x \in [0, 1]$. Since this holds for every n it contradicts $p_n \downarrow 0$. Hence, f_n does not converge in order.

For monotone sequences the situation is different, as will be shown in the first part of the next theorem. In the second part we show that a u-uniform Cauchy sequence has a u-uniform limit if and only if the sequence converges in norm. This is analogous to Theorem 10.3(iii), where it was shown that if a sequence is u-uniformly Cauchy as well as v-uniformly Cauchy, then the sequence converges u-uniformly if and only if it converges v-uniformly.

Theorem 15.3. *(i) If $f_n \uparrow$ in the normed Riesz space E and $f_n \to f$ (norm), then $f_n \uparrow f$. Similarly for decreasing sequences. In other words, if f is the norm limit of a monotone sequence, then f is also the order limit of the sequence.*

(ii) If for some $u \in E^+$ the sequence $(f_n : n = 1, 2, \ldots)$ is u-uniformly Cauchy, then the sequence is also Cauchy in norm. In this case the sequence has a u-uniform limit if and only if the sequence has a norm limit, and these limits are then the same. Any Banach lattice is therefore uniformly complete.

Proof. (i) Let $f_n \uparrow$ and $f_n \to f$ (norm). Since $f_n \geq f_m$ for all $n \geq m$, we have $f \geq f_m$ by Theorem 15.1(ii) in the present section. This holds for all m, so f is an upper bound of the sequence. Let g be another upper bound. Then $f_n \leq g$ for all n, so $f \leq g$ (once more by the same theorem). It follows that f is the least upper bound of the sequence, i.e., $f_n \uparrow f$.

(ii) It is evident that any u-uniform Cauchy sequence is Cauchy in norm. It is also evident that if in this case the sequence has a u-uniform limit f, then f is also the norm limit. It remains to prove that if the u-uniform Cauchy sequence $(f_n : n = 1, 2, \ldots)$ has the norm limit f, then f is also the u-uniform limit of f_n. For any given $\epsilon > 0$ we have $\mid f_m - f_n \mid \leq \epsilon u$ for $m, n \geq n(\epsilon)$, and also (n fixed, $m \to \infty$) the sequence

$$(\mid f_m - f_n \mid : m = n + 1, n + 2, \ldots)$$

converges in norm to $\mid f - f_n \mid$. Hence, once more by Theorem 15.1(ii), we get $\mid f - f_n \mid\leq \epsilon u$ for $n \geq n(\epsilon)$. This shows that f is the u-uniform limit of f_n. ∎

We have seen that norm convergence and order convergence are to a great extent independent. Fortunately, as we shall prove in the next theorem, it is so that if a sequence is convergent in norm as well as in order, then the limits are the same.

Theorem 15.4. *If $f_n \to f$ (norm) and $f_n \to g$ in the normed Riesz space E, then $f = g$.*

Proof. We may assume that $g = 0$. Then $\mid f_n \mid$ converges in norm to $\mid f \mid$ and in order to zero, so there exists a sequence $p_n \downarrow 0$ such that $\mid f_n \mid\leq p_n \downarrow 0$. We have to prove that $\mid f \mid= 0$. Since $\mid f_m \mid\leq p_m \leq p_n$ for all $m \geq n$ and $\mid f_m \mid$ converges in norm to $\mid f \mid$ as $m \to \infty$, we have $\mid f \mid\leq p_n$ by Theorem 15.1(ii). This holds for every n, so $\mid f \mid\leq \inf p_n = 0$. It follows that $\mid f \mid= 0$. ∎

Example 15.5. (i) The Riesz space $C([a, b])$ of all real continuous functions on the closed interval $[a, b]$ is a normed Riesz space with respect to the *uniform norm*

$$\parallel f \parallel= \max(\mid f(x) \mid: x \in [a, b]).$$

The sequence $(f_n : n = 1, 2, \ldots)$ in $C([a, b])$ converges in norm to f whenever $\parallel f - f_n \parallel\to 0$, i.e., whenever

$$\max(\mid f(x) - f_n(x) \mid: x \in [a, b]) \to 0 \text{ as } n \to \infty.$$

This is equivalent to saying that for any $\epsilon > 0$ there exists $n(\epsilon)$ such that $\mid f(x) - f_n(x) \mid\leq \epsilon$ for all $n \geq n(\epsilon)$ and all $x \in [a, b]$ simultaneously, i.e., it is equivalent to e-uniform convergence of f_n to f, where e is the unit function defined by $e(x) = 1$ for all $x \in [a, b]$. Hence, norm convergence is the same as e- uniform convergence. Note that e - uniform convergence is exactly what is called uniform convergence in classical analysis. It is a well-known theorem that any e-uniform Cauchy sequence converges e-uniformly to a continuous function, i.e., $C([a, b])$ is e-uniformly complete, and hence norm complete. This shows that $C([a, b])$ is a Banach lattice with respect to the uniform norm. Furthermore, the unit function e is a strong order unit, that is to say, the order ideal A_e generated by e is the whole space $C([a, b])$. In other words, if u is any non- negative function in the space, then $u \leq \alpha e$ for some positive number α (depending on u). It follows that every u-uniform Cauchy sequence is an e-uniform Cauchy sequence, which implies (since any e-uniform Cauchy sequence has an e-uniform limit) that any u-uniform Cauchy sequence has an u-uniform limit (see Theorem 10.3(iii)). In other words, the space $C([a, b])$ is uniformly complete.

It remains to consider order convergence in $C([a, b])$. At first it might be thought that $f_n \uparrow f$ in $C([a, b])$ is equivalent to pointwise (monotone) convergence of f_n to f. This is true, however, only in one direction. If $f_n(x) \uparrow f(x)$ for every $x \in [a, b]$, with all f_n and f continuous, then $f_n \uparrow f$ holds in $C([a, b])$. The converse is false, as shown by the following example.

Let $(r_n : n = 1, 2, \ldots)$ be the set of all rational numbers in $[a, b]$ and let ϵ be a number satisfying $0 < \epsilon < b - a$. Furthermore, let $\Delta_n (n = 1, 2, \ldots)$ be the intersection of $[a, b]$ and the interval

$$[r_n - (\epsilon/2^{n+1}), r_n + (\epsilon/2^{n+1})].$$

Now, let g_n be a non-negative continuous function on $[a, b]$, vanishing outside Δ_n and such that $0 \le g_n(x) \le 1$ and $g_n(r_n) = 1$. Finally, let $f_n = g_1 \vee \ldots \vee g_n$ for $n = 1, 2, \ldots$. Then $f_n \uparrow$ in $C([a, b])$ and the unit function e is an upper bound of the sequence. If h is another upper bound, then $h(r_n) \ge 1$ for all n, and so $h \ge e$ by continuity. Hence e is the least upper bound, i.e., $f_n \uparrow e$. But, as we show now, the convergence is not pointwise. Of course, the pointwise limit s of the monotone sequence f_n exists, i.e., $f_n(x) \uparrow s(x)$ for every $x \in [a, b]$, and we have $s(r_n) = 1$ for all r_n and $s(x) = 0$ for every x outside $\cup_{n=1}^{\infty} \Delta_n$. The Lebesgue measure of $\cup_1^{\infty} \Delta_n$ is at most $\Sigma_1^{\infty} \epsilon/2^n = \epsilon$, so s vanishes on a set of measure at least $(b - a) - \epsilon$. Hence, although the functions e and s have the same value at each rational point in $[a, b]$, they have different values at many irrational points.

In the case of the normed Riesz space $C(X)$ of all real continuous functions on a compact (Hausdorff) topological space X, similar assertions hold (with the possible exception of the last counterexample which may not admit a generalization to the compact set X under consideration). Hence, $C(X)$ is a Banach lattice with respect to the uniform norm.

If we have an open interval (a, b), where $a = -\infty$ and (or) $b = +\infty$ is allowed, the Riesz space $C_b(a, b)$ of all bounded real continuous functions on (a, b) may likewise be equipped with the uniform norm

$$\| f \| = \sup(| f(x) |: x \in (a, b)),$$

and again similar assertions hold as in the case of $C([a, b])$. Everything (except perhaps for the counterexample) may be generalized to $C_b(X)$, where X is an arbitrary (Hausdorff) topological space.

(ii) As in Example 9.5, let μ be a countably additive (non-negative and not identically zero) σ-finite measure in the non-empty point set X and let $L_0 = L_0(X, \mu)$ be the Archimedean Riesz space of all real μ-measurable functions on X, with the understanding that functions in L_0 differing only on a set of measure zero are identified (i.e., the elements of L_0 are actually equivalence classes of functions, two functions being in the same equivalence class if and only if they differ only on a set of measure zero). For $1 \le p < \infty$, the subset of L_0 consisting of all $f \in L_0$ for which $\int_X | f(x) |^p \, d\mu$ is finite is denoted by $L_p = L_p(X, \mu)$. The set of all bounded $f \in L_0$ is denoted by

$L_\infty = L_\infty(X, \mu)$. More precisely, we have $f \in L_\infty$ whenever there exists a non-negative (finite) number M such that $|\, f(x)\, |\leq M$ holds for μ-almost every $x \in X$ (the exceptional set of measure zero possibly depending on the choice of f from its equivalence class). The spaces $L_p(1 \leq p < \infty)$ and the space L_∞ are ideals in L_0 having X as carrier. To see that X is the carrier, observe that the characteristic function of any subset of X of finite measure is a member of every $L_p(1 \leq p \leq \infty)$; for the notion of a carrier, see Example 9.5. Since L_0 is Dedekind complete (see Example 12.5(iii)), every space L_p is Dedekind complete as well.

For $1 \leq p < \infty$ the space L_p is a normed Riesz space with respect to

$$\| \, f \, \|_p = (\int_X |\, f(x)\, |^p \, d\mu)^{1/p}$$

as the norm of $f \in L_p$. This is a well-known theorem, but for the sake of completeness we present a proof in the exercises. Note that different functions in the same equivalence class give the same value for $\int |\, f\, |^p \, d\mu$, so $\| \, f \, \|_p$ is uniquely determined. The space L_∞ is a normed Riesz space with respect to

$$\| \, f \, \|_\infty = \mathrm{esssup}(|\, f(x)\, | : x \in X)$$

as norm, where esssup stands for *essential supremum*. This has to be explained. As stated above, there exists for any $f \in L_\infty$ a non-negative number M such that $|\, f(x)\, |\leq M$ for almost every $x \in X$. In other words, the set $(x : |\, f(x)\, |> M)$ is of measure zero. Let M_0 be the infimum of all M for which this holds. Then $(x : |\, f(x)\, |> M_0)$ is still of measure zero (because $M_n \downarrow M_0$ implies

$$(x : |\, f(x)\, |> M_0) = \cup_{n=1}^\infty (x : |\, f(x)\, |> M_n)),$$

but for any $M^* < M_0$ the set $(x : |\, f(x)\, |> M^*)$ is of positive measure. The number M_0 is called the *essential supremum* of $|\, f\, |$. The proof that $\| \, f \, \|_\infty$ is a norm in L_∞ is immediate by observing that

$$|\, f_1(x) + f_2(x)\, |\leq|\, f_1(x)\, | + |\, f_2(x)\, |\leq\| \, f_1 \, \|_\infty + \| \, f_2 \, \|_\infty$$

for almost every $x \in X$.

For monotone sequences in $L_p(1 \leq p \leq \infty)$ order convergence is the same as pointwise convergence, i.e., if $f_n \uparrow f$ or $f_n \downarrow f$ and f and all f_n are members of L_p, then f_n converges to f pointwise almost everywhere if and only if f_n converges to f in order. It follows immediately that if f_n converges to f in order (not necessarily monotonely), then f_n converges to f pointwise almost everywhere. The converse does not hold, as shown by the following example. Let X be the interval $[0,1]$ with Lebesgue measure and, for $n = 1, 2, \ldots$, let Δ_n be the interval $[2^{-n}, 2^{-(n-1)}]$. For $1 \leq p < \infty$, let f_n be equal to $2^{n/p}$ on Δ_n and zero outside Δ_n, so $\int |\, f_n\, |^p \, dx = 1$. For $p = \infty$, let f_n be equal to n on Δ_n and zero outside Δ_n, so $\parallel f_n \parallel_\infty = n$. Obviously, f_n converges to zero pointwise on $[0,1]$, but not in order. Indeed, assume that $1 \leq p < \infty$ and $0 \leq f_n \leq q_n$ for some sequence $q_n \downarrow 0$ in L_p. Then $q_n \geq f_n$ for all n, and so $q_n \geq q_m \geq f_m$ for all $m \geq n$, which implies (since the functions $f_n; n = 1, 2, \ldots$, are mutually disjoint) that

$$\int (q_n)^p dx \geq \Sigma_n^\infty \int (f_m)^p dx = \infty.$$

Contradiction. For $p = \infty$ we get similarly that $\parallel q_n \parallel_\infty \geq \parallel f_m \parallel_\infty = m$ for all $m \geq n$, so $\parallel q_n \parallel_\infty = \infty$, which again is contradictory. Hence, the sequence $(f_n : n = 1, 2, \ldots)$ does not converge in order. The situation improves if the sequence (f_n) has a majorant in absolute value in L_p, because then pointwise convergence of f_n to zero implies convergence in order to zero. In this case, defining u_n to be the pointwise supremum of all $\mid f_m \mid; m \geq n$, we have $u_n \downarrow 0$ pointwise, and so $0 \leq \mid f_n \mid \leq u_n$ for all n and $u_n \downarrow 0$ in order.

For $1 \leq p < \infty$ norm convergence in L_p does not always imply convergence in order. This was proved for $p = 1$ in Example 15.2(ii); the proof for $1 < p < \infty$ is similar. On the other hand, if f_n converges in order to f in L_p, then $\mid f - f_n \mid \leq q_n \downarrow 0$ for an appropriate sequence (q_n) in L_p. But then $\int (q_n)^p d\mu \downarrow 0$ by a well-known theorem in integration theory, so $\parallel f - f_n \parallel_p \to 0$.

For $p = \infty$ the situation is different because (similarly as for $C(X)$) norm convergence is the same as e-uniform convergence, where again e is the unit function identically equal to one. Hence, norm convergence (i.e., e-uniform convergence) implies order convergence. On the other hand, order convergence does not always imply norm convergence, as shown in Example 15.2(i). Finally, observe that every e-uniform Cauchy sequence in L_∞ has an e-uniform limit and e is a strong order unit in L_∞, i.e., every $u \geq 0$ in L_∞ is majorized by a positive multiple of e. Hence, exactly as for $C(X)$, every u-uniform Cauchy sequence has an u-uniform limit, i.e., L_∞ is relatively uniformly complete.

It was shown in Example 15.2(ii) that a norm convergent sequence is not necessarily order convergent. In L_p -spaces $(1 \leq p < \infty)$ it is well-known that every norm convergent sequence has a subsequence converging pointwise (and since the subsequence is majorized in absolute value, it converges in order). It is reasonable, therefore, to conjecture that in a Banach lattice every norm convergent sequence has a subsequence converging in order. Even more is true. We shall prove that in a Banach lattice every norm convergent sequence has a subsequence converging relatively uniformly.

Theorem 15.6. *Every norm convergent sequence in a Banach lattice has a subsequence converging relatively uniformly (and hence the subsequence converges in order).*

Proof. To begin with, note that if E is a Banach lattice and $v_n \in E^+$ for $n = 1, 2, \ldots$ with $\| v_n \| \leq n^{-3}$ for all n, then $\Sigma_1^\infty \| n v_n \|$ is finite, so (as shown in section 14) the series $\Sigma_1^\infty n v_n$ converges in norm in E to some element $s \in E$, i.e., the partial sums $s_n = \Sigma_{k=1}^n k v_k$ satisfy $0 \leq s_n$ and $s_n \to s$ in norm. But then $s_n \uparrow s$ by Theorem 15.3(i), which implies that $n v_n \leq s$ for all n.

Assume now that f_n converges in norm to f, so $\| f - f_n \| \to 0$ as $n \to \infty$. Choose a subsequence $g_k = f_{n_k} (k = 1, 2, \ldots)$ such that $\| f - g_k \| \leq k^{-3}$ for all k. Then, as shown above, there exists an element $s \in E^+$ such that $k \mid f - g_k \mid \leq s$ for all k, so $\mid f - g_k \mid \leq k^{-1} s$ for all k. This shows that the sequence $g_k = f_{n_k} (k = 1, 2, \ldots)$ converges s-uniformly to f. $\blacksquare$

In Theorem 15.3(i) we have proved that if the increasing sequence (f_n) converges in norm to f_0, then f_0 is the supremum of the series, i.e., $f_n \uparrow f_0$. A similar result holds for an upwards directed set. We recall that, according to Definition 10.9, the (non-empty) set D in the Riesz space E is upwards directed whenever for every two elements f and g in D there exists an element $h \in D$ such that $h \geq f \vee g$. The definition of a downwards directed set is similar. If D is an upwards directed set in the normed Riesz space E, we say that D converges in norm to $f_0 \in E$ if for any $\epsilon > 0$ there exists an element $f(\epsilon) \in D$ such that $\| f - f_0 \| \leq \epsilon$ for all $f \in D$ satisfying $f \geq f(\epsilon)$.

Theorem 15.7. *If D is an upwards directed set in the normed Riesz space E such that D converges in norm to f_0, then $f_0 = \sup D$.*

Proof. We show first that f_0 is an upper bound of D. For that purpose, let $f^* \in D$ be given. We prove that $f^* \leq f_0$. Since D converges in norm to f_0, there exists an element $f(1) \in D$ such that $\| f_0 - f \| \leq 1$ for all $f \in D$ satisfying $f \geq f(1)$. Now, choose $f_1 \in D$ such that $f_1 \geq f^* \vee f(1)$. This is possible because D is upwards directed. Then $f_1 \geq f^*$ and $\| f_0 - f_1 \| \leq 1$. Next, there exists an element $f(2) \in D$ such that $\| f_0 - f \| \leq 2^{-1}$ for all $f \in D$ satisfying $f \geq f(2)$. Choose $f_2 \in D$ such that $f_2 \geq f_1 \vee f(2)$. Then $f_2 \geq f_1$ and $\| f_0 - f_2 \| \leq 2^{-1}$. Continuing in this manner, we obtain a sequence $f^* \leq f_1 \leq f_2 \leq \cdots$ in D such that $\| f_0 - f_n \| \leq n^{-1}$ for $n = 1, 2, \ldots$. Hence, the sequence is increasing and converging in norm to f_0. But then f_0 is the supremum of the sequence, i.e., $f^* \leq f_n \uparrow f_0$. Since f^* is arbitrary in D, this shows that f_0 is an upper bound of D. For any other upper bound g of D we have $f_n \leq g$ for all n, so $\sup f_n = f_0 \leq g$ by Theorem 15.1(ii). It follows that f_0 is the least upper bound of D, i.e., $f_0 = \sup D$. $\blacksquare$

Exercise 15.8. Show that in a Banach lattice every order closed set is norm closed.

Hint: Use that in a Banach lattice every norm convergent sequence has a subsequence converging in order.

Exercise 15.9. If E is a normed Riesz space but not a Banach lattice, it may happen that a norm convergent sequence does not have any subsequence converging in order, as shown by the following example. Let E be the normed Riesz space of all real continuous functions f on $[0,1]$ with norm $\| f \| = \int_0^1 | f(x) | \, dx$ and let the sequence (f_n) in E be such that $0 \le f_n(x) \le 1$ on $[0,1]$ for all n, and furthermore

$$f_1(1/2) = 1 \text{ and } \| f_1 \| = 2^{-1},$$

$$f_2(1/3) = f_2(2/3) = 1 \text{ and } \| f_2 \| = 2^{-2},$$

generally

$$f_n = 1 \text{ in } {}^1/_{(n+1)}, \dots, {}^n/_{(n+1)} \text{ and } \| f_n \| = 2^{-n}.$$

Show that f_n converges to zero in norm, but there is no subsequence converging to zero in order. Now let D be the set consisting of all f_n. Show that D is order closed but not norm closed.

Hint: If p is a majorant of a subsequence, then $p(x) \ge 1$ on a dense subset of $[0,1]$, and so $p(x) \ge 1$ for all $x \in [0,1]$. Assuming now that $0 \le f_{n_k} \le p_k \downarrow 0$, the function p_k is a majorant for all $f_{n_j} (j \ge k)$.

Exercise 15.10. Let E be a normed Riesz space. It was proved in Theorem 15.1 that if $f_n \to f$ in norm and $g_n \to g$ in norm in E, then $f_n \vee g_n \to f \vee g$ in norm and $f_n \wedge g_n \to f \wedge g$ in norm. In particular, $f_n^+ \to f^+$ in norm, $f_n^- \to f^-$ in norm and $| f_n | \to | f |$ in norm. Show that if $f_n \to f$ in norm and $f_n \perp h$ for all n, then $f \perp h$. Show now that if $f_n \to f$ in norm and all f_n are contained in a band B, then $f \in B$. Hence, every band in a Banach lattice is a Banach lattice on its own.

Exercise 15.11. Let E be a uniformly complete normed Riesz space (observe that in view of Theorem 15.3(ii) every Banach lattice satisfies this condition). As shown in Theorem 13.4, every element $h = f + ig$ (with f and g in E) in $E + iE$ has an absolute value $| h |$ in E. For this element h we define the number $\| h \|$ by $\| h \| = \| | h | \|$.

(i) Show that the thus defined $\| h \|$ is a norm in the complex vector space $E + iE$, extending the existing norm in E and such that $| h_1 | \le | h_2 |$ implies $\| h_1 \| \le \| h_2 \|$.

(ii) Show that $E + iE$ is a Banach space if and only if E is a Banach lattice. In this case $E + iE$ is called a *complex Banach lattice*.

Exercise 15.12. Let E be as in the preceding exercise. Order convergence, u-uniform convergence (for any $u \in E^+$) and relatively uniform convergence are defined in $E + iE$ as in E.

(i) Show that if $h_n \to h$ in norm and $h_n \to k$ in order in $E + iE$, then $h = k$.

(ii) Show that if $h_n \to h$ relatively uniformly, then $h_n \to h$ in norm as well as in order.

(iii) Show that if (h_n) is a u-uniform Cauchy sequence in $E + iE$, then the sequence has a u-uniform limit if and only if it has a norm limit. Hence, every complex Banach lattice is uniformly complete.

(iv) Show that in a complex Banach lattice $E + iE$ every norm convergent sequence has a subsequence which is relatively uniformly convergent. Hence, every order closed set in $E + iE$ is norm closed.

(v) Show that if A_r is an ideal in E, then the ideal $A = A_r + iA_r$ in $E + iE$ is closed (in order, relatively uniformly or in norm) if and only if A_r is closed in E.

Exercise 15.13. Show that if (X, μ) is a σ-finite measure space and $E = L_p(X, \mu)$ for some p satisfying $1 \leq p \leq \infty$, then $E + iE$ admits an absolute value (i.e., if $h = f + ig \in E + iE$, then $\mid h \mid = \sup_\theta Re(he^{-i\theta})$ exists), and $\mid h \mid$ satisfies

$$\mid h \mid (x) = (f^2(x) + g^2(x))^{1/2}$$

for μ-almost every x, i.e., $\mid h \mid$ is the familiar pointwise absolute value of h.

Show that the same holds if E is the space $C(X)$ of all real continuous functions on a (Hausdorff) topological space, so in particular if $E = C(\Delta)$, where Δ is a closed or open interval in $\mathbb{R}$.

Hint: See Example 13.2.

Exercise 15.14. (i) Let $v = \phi(u)$, for $u \geq 0$, be non-negative, continuous and strictly increasing with $\phi(0) = 0$. Note that the inverse function $u = \psi(v)$ has the same properties. For $a \geq 0, b \geq 0$, let

$$F(a) = \int_0^a \phi(u)du \text{ and } G(b) = \int_0^b \psi(v)dv.$$

Show that $ab \leq F(a) + G(b)$.

(ii) Let $1 < p < \infty$ and let q be given by $p^{-1} + q^{-1} = 1$. Now, let $v = u^{p-1}$ in part (i). Show that $u = v^{q-1}$ and derive from (i) that

$$ab \leq p^{-1}a^p + q^{-1}b^q \text{ for all } a \geq 0, b \geq 0.$$

Hint: For part (i), draw the graph of ϕ and observe that $F(a)$ is the "area under the curve" between zero and a.

Exercise 15.15. As in Exercise 15.5(ii), let μ be a countably additive and σ-finite measure in the non-empty point set X. Furthermore, let p and q be positive numbers such that $p^{-1} + q^{-1} = 1$ and let f and g be complex μ-measurable functions on X such that the integrals $\int_X |f|^p \, d\mu$ and $\int_X |g|^q \, d\mu$ exist as finite numbers. Then

$$|\int_X fg d\mu| \leq \int_X |fg| \, d\mu \leq (\int_X |f|^p \, d\mu)^{1/p} \cdot (\int_X |g|^q \, d\mu)^{1/q}.$$

This is known as *Hölder's inequality* for integrals. Note that for $p = q = 2$ we obtain the familiar *Schwarz-Buniakovsky inequality*. We indicate the proof. If $\int |f|^p \, d\mu = 0$, then $f(x) = 0$ for almost every $x \in X$, so $\int |fg| \, d\mu = 0$. Similarly if $\int |g|^q \, d\mu = 0$. Assume, therefore, that

$$A = (\int |f|^p \, d\mu)^{1/p} > 0 \text{ and } B = (\int |g|^q \, d\mu)^{1/q} > 0$$

We may also assume that $|f(x)|$ and $|g(x)|$ are finite for every $x \in X$. Substitute now $a = A^{-1} |f(x)|$ and $b = B^{-1} |g(x)|$ in $ab \leq p^{-1}a^p + q^{-1}b^q$ and integrate. Show that we thus obtain

$$(AB)^{-1} \int |fg| \, d\mu \leq$$

$$(pA^p)^{-1} \int |f|^p \, d\mu + (qB^q)^{-1} \int |g|^q \, d\mu =$$

$$p^{-1} + q^{-1} = 1,$$

so $\int |fg| \, d\mu \leq AB$, as desired.

Exercise 15.16. If $E = L_p(X, \mu)$ for some σ-finite μ and some p satisfying $1 \leq p \leq \infty$, we extend the definition of the norm $\| f \|_p$ to $E + iE$ by $\| f \|_p = \| |f| \|_p$ for any $f \in E + iE$, as indicated in Exercise 15.10. Let now $p^{-1} + q^{-1} = 1$ (with the understanding that $q = \infty$ for $p = 1$ and $q = 1$ for $p = \infty$). Show that if the complex μ-measurable functions f and g satisfy $\| f \|_p < \infty$ and $\| g \|_q < \infty$, then Hölder's inequality holds, i.e.,

$$|\int fg d\mu| \leq \int |fg| \, d\mu \leq \| f \|_p \cdot \| g \|_q .$$

Hint: The cases $p = 1$ and $p = \infty$ are easy.

Exercise 15.17. Let $E = L_p(X, \mu)$ as in the preceding exercise and let f and g be members of $E + iE$. Show that the triangle inequality holds for the norm, i.e.,

$$\| f + g \|_p \leq \| f \|_p + \| g \|_p .$$

Hint: The proof is easily reduced to the case that $1 < p < \infty$, $f(x) \geq 0$ and $g(x) \geq 0$ for all $x \in X$ and $\| f + g \|_p > 0$. In this case we have

$$\int (f + g)^p d\mu = \int f(f + g)^{p-1} d\mu + \int g(f + g)^{p-1} d\mu.$$

Define q by $p^{-1} + q^{-1} = 1$ and apply Hölder's inequality to each of the last two terms. Since $(p - 1)q = p$, we then get

$$\int (f + g)^p d\mu \leq$$

$$\| f \|_p \left(\int (f + g)^{(p-1)q} d\mu \right)^{1/q} + \| g \|_p \left(\int (f + g)^{(p-1)q} d\mu \right)^{1/q} =$$

$$\| f \|_p \left(\int (f + g)^p d\mu \right)^{1/q} + \| g \|_p \left(\int (f + g)^p d\mu \right)^{1/q}.$$

Dividing by the positive number $(\int (f + g)^p d\mu)^{1/q}$, it follows that

$$\left(\int (f + g)^p d\mu \right)^{1 - q^{-1}} \leq \| f \|_p + \| g \|_p .$$

Since $1 - q^{-1} = p^{-1}$, this is the desired result.

To end our discussion in this section we present two theorems that will be of use when we discuss Banach lattices with an order continuous norm (in section 17). In Theorem 7.6 it was proved that if A_1 and A_2 are ideals in the Riesz space E, then the algebraic sum $A_1 + A_2$ is an ideal in E. It may be asked what happens if A_1 and A_2 are norm closed ideals in the Banach lattice E.

Theorem 15.18. *If A_1 and A_2 are norm closed ideals in the Banach lattice E, then $A_1 + A_2$ is a norm closed ideal in E.*

Proof. We first recall the proof that $A_1 + A_2$ is an ideal (with notations a little different from those in Theorem 7.6). Let $\mid g \mid \leq \mid f_1 + f_2 \mid$ with $f_1 \in A_1$ and $f_2 \in A_2$. Then $g^+ \leq \mid g \mid \leq \mid f_1 \mid + \mid f_2 \mid$, so (by the Riesz decomposition theorem) g^+ can be written as $g^+ = h_1^+ + h_2^+$ with $0 \leq h_1^+ \leq \mid f_1 \mid$ and $0 \leq h_2^+ \leq \mid f_2 \mid$, which implies that $h_1^+ \in A_1$ and $h_2^+ \in A_2$. Similarly, $g^- = h_1^- + h_2^-$ with $0 \leq h_1^- \in A_1$ and $0 \leq h_2^- \in A_2$. It follows that $g = h_1 + h_2$ with $h_1 = h_1^+ - h_1^-$ and $h_2 = h_2^+ - h_2^-$. Note that the notations are correct (i.e., $h_1^+ \wedge h_1^- = h_2^+ \wedge h_2^- = 0$). Also note that $\mid h_1 \mid = h_1^+ + h_1^- \leq g^+ + g^- = \mid g \mid$, and similarly $\mid h_2 \mid \leq \mid g \mid$. To summarize, we have proved that if $\mid g \mid \leq \mid f_1 + f_2 \mid$ with $f_1 \in A_1$ and $f_2 \in A_2$, then $g = h_1 + h_2$ with $h_1 \in A_1$ and $h_2 \in A_2$, and furthermore $\mid h_1 \mid$ and $\mid h_2 \mid$ are majorized by $\mid g \mid$. Hence, in the case that E is normed, we have $\| h_1 \| \leq \| g \|$ and $\| h_2 \| \leq \| g \|$.

Assume now that A_1 and A_2 are closed ideals in the Banach lattice E and $(f_n : n = 1, 2, \ldots)$ is a sequence in $A_1 + A_2$ converging in norm to $f \in E$.

We have to show that $f \in A_1 + A_2$. Passing to a subsequence if necessary we may assume that $\| f_n - f_{n-1} \| \leq n^{-2}$ for $n = 2, 3, \ldots$. Write $g_1 = f_1$ and $g_n = f_n - f_{n-1}$ for $n = 2, 3, \ldots$. Then the partial sums of Σg_n converge in norm to f and $\Sigma \| g_n \|$ converges. Now decompose each g_n as above, so $g_n = h_{n1} + h_{n2}$ with $h_{n1} \in A_1, h_{n2} \in A_{n2}$ and furthermore $\| h_{n1} \| \leq \| g_n \|$ as well as $\| h_{n2} \| \leq \| g_n \|$. Then $\Sigma \| h_{n1} \|$ and $\Sigma \| h_{n2} \|$ converge, so (since E is a Banach lattice, and therefore a Banach space) the partial sums of Σh_{n1} and Σh_{n2} converge in norm to elements $h_1 \in A_1$ and $h_2 \in A_2$ respectively (we use here that A_1 and A_2 are closed). Hence, since $h_{n1} + h_{n2} = g_n$ holds for each n, the partial sums of Σg_n converge in norm to $h_1 + h_2 \in A_1 + A_2$. But, as observed above, the partial sums of Σg_n converge in norm to f. Therefore, $f = h_1 + h_2 \in A_1 + A_2$.   ∎

Theorem 15.19. *The norm closure S^- of a subset S of the normed vector space V is (by definition) the smallest norm closed set containing S as a subset. It is easy to see that S^- is the set of all norm limits of sequences in S that converge in norm. It is evident that the norm closure of a linear subspace is a linear subspace.*

(i) If A is an ideal in the normed Riesz space E, then the norm closure A^- is also an ideal in E.

(ii) If A_1 and A_2 are ideals in the Banach lattice E, then $(A_1 + A_2)^- = A_1^- + A_2^-$.

(iii) If A is the ideal in the Banach lattice E generated by the subset D of E, then the norm closure A^- is called the closed ideal generated by D. Now, if A_1 and A_2 are the ideals generated by the subsets D_1 and D_2 respectively, then $A_1^- + A_2^-$ is the closed ideal generated by $D_1 \cup D_2$.

Proof. (i) If the sequence (f_n) in A converges in norm to $f \in A^-$, then the sequence $(| f_n |)$ in A converges in norm to $| f |$, so $| f | \in A^-$. It remains to prove that if $0 \leq f \in A^-$ and $0 \leq g \leq f$, then $g \in A^-$. There exists a sequence (f_n) in A converging to f, and we may assume that $f_n \geq 0$ for all n. Then $g_n = f_n \wedge g \in A$ holds for all n and g_n converges to $f \wedge g = g$. This shows that $g \in A^-$.

(ii) It is evident that $A_1^- + A_2^-$ is a subset of $(A_1 + A_2)^-$. To see that $(A_1 + A_2)^-$ is a subset of $A_1^- + A_2^-$ observe that (in view of the preceding theorem) $A_1^- + A_2^-$ is a closed set containing $A_1 + A_2$, whereas $(A_1 + A_2)^-$ is the smallest closed set containing $A_1 + A_2$.

(iii) Observe that $A_1 + A_2$ is the ideal generated by $D_1 \cup D_2$ (see Exercise 7.11).   ∎

CHAPTER 8
The Riesz-Fischer Property and Order Continuous Norms

16. The Riesz-Fischer Property

As we have seen in section 14, the normed vector space V is norm complete (in other words, V is a Banach space) if and only if every absolutely convergent series in V is convergent in norm. More precisely, V is a Banach space if and only if it follows from $\Sigma_1^\infty \parallel f_n \parallel < \infty$ (all f_n in V) that the partial sums $s_n = \Sigma_1^n f_k$ have a norm limit in V (as $n \to \infty$). The norm limit is then often written as $\Sigma_1^\infty f_n$ (as we did in section 14), but this may cause confusion if $V = E$ is a normed Riesz space because then $\Sigma_1^\infty u_n$ (all u_n in E^+) is also used to denote the order limit (i.e., the supremum) of the increasing sequence of partial sums $s_n = \Sigma_1^n u_k$, if this limit exists. If the norm limit for an increasing sequence of positive elements exists, then norm limit and order limit are identical (see Theorem 15.3(i)), so that in this case there can arise no misunderstanding. It may happen, however, that the order limit exists but the norm limit does not exist. For this reason we shall use Σ_1^∞ only to denote an order limit of partial sums of a series with positive terms (if existing).

The first point to observe is that the normed Riesz space E is a Banach lattice if and only if convergence of $\Sigma \parallel u_n \parallel$, with all u_n in E^+, implies norm convergence of the sequence $(\Sigma_1^n u_k : n = 1, 2, \ldots)$. To see this, assume that the last condition is satisfied. We have to show only that convergence of the series $\Sigma \parallel f_n \parallel$ implies norm convergence of the sequence

$$(s_n = \Sigma_1^n f_k : n = 1, 2, \ldots).$$

Convergence of $\Sigma \parallel f_n \parallel$ implies convergence of $\Sigma \parallel f_n^+ \parallel$ and $\Sigma \parallel f_n^- \parallel$, so by hypothesis $s_n' = \Sigma_1^n f_k^+$ and $s_n'' = \Sigma_1^n f_k^-$ converge in norm. It follows that $s_n = s_n' - s_n''$ converges in norm. It is a remarkable result, due to I. Halperin and W.A.J. Luxemburg (1956), that it is also true that E is a Banach lattice if and only if it follows from convergence of $\Sigma \parallel u_n \parallel$, with all u_n in E^+, that the order limit of the sequence $(s_n = \Sigma_1^n u_k)$ exists in E. The property that convergence of $\Sigma_1^\infty \parallel u_n \parallel$ implies the existence of $\Sigma_1^\infty u_n$ is called the *Riesz-Fischer property*. To establish the Halperin-Luxemburg result we first prove a lemma.

Lemma 16.1. *If E has the Riesz-Fischer property, i.e., if for any sequence $(u_n : n = 1, 2, \ldots)$ of positive elements in E for which $\Sigma \parallel u_n \parallel$ converges the order limit $\Sigma_1^\infty u_n$ exists, then $\parallel \Sigma u_n \parallel \leq \Sigma \parallel u_n \parallel$ for any such sequence.*

Proof. Assume that for some sequence (u_n) in E^+ this fails to be true. Then there exists a number $\epsilon > 0$ such that

$$\parallel \Sigma_1^\infty u_k \parallel > \epsilon + \Sigma_1^\infty \parallel u_k \parallel .$$

This inequality holds as well for every tail of the series, i.e.,

$$\parallel \Sigma_{n+1}^\infty u_k \parallel = \parallel \Sigma_1^\infty - \Sigma_1^n \parallel \geq \parallel \Sigma_1^\infty \parallel - \parallel \Sigma_1^n \parallel \geq$$

$$\parallel \Sigma_1^\infty u_k \parallel - \Sigma_1^n \parallel u_k \parallel > \epsilon + \Sigma_{n+1}^\infty \parallel u_k \parallel$$

for all $n = 1, 2, \ldots$. Let now the the natural numbers $n_k (k = 1, 2, \ldots)$ be such that $n_1 < n_2 < \cdots$ and $\Sigma_{n=n_k}^\infty \parallel u_n \parallel \leq k^{-2}$ for all k. The corresponding elements $s_k = \Sigma_{n_k}^\infty u_n$ form a decreasing sequence in E^+. Arranging all u_n occurring in all s_k into a single sequence, we obtain a sequence $(w_n : n = 1, 2, \ldots)$ in which each u_n occurs at least k times if $n > n_k$. Note that $\Sigma \parallel w_n \parallel \leq \Sigma k^{-2}$, so $s = \Sigma w_n$ exists by hypothesis. It is evident that each sum $\Sigma_1^k s_m$ is less than or equal to s, and so $s \geq \Sigma_1^k s_m \geq k s_k$ for $k = 1, 2, \ldots$. It follows that

$$\parallel s \parallel \geq k \parallel s_k \parallel \geq k\epsilon$$

for $k = 1, 2, \ldots$. This is impossible. Hence, if E has the Riesz-Fischer property, then the infinite triangle inequality $\parallel \Sigma u_n \parallel \leq \Sigma \parallel u_n \parallel$ holds in E^+.   ∎

Theorem 16.2. *The normed Riesz space E is a Banach lattice if and only if E has the Riesz-Fischer property.*

Proof. Assume first that E is a Banach lattice. Let $u_n \in E^+$ for $n = 1, 2, \ldots$ and $\Sigma \parallel u_n \parallel < \infty$. Then the sequence $(s_n = \Sigma_1^n u_k : n = 1, 2, \ldots)$ is Cauchy in norm, so the norm limit s of s_n exists. Since s_n is increasing as n increases, it follows that $s_n \uparrow s$ (see Theorem 15.3(i)), i.e., $\Sigma_1^\infty u_n$ exists and is equal to s. This shows that E has the Riesz-Fischer property.

Assume now, conversely, that E has the Riesz-Fischer property. We have to show that E is a Banach lattice. In view of what was observed above it is sufficient to show that if $\Sigma_1^\infty \parallel u_n \parallel$ converges with all u_n in E^+, then the sequence $(s_m = \Sigma_1^m u_k : m = 1, 2, \ldots)$ converges in norm. For this purpose, note that $s = \Sigma_1^\infty u_n$ exists by hypothesis, i.e., $s = \sup_n s_n$ for $s_n = u_1 + \cdots + u_n$. Then

$$s - s_m = \sup_n (s_n - s_m) = \Sigma_{k=m+1}^\infty u_k,$$

and so (by the infinite triangle inequality, as proved in the last lemma)

$$\| s - s_m \| \leq \Sigma_{m+1}^\infty \| u_k \| \to 0 \text{ as } m \to \infty.$$

This is the desired result. The reader will have seen that the essential part of the proof is contained in the lemma about the infinite triangle inequality. ∎

Example 16.3. We return to the spaces $L_p = L_p(X,\mu), 1 \leq p \leq \infty$, investigated already in the Examples 9.5, 12.5(iii) and 15.5(ii). These spaces are normed Riesz spaces; we shall prove now that they are in fact Banach lattices. This will be done by showing that L_p has the Riesz-Fischer property. For this purpose, assume first that $(s_n : n = 1, 2, \ldots)$ is an increasing sequence of nonnegative functions in L_p. The limit $s(x)$ of the sequence $(s_n(x) : n = 1, 2, \ldots)$ exists for every $x \in X$, where it is possible that $s(x) = +\infty$ at several (or even all) points $x \in X$. It is true now that either $s \notin L_p$ and $\| s_n \|_p \uparrow \infty$ as $n \to \infty$ or $s \in L_p$ and $\| s_n \|_p \uparrow \| s \|_p$ as $n \to \infty$. For $1 \leq p < \infty$ this follows immediately from the well-known theorem of Beppo Levi on the integration of increasing sequences, asserting that $\int s_n^p d\mu$ tends either to $+\infty$ or to $\int s^p d\mu$. For $p = \infty$, if $\lim \| s_n \|_\infty$ is finite, say $\lim \| s_n \| = M$, then $s_n(x) \leq M$ holds for almost every x for all n simultaneously, so $s(x) \leq M$ holds for almost every x, i.e., $\| s \|_\infty \leq M$. Since $\| s \|_\infty \geq \| s_n \|_\infty$ for all n, it is evident that $\| s \|_\infty \geq M$. Hence $\| s \|_\infty = M = \lim \| s_n \|_\infty$.

We prove now that L_p has the Riesz-Fischer property. Assume that $0 \leq u_n \in L_p$ for $n = 1, 2, \ldots$ and $\Sigma_1^\infty \| u_k \|$ converges. Write $\alpha = \Sigma_1^\infty \| u_k \|$ and $s_n = \Sigma_1^n u_k$ for $n = 1, 2, \ldots$. Then $\| s_n \| \leq \Sigma_1^\infty \| u_k \| = \alpha$ for all n, so the pointwise limit (i.e., the order limit) $s = \Sigma_1^\infty u_k$ satisfies $\| s \| = \lim \| s_n \| \leq \alpha$. This shows that $s = \Sigma_1^\infty u_k$ is a member of L_p. In other words, the Riesz-Fischer condition is satisfied. Hence, L_p is a Banach lattice.

In particular, choosing for X the set $(1, 2, \ldots)$ of natural numbers and for μ the counting measure (i.e., each point has measure one), it has been shown thus that the sequence spaces $\ell_p (1 \leq p \leq \infty)$ are Banach lattices.

As indicated in Exercise 15.10(ii), the complexification $L_p + iL_p$ of L_p is now also a Banach space, i.e., every norm Cauchy sequence in $L_p + iL_p$ has a norm limit. In many cases, if it is understood from the beginning that we have to deal with complexvalued functions, the space $L_p + iL_p$ is simply denoted by L_p again. A similar remark holds for $C([a, b]) + iC([a, b])$ or more generally for $C(X) + iC(X)$, where X is a compact Hausdorff space.

Exercise 16.4. There is a property similar to the Riesz-Fischer property, but somewhat weaker. This property is called the weak Riesz-Fischer property and can be formulated in two ways which at first seem different, although they are in fact equivalent. We shall say that the normed Riesz space E has the *weak Riesz-Fischer property* if it follows from $0 \leq u_n \in E$ for $n = 1, 2, \ldots$ and $\Sigma_1^\infty \| u_n \|$ convergent that the partial sums $s_n = u_1 + \cdots + u_n$ are bounded above, i.e., there exists an element $v \in E^+$ such that $s_n \leq v$ holds for all n.

We indicate how to show that E has the weak Riesz-Fischer property if and only if the following condition (*) is satisfied:

(*) If $0 \le u_n \in E$ for $n = 1, 2, \ldots$ and $\Sigma_1^\infty \parallel u_n \parallel$ is finite, then the elements u_n are bounded above, i.e., there exists an element $w \in E^+$ such that $u_n \le w$ for all n.

It is sufficient to show that (*) implies the weak Riesz-Fischer property. Hence, assume (*) and let $0 \le u_n \in E$ with $\Sigma \parallel u_n \parallel$ finite. There exist natural numbers $n_1 < n_2 < \ldots$ such that $n_1 = 1$ and

$$\Sigma_{j=n_k}^{n_{k+1}} \parallel u_j \parallel \, \le C \cdot 4^{-k} \text{ for } k = 1, 2, \ldots,$$

where C is a constant satisfying $C \ge 4\Sigma_1^\infty \parallel u_j \parallel$. This condition for C is to make it possible to choose $n_1 = 1$. Now, let

$$v_k = \Sigma_{j=n_k}^{n_{k+1}} u_j \text{ for } k = 1, 2, \ldots.$$

Then

$$\Sigma_1^\infty \parallel 2^k v_k \parallel \, \le C\Sigma_1^\infty 2^{-k} < \infty,$$

so in view of condition (*) there exists an element $w \in E^+$ such that $2^k v_k \le w$ for all k. Observe next that for any given natural number n there exists a natural number k such that

$$u_1 + \cdots + u_n \le v_1 + \cdots + v_k,$$

so

$$0 \le u_1 + \cdots + u_n \le v_1 + \cdots + v_k \le (2^{-1} + 2^{-2} + \cdots + 2^{-k})w \le w.$$

This shows that E has the weak Riesz-Fischer property.

Exercise 16.5. Show that if the normed Riesz space E is Dedekind σ–complete and has the weak Riesz-Fischer property, then E is a Banach lattice.

Hint: Show that E has the Riesz-Fischer property.

Exercise 16.6. The conditions mentioned in the preceding exercise (Dedekind σ–completeness and the weak Riesz-Fischer property together) are sufficient but not necessary for norm completeness, because for example $C([0, 1])$ is a Banach lattice which fails to be Dedekind σ-complete. We indicate how to show that E is a Banach lattice if and only if E is uniformly complete and has the weak Riesz-Fischer property. The proof in one direction is easy (since every Banach lattice is uniformly complete; see Theorem 15.3(ii)). Assume, therefore, that E is uniformly complete and has the weak Riesz-Fischer property. We show that E has the Riesz-Fischer property. To this end, let $0 \le u_n \in E$ for $n = 1, 2, \ldots$ such that $\Sigma \parallel u_n \parallel$ converges. Let $(v_k : k = 1, 2, \ldots)$ be the same sequence as in Exercise 16.4, so $v_k \le 2^{-k}w$

for some $w \in E^+$ and all k. Write $s_n = v_1 + \cdots + v_n$ for $n = 1, 2, \ldots$. The sequence $(s_n : n = 1, 2, \ldots)$ is a w-uniform Cauchy sequence, so in view of the uniform completeness s_n converges w-uniformly to some element s. Then $\| s - s_n \| \to 0$ and, since s_n is increasing, it follows that $s_n \uparrow s$, i.e., $s = \Sigma v_k$. But then s is also the supremum of the partial sums of the series Σu_n, so $s = \Sigma u_n$.

Exercise 16.7. (i) Show that the following normed Riesz spaces have the weak Riesz-Fischer property, but not the Riesz-Fischer property.

(a) E consists of all continuous functions f on $[0,1]$ such that the graph of f consists of a finite number of straight line segments; the norm is the uniform norm. Observe that every non-negative continuous function on $[0,1]$ has a majorant in E.

(b) E is the linear subspace of $L_1([0, 1], \mu)$, where μ is Lebesgue measure, consisting of all $f \in L_1([0, 1], \mu)$ assuming finitely or countably many different values. Observe that every non-negative function in L_1 has a majorant in E.

(ii) Let μ be Lebesgue measure in $X = [0, 1]$. Show that the space

$$E = L_1(X, \mu) \cap L_\infty(X, \mu),$$

equipped with the L_1-norm, does not have the weak Riesz-Fischer property. Note that E^+ contains many increasing Cauchy sequences that fail to be bounded above. Show also that E contains a sequence converging in norm to zero such that the sequence is not bounded, neither above nor below.

Hint: For the weak Riesz-Fischer property in (ii), observe that E is an ideal in L_1, so E is Dedekind complete. The space E is not a Banach lattice, however. Now use Exercise 16.5.

17. Order Continuous Norms

Before dealing with the main subject in this section, that is to say, with the properties of Banach lattices having order continuous norm (the definition will follow), we return for some moments to directed sets. Since these sets will occur on several further occasions in the sequel, it is convenient to fix some notations similar to those for monotone sequences. If D is an upwards (or downwards) directed set in the Riesz space E, we shall write $D \uparrow$ (or $D \downarrow$). If $D \uparrow$ and D has the supremum f_0, we shall write $D \uparrow f_0$. Similarly, if $D \downarrow$ and D has infimum f_0, we write $D \downarrow f_0$. It was proved in Theorem 15.7 that if $D \uparrow$ in the normed Riesz space E and D converges in norm to f_0, then $D \uparrow f_0$. Similarly, of course, if $D \downarrow$ and D converges in norm to f_0, then $D \downarrow f_0$. Recall that, by definition, the upwards directed set D converges in norm to f_0 if for any $\epsilon > 0$ there exists an element $f(\epsilon) \in D$ such that

$\| f - f_0 \| \leq \epsilon$ for all $f \in D$ satisfying $f \geq f(\epsilon)$. It is not surprising that for a directed set in a normed Riesz space there is also a notion similar to that of a Cauchy sequence. The upwards directed set D in the normed Riesz space E is called a *norm Cauchy system* if for any $\epsilon > 0$ there exists an element $f(\epsilon) \in D$ such that $\| f_1 - f_2 \| \leq \epsilon$ for all f_1, f_2 in D satisfying $f_1 \geq f(\epsilon)$ and $f_2 \geq f(\epsilon)$. The definition for a downwards directed set is similar. It is evident that if the directed set D converges in norm, then D is a Cauchy system. In the converse direction, we shall prove that in a Banach lattice every Cauchy system converges (in norm). For this purpose we first present a lemma.

Lemma 17.1. *Let D be an upwards directed Cauchy system in the normed Riesz space E. Then there exists an increasing sequence in D such that the sequence and D have the same upper bounds (where it is possible that the set of these upper bounds is empty). A similar assertion holds for downwards directed Cauchy systems.*

Proof. Let $(\epsilon_n : n = 1, 2, \ldots)$ be a sequence of positive numbers such that $\epsilon_n \downarrow 0$. By hypothesis there exists an element f_1 in D such that $\| f - f_1 \| \leq \epsilon_1$ for all $f \in D$ satisfying $f \geq f_1$. Also, there exists f_2^* in D such that $\| f - f_2^* \| \leq \epsilon_2$ for all $f \in D$ satisfying $f \geq f_2^*$. Now, choose $f_2 \in D$ such that $f_2 \geq f_1 \vee f_2^*$. Then $f_2 \geq f_1$ and all $f \geq f_2$ in D satisfy

$$0 \leq f - f_2 \leq f - f_2^*, \text{ so } \| f - f_2 \| \leq \epsilon_2.$$

Similarly, it is seen that there exists f_3 in D such that $f_3 \geq f_2$ and $\| f - f_3 \| \leq \epsilon_3$ for all $f \in D$ satisfying $f \geq f_3$. Continuing, we obtain an increasing sequence $(f_n : n = 1, 2, \ldots)$ in D such that $\| f - f_n \| \leq \epsilon_n$ for all $f \in D$ satisfying $f \geq f_n$. In particular we have $\| f_m - f_n \| \leq \epsilon_n$ for $m \geq n$. This shows that the sequence is a Cauchy sequence.

To show that $(f_n : n = 1, 2, \ldots)$ has the required property concerning its upper bounds, it is sufficient to prove that if g is any upper bound of the sequence, then g is an upper bound of D, i.e., $g \vee f = g$ for every $f \in D$. To this end, let $f \in D$ and f_n in the sequence be given. There exists an element $f' \in D$ such that $f' \geq f \vee f_n$, and so

$$\| (f \vee f_n) - f_n \| \leq \| f' - f_n \| \leq \epsilon_n.$$

By one of the Birkhoff inequalities it follows that

$$0 \leq (g \vee f) - g = g \vee (f \vee f_n) - g \vee f_n \leq (f \vee f_n) - f_n,$$

and so

$$\| (g \vee f) - g \| \leq \| (f \vee f_n) - f_n \| \leq \epsilon_n.$$

This holds for every n. Since $\epsilon_n \downarrow 0$, it follows that $\| (g \vee f) - g \| = 0$, so $(g \vee f) - g = 0$, i.e., $g \vee f = g$. ∎

Theorem 17.2. *Every norm Cauchy system in a Banach lattice converges (in norm). If the system is upwards directed, the norm limit is the supremum of the system. In other words, if $D \uparrow$ and D is a norm Cauchy system, then D converges in norm to some f_0, and $D \uparrow f_0$. Similarly, if D is downwards directed.*

Proof. Let D be an upwards directed Cauchy system in the Banach lattice E. Then, by the lemma, D contains an increasing Cauchy sequence $(f_n : n = 1, 2, \ldots)$ such that $\| f - f_n \| \leq \epsilon_n \downarrow 0$ for all $f \in D$ satisfying $f \geq f_n$. Since E is a Banach lattice, the norm limit f_0 of the sequence exists and, because f_n is increasing with n, the norm limit is also the supremum, i.e., $f_n \uparrow f_0$. We prove that D converges in norm to f_0. Given $\epsilon > 0$, there exists a natural number n_0 such that $\| f_0 - f_n \| \leq \epsilon/2$ for all $n \geq n_0$. Now choose $n \geq n_0$ such that the corresponding ϵ_n satisfies $\epsilon_n \leq \epsilon/2$. Then, for any $f \in D$ satisfying $f \geq f_n$, we have

$$\| f - f_0 \| \leq \| f - f_n \| + \| f_n - f_0 \| \leq \epsilon_n + \epsilon/2 \leq \epsilon.$$

This shows that D converges in norm to f_0. In view of $f_n \uparrow f_0$ it follows from the lemma that $D \uparrow f_0$. Note that $D \uparrow f_0$ follows also from Theorem 15.7. ∎

Exercise 17.3. Recall that there exist normed Riesz spaces containing upwards directed Cauchy systems that fail to be bounded above (see Exercise 16.7(ii)). Show that every upwards directed Cauchy system in the normed Riesz space E is bounded above if and only if E has the weak Riesz-Fischer property.

If D is an upwards directed set in the Riesz space E and D is bounded above, then the set G of all upper bounds of D is not empty and it is evident that G is downwards directed. Hence, the set

$$G - D = (g - f : g \in G, f \in D)$$

is downwards directed because if $g_1 - f_1$ and $g_2 - f_2$ are given in $G - D$ and we choose $g_3 \leq g_1 \wedge g_2$ in G and $f_3 \geq f_1 \vee f_2$ in D, then

$$g_3 - f_3 \leq (g_1 - f_1) \wedge (g_2 - f_2).$$

The set $G - D$ is a subset of E^+ and sometimes it is important to know whether or not $G - D$ has an infimum. The following lemma provides an answer for the case that E is Archimedean.

Lemma 17.4. *If $D \uparrow$ in the Archimedean Riesz space E and D is bounded above with G as the set of its upper bounds, then $G - D \downarrow 0$.*

Proof. Assume that $0 \leq w \leq g - f$ for all $g \in G$ and all $f \in D$. We have to prove that $w = 0$. Fix $g \in G$. It follows from $g - w \geq f$ for all $f \in D$ that $g - w$ is an upper bound of D, i.e., $g - w \in G$. Thus, $g \in G$ implies that $g - w \in G$. By induction we get $g - nw \in G$ for $n = 1, 2, \ldots$. But then $g - nw \geq f$ for every $f \in D$, so (fixing $f \in D$ as well) $nw \leq g - f$ for $n = 1, 2, \ldots$. Since E is Archimedean, it follows that $w = 0$. ∎

We now introduce a notion, somewhat stronger than Dedekind completeness.

Definition 17.5. (i) The Riesz space E is said to be *order separable* if every set in E possessing a supremum contains a finite or countable subset having the same supremum (equivalently, every set possessing an infimum contains a finite or countable subset having the same infimum).

(ii) The Riesz space E is said to be *super Dedekind complete* if E is order separable and Dedekind complete, i.e., if every set in E which is bounded above has a supremum and contains a finite or countable subset having the same supremum.

Theorem 17.6. *(i) The Riesz space E is order separable if and only if every upwards directed set in E^+ possessing a supremum contains a finite or countable subset having the same supremum. In this case the subset of the upwards directed set in E^+ may be assumed to be a monotonely increasing sequence.*

(ii) The Riesz space E is super Dedekind complete if and only if every upwards directed set in E^+ which is bounded above possesses a supremum and contains a finite or countable subset having the same supremum.

(iii) If E is order separable or super Dedekind complete, then the same holds for any ideal in E.

Proof. (i) Assume that every upwards directed set in E^+ possessing a supremum has a finite or countable subset having the same supremum. Let D be a subset of E possessing a supremum, say $\sup D = f_0$, and assume first that D contains finite suprema (i.e., D contains all finite suprema of its own elements). This implies that $D \uparrow f_0$. Choose $f_1 \in D$ and denote the set $(f \vee f_1 : f \in D)$ by D_1. Then the subset D_1 of D satisfies $D_1 \uparrow f_0$ and $f \geq f_1$ holds for all $f \in D_1$. Let $D_2 = (f - f_1 : f \in D_1)$. Then D_2 is a subset of E^+ and $D_2 \uparrow f_0 - f_1$. Hence, by hypothesis, D_2 contains a finite or countable subset having the same supremum as D_2. It follows immediately that D_1 (and hence D) has the same property.

Now let it be given only that D is a subset of E having the supremum f_0. Adjoining to D all finite suprema of its elements we obtain a set D_3 still having supremum f_0. Hence, by what has been proved already, D_3 contains a finite or countable subset $(g_1, g_2, \ldots)$ having supremum f_0. The elements $g_n (n = 1, 2, \ldots)$ are not necessarily members of D, but each g_n is a finite supremum of members of D. Replacing each g_n by these members of D we obtain a finite or countable subset of D having f_0 as supremum. Finally, let D_4 be any upwards directed subset of E^+ having supremum f_0 and let $(h_1, h_2, \ldots)$ be a subset of D_4 such that $\sup h_n = f_0$. We shall define a sequence $(k_1, k_2, \ldots)$ in D_4 such that $k_1 \leq k_2 \leq \cdots$ and $k_n \geq h_n$ for all n (so $\sup k_n = f_0$). Let $k_1 = h_1$ and let $k_2 \in D_4$ be such that $k_2 \geq h_1 \vee h_2 = k_1 \vee h_2$. Next, let $k_3 \in D_4$ be such that $k_3 \geq k_2 \vee h_3$. Generally, if $k_1, \ldots, k_n$ are already defined, let $k_{n+1} \in D_4$ be such that $k_{n+1} \geq k_n \vee h_{n+1}$.

(ii) The proof follows by combining the results in part (i) and in Theorem 12.1(i).

(iii) Follows from (i) and (ii).   ∎

As an example, let μ be a σ-finite measure in the point set X. The Riesz space $L_0 = L_0(X, \mu)$ of all (real) μ-measurable functions on X is now super Dedekind complete. This is seen by observing that the proof for Dedekind completeness of L_0 (in Example 12.5(iii)) is based on the fact that if D is a bounded subset of L_0^+, then D contains a countable subset the (pointwise) supremum of which is the supremum of D. It follows that every ideal in L_0 is super Dedekind complete; in particular, all spaces $L_p(X, \mu)$ for $0 < p \leq \infty$ are super Dedekind complete. Looking again at the proof for L_0, it may be of some interest to observe that the proof begins by showing that the space $L_1(X, \mu)$, with $\mu(X)$ finite, is super Dedekind complete.

The Riesz space E of all (real) bounded functions on an uncountable point set X, with pointwise ordering, is Dedekind complete, but not super Dedekind complete. To see this, let for example D be the set of all functions in E assuming the value one at one point of X and vanishing at all other points. Then $\sup D$ is the function identically one and D does not have a finite or countable subset with the same supremum. Observe that if μ is the counting measure in X (i.e., each point of X has measure one), then E is exactly the space $L_\infty(X, \mu)$. Of course, the measure μ is not σ-finite in this case.

All preparations are made to introduce now the notions of order continuous and σ-order continuous norms.

Definition 17.7. The normed Riesz space E is said to have *order continuous norm* if, for any subset $D \downarrow 0$ in E, we have $\inf(\| f \| : f \in D) = 0$. The norm is said to be *σ-order continuous* if, for any sequence $f_n \downarrow 0$ in E, we have $\| f_n \| \downarrow 0$.

Some comments are appropriate. Evidently, the norm in E is σ-order continuous if and only if $f_n \uparrow f_0$ (or $f_n \downarrow f_0$) implies that $\| f_0 - f_n \| \to 0$, i.e., f_n converges in norm to f_0. Conversely, it follows in any normed Riesz space from $f_n \uparrow$ (or $f_n \downarrow$) and $\| f_0 - f_n \| \to 0$ that $f_n \uparrow f_0$ (or $f_n \downarrow f_0$). Hence, order convergence and norm convergence for monotone sequences are equivalent if and only if the norm in E is σ-order continuous. Similarly, the norm in E is order continuous if and only if $D \uparrow f_0$ (or $D \downarrow f_0$) implies that

$$\inf(\| f - f_0 \|: f \in D) = 0,$$

i.e., D converges in norm to f_0. Conversely, it follows in any normed Riesz space from $D \uparrow$ (or $D \downarrow$) and norm convergence of D to f_0 that $f_0 = \sup D$ (or $f_0 = \inf D$). Hence, norm convergence of an upwards (or downwards) directed set D to f_0 is equivalent to $f_0 = \sup D$ (or $f_0 = \inf D$) if and only if the norm in E is order continuous.

It is evident that if E is order separable and the norm in E is σ-order continuous, then the norm is order continuous. This holds in particular for the spaces $L_p(X, \mu), 1 \leq p < \infty$, where μ is a σ-finite measure in the point set X. The sequence space ℓ_∞ and the space $L_\infty([0, 1], \mu)$, with μ Lebesgue measure, are examples of Banach lattices with norm failing to be σ-order continuous.

Theorem 17.8. *Any Banach lattice having order continuous norm is super Dedekind complete.*

Proof. Let E be a Banach lattice having order continuous norm and let D be an upwards directed set in E such that D is bounded above. Denote the set of all upper bounds of D by G. In view of Lemma 17.4 we have $G - D \downarrow 0$, so

$$\inf(\| g - f \|: g \in G, f \in D) = 0,$$

since the norm in E is order continuous. Therefore, given $\epsilon > 0$, there exist $g \in G$ and $f_1 \in D$ such that $\| g - f_1 \| < \epsilon$. Since $0 \leq f_2 - f_1 \leq g - f_1$ for all $f_2 \in D$ satisfying $f_2 \geq f_1$, it follows that $\| f_2 - f_1 \| < \epsilon$ for all $f_2 \geq f_1$ in D. This shows that D is a norm Cauchy system. The space E is a Banach lattice, so the Cauchy system D converges in norm to some $f_0 \in E$ and f_0 is now the supremum of D (see Theorem 17.2). Furthermore, D contains an increasing sequence having the same supremum (see Lemma 17.1). These results show that E is super Dedekind complete. ∎

In the next theorem we present the first characterization of Banach lattices having order continuous norm.

Theorem 17.9. *For a Banach lattice E the following conditions are equivalent.*

(i) E has order continuous norm.
(ii) E has σ-order continuous norm and E is Dedekind σ-complete.
(iii) Every sequence in E which is increasing and bounded above converges in norm.

Proof. (i)$\rightarrow$ (ii) Evident, since in view of the preceding theorem E is super Dedekind complete.

(ii)$\rightarrow$ (iii) Let $(f_n : n = 1, 2, \ldots)$ be an increasing sequence in E which is bounded above. Then, since E is Dedekind σ-complete, the supremum of the sequence exists, so $f_n \uparrow f_0$. It follows that $f_0 - f_n \downarrow 0$, so $\parallel f_0 - f_n \parallel \downarrow 0$ since E has σ-order continuous norm. This shows that f_n converges in norm to f_0.

(iii)$\rightarrow$ (i) Let $D \downarrow 0$ in E. We have to show that $\inf(\parallel f \parallel : f \in D) = 0$, and for this purpose we may just as well assume that D contains finite infima (i.e., D contains all finite infima of its own elements), since this does not make any difference for the number $\inf(\parallel f \parallel : f \in D)$. Now fix some $f_0 \in D$ and define D_1 to be the set of all $(f \wedge f_0 : f \in D)$. Then $D_1 \downarrow 0$ and

$$\inf(\parallel f \parallel : f \in D_1) = \inf(\parallel f \parallel : f \in D),$$

so we may just as well assume that we have to do with D_1 from the beginning on. In other words, we may assume that $D \downarrow 0$ and D has an upper bound f_0. The set $D_2 = (f_0 - f : f \in D)$ satisfies $D_2 \uparrow f_0$, from which it follows that D_2 is a norm Cauchy system, because otherwise there would exist some $\epsilon > 0$ and an increasing sequence $f_1 \leq f_2 \leq \cdots$ in D_2 such that $\parallel f_{n+1} - f_n \parallel > \epsilon$ for $n = 1, 2, \ldots$, contradicting our hypothesis (iii). Hence, since E is a Banach lattice, D_2 converges in norm to some $g \in E$ and $g = \sup D_2$. But $\sup D_2 = f_0$, and so $g = f_0$. It has been proved thus that D_2 converges in norm to f_0. Since $D_2 = (f_0 - f : f \in D)$ by definition, it follows that D converges in norm to the null element, i.e., given $\epsilon > 0$, there exists an element $f(\epsilon) \in D$ such that $\parallel f \parallel \leq \epsilon$ for all $f \in D$ satisfying $f \leq f(\epsilon)$. Hence, $\inf(\parallel f \parallel : f \in D) = 0$ ∎

The results in the following exercise are essentially due to H. Nakano (1943).

Exercise 17.10. let E be a normed Riesz space, but not necessarily a Banach lattice. Show that the conditions (ii), (iii) in the theorem above are still equivalent and each of these is equivalent to

(i') E has order continuous norm and E is super Dedekind complete.

Hint: The proofs for (i') $\rightarrow$ (ii) $\rightarrow$ (iii) are as in the theorem. In the proof that (iii) implies (i'), use Lemma 17.1 to see that the Cauchy system D_2 contains an increasing Cauchy sequence having supremum f_0, and therefore (by hypothesis (iii)) converging in norm to f_0. To show that E is super Dedekind complete, let $D \uparrow$ and let D be bounded above. Then again D is a Cauchy system since otherwise (iii) would be violated. It follows that D contains an increasing Cauchy sequence having the same upper bounds as D. By hypothesis (iii) the sequence converges in norm to some f_0 and, once more by Lemma 17.1, f_0 is the supremum of the sequence as well as of D.

As defined in section 8, the elements f, g in the Riesz space E are said to be disjoint if $\inf(\mid f \mid, \mid g \mid) = 0$, i.e., $\mid f \mid \wedge \mid g \mid = 0$. The sequence

$(f_n : n = 1, 2, \ldots)$ in E is said to be a *disjoint sequence* if every two elements in the sequence are disjoint, i.e., $\inf(|\, f_m \,|, |\, f_n \,|) = 0$ for $m \neq n$. It is admitted here that $f_n = 0$ for one or more values of n. If $k \geq 2$ is a natural number, the sequence (f_n) in E is said to be a *k-disjoint sequence* if for every k-tuple of elements in the sequence the infimum of the absolute values of these elements is the null element of E. A disjoint sequence, therefore, is the same as a 2-disjoint sequence. Evidently, every k-disjoint sequence is $(k + 1)$-disjoint, but the converse does not hold in general. Disjoint sequences in a Banach lattice possessing order continuous norm have a special property. As shown in the last theorem, it is true in such a space E that every order bounded increasing sequence converges in norm. This implies easily that every order bounded disjoint sequence in E^+ converges in norm to the null element. The proof is almost immediate. Let $(u_n : n = 1, 2, \ldots)$ be such a sequence, say $0 \leq u_n \leq u_0$ for all n and $u_m \wedge u_n = 0$ for $m \neq n$. Then the partial sums $s_n = u_1 + \cdots + u_n$ satisfy $s_n \uparrow$ and

$$s_n = u_1 \vee \cdots \vee u_n \leq u_0,$$

so s_n converges in norm. But then $\|\, u_n \,\| = \|\, s_n - s_{n-1} \,\| \to 0$ as $n \to \infty$. It is a remarkable result that this property of disjoint sequences characterizes Banach lattices having order continuous norm, that is to say, the Banach lattice E has order continuous norm if and only if every order bounded disjoint sequence in E^+ converges in norm to the null element. The proof is not trivial and some preliminary discussion is needed. The first point we shall discuss is a certain method to obtain a disjoint sequence from a given sequence. There exist two variants of the method, one of which can be used for order bounded sequences in any arbitrary Riesz space and the other one works for norm bounded sequences in a Banach lattice.

Theorem 17.11. *(i) Let E be a Riesz space and assume that $(u_n : n = 1, 2, \ldots)$ is an order bounded sequence in E^+, say $0 \leq u_n \leq u$ for all n. Define the sequence $(v_n : n = 1, 2, \ldots)$ by $v_1 = 0$ and*

$$v_n = (u_n - n\Sigma_{k=1}^{n-1} u_k - n^{-1}u)^+ \text{ for } n = 1, 2, \ldots$$

Then (v_n) is a disjoint sequence such that $0 \leq v_n \leq u_n$ holds for all n.

(ii) Let E be a Banach lattice and assume that $(u_n : n = 1, 2, \ldots)$ is a norm bounded sequence in E^+. Define the sequence $(v_n : n = 1, 2, \ldots)$ by

$$v_1 = (u_1 - \Sigma_{k=2}^{\infty} 2^{-k} u_k)^+,$$

$$v_n = (u_n - 2^n \Sigma_{k=1}^{n-1} u_k - \Sigma_{k=n+1}^{\infty} 2^{-k} u_k)^+ \text{ for } n = 2, 3, \ldots$$

Then (v_n) is a disjoint sequence such that $0 \leq v_n \leq u_n$ holds for all n.

Proof. (i) Let $2 \leq m < n$. Then

$$v_n \leq (u_n - nu_m)^+,$$

$$v_m = (u_m - m\Sigma_1^{m-1}u_k - m^{-1}u)^+ \leq (u_m - m^{-1}u_n)^+ \leq$$

$$(u_m - n^{-1}u_n)^+ = n^{-1}(nu_m - u_n)^+ = n^{-1}(u_n - nu_m)^-,$$

so $v_n \wedge v_m = 0$, since $(u_n - nu_m)^+ \wedge (u_n - nu_m)^- = 0$. It is evident that $0 \leq v_n \leq u_n$ for all n.

(ii) Note first that $\Sigma_1^\infty 2^{-k}u_k$ converges absolutely because the sequence (u_n) is bounded in norm. Hence, the partial sums converge in norm and their limit is also their supremum. Thus, according to our conventions about notations, $\Sigma_1^\infty 2^{-k}u_k$ exists. The same holds for $\Sigma_{n+1}^\infty 2^{-k}u_k$. Now, if $1 \leq m < n$, we have

$$v_n \leq (u_n - 2^n u_m)^+,$$

$$v_m \leq (u_m - 2^{-n}u_n)^+ = 2^{-n}(2^n u_m - u_n)^+ = 2^{-n}(u_n - 2^n u_m)^-,$$

so $v_n \wedge v_m = 0$.

Finally, observe that in part (i) the same result holds if we define v_n by

$$v_n = (u_n - \alpha_n \Sigma_{k=1}^{n-1} u_k - \alpha^{-1}u)^+,$$

where $\alpha_n \uparrow \infty$. A similar remark holds for (ii), provided $\Sigma_1^\infty \alpha_n^{-1}$ converges. $\blacksquare$

It is of importance for what follows to estimate $u_n - v_n$ in part (i) of the last theorem. Using that $f - (f - g)^+ = f \wedge g$ for arbitrary f and g (formula (2) in the proof of Theorem 5.2), we obtain for $n \geq 2$ that

$$0 \leq u_n - v_n = u_n - (u_n - n\Sigma_1^{n-1}u_k - n^{-1}u)^+ =$$

$$u_n \wedge (n\Sigma_1^{n-1}u_k + n^{-1}u) \leq u_n \wedge (n\Sigma_1^{n-1}u_k) + n^{-1}u$$

and

$$u_n - v_n \geq u_n \wedge (n\Sigma_1^{n-1}u_k).$$

Hence, writing $w_1 = 0$ and $w_n = u_n \wedge (n\Sigma_1^{n-1}u_k)$ for $n = 2, 3, \ldots$, we have

$$v_n + w_n \leq u_n \leq v_n + w_n + n^{-1}u \text{ for } n = 1, 2, \ldots. \tag{1}$$

For further information we have to investigate, therefore, the behaviour of the sequence (w_n).

Theorem 17.12. *Let E be an arbitrary Riesz space. If $(u_n : n = 1, 2, \ldots)$ is a $(k+1)$-disjoint sequence in E^+ and $(w_n : n = 1, 2, \ldots)$ is defined by $w_1 = 0$ and*

$$w_n = u_n \wedge (n\Sigma_{j=1}^{n-1} u_j) \ for \ n = 2, 3, \ldots,$$

then (w_n) is a k-disjoint sequence in E^+. Evidently, if (u_n) is order bounded, then (w_n) is order bounded.

Proof. Let $J = \{n(1), \ldots, n(k)\}$ be a set of k natural numbers ($k \geq 2$) and assume that $n(1)$ is the smallest number in J. If $n(1) = 1$, then $w_{n(1)} = w_1 = 0$, and so $\inf(w_n : n \in J) = 0$. If $n(1) > 1$, we have

$$0 \leq \inf(w_n : n \in J) \leq u_{n(1)} \wedge \cdots \wedge u_{n(k)} \wedge n(1)\Sigma_{j=1}^{n(1)-1} u_j \leq$$

$$n(1)\Sigma_{j=1}^{n(1)-1} u_{n(1)} \wedge \cdots \wedge u_{n(k)} \wedge u_j = 0,$$

since (u_n) is $(k+1)$-disjoint. This shows that (w_n) is a k-disjoint sequence.∎

Speaking somewhat unprecisely, we may derive now from formula (1) above that every positive order bounded and $(k+1)$-disjoint sequence in a Riesz space is almost equal to the sum of a positive disjoint sequence and a positive k- disjoint sequence. In particular, it follows from (1) that if E is a normed Riesz space having the property that, for some $k \geq 3$, every order bounded $(k-1)$- disjoint sequence tends in norm to zero, then the same holds for every order bounded k-disjoint sequence. Hence, it is easy now to prove the following theorem.

Theorem 17.13. *If every order bounded disjoint sequence in the normed Riesz space E tends in norm to zero, then, for every $k = 2, 3, \ldots$, every order bounded k-disjoint sequence tends in norm to zero.*

Proof. The theorem holds for $k = 2$, so by induction it holds for every $k = 2, 3, \ldots$.   ∎

We shall now formulate and prove the main theorem in this section containing (as already announced) another characterization of Banach lattices possessing order continuous norm.

Theorem 17.14. *Let E be a Banach lattice. Then E has order continuous norm if and only if every order bounded disjoint sequence in E converges in norm to zero.*

Proof. We have seen already that if E has order continuous norm, then every order bounded disjoint sequence converges in norm to zero. For the converse, assume now that every order bounded disjoint sequence in E has this property. Equivalently, assume that, for $k = 2, 3, \ldots$, every order bounded k-disjoint sequence in E converges in norm to zero. We have to show that E has order continuous norm or, equivalently, we have to show that every order bounded increasing sequence in E converges in norm (see Theorem 17.9). Assuming that the last condition is not satisfied, there exists an order bounded increasing sequence in E that does not converge in norm. We may assume that the sequence is contained in E^+ and, passing to a subsequence if necessary, we obtain an increasing sequence (u_n) such that $0 \leq u_n \leq u$ for some $u \in E^+$ and $\| u_{n+1} - u_n \| > \epsilon$ for some ϵ and all n. Let $k(k \geq 2)$ be a natural number such that $\| k^{-1}u \| < \epsilon/2$ and define v_n by

$$v_n = (u_{n+1} - u_n - k^{-1}u)^+ \text{ for } n = 1, 2, \ldots.$$

We prove that (v_n) is a k-disjoint sequence (and so $\| v_n \| \to 0$ as $n \to \infty$). For the proof, let $J = \{n(1), \ldots, n(k)\}$ be an arbitrary set of k different natural numbers with $n(k)$ the largest among these numbers. Then

$$0 \leq \Sigma_{j=1}^{k}\bigl(u_{n(j)+1} - u_{n(j)}\bigr) \leq \Sigma_{n=1}^{n(k)}\bigl(u_{n+1} - u_n\bigr) =$$

$$u_{n(k)+1} - u_1 \leq u,$$

so $w = \inf(u_{n+1} - u_n : n \in J)$ satisfies

$$kw \leq \Sigma_{j=1}^{k}\bigl(u_{n(j)+1} - u_{n(j)}\bigr) \leq u.$$

It follows that

$$\inf(u_{n+1} - u_n - k^{-1}u : n \in J) = w - k^{-1}u \leq 0,$$

which implies in view of one of the distributive laws (Corollary 6.2) that $\inf(v_n : n \in J) = 0$. Hence, (v_n) is an order bounded k-disjoint sequence. As observed already, it follows that $\| v_n \| \to 0$ as $n \to \infty$. But then, since

$$(u_{n+1} - u_n) - v_n = (u_{n+1} - u_n) - (u_{n+1} - u_n - k^{-1}u)^+ =$$

$$(u_{n+1} - u_n) \wedge k^{-1}u,$$

we get

$$\| u_{n+1} - u_n \| = \| v_n + (u_{n+1} - u_n) \wedge k^{-1}u \| \leq \| v_n \| + \| k^{-1}u \| <$$

$$\| v_n \| + \frac{\epsilon}{2} < \epsilon$$

for sufficiently large n. This contradicts the assumption that $\| u_{n+1} - u_n \| > \epsilon$ for all n. Hence, every order bounded increasing sequence in E converges in norm. ∎

The last theorem is due to P. Meyer-Nieberg (1972) and, independently, to D.H.Fremlin (1974). Their proofs are different. The proof presented here is a somewhat modified version of Fremlin's proof (which works for the more general case of a topological Riesz space).

Exercise 17.15. Let μ be a countably additive σ-finite (and non-negative) measure in the point set X and let $1 \leq p < \infty$. Show that the norm in $L_p(X, \mu)$ is order continuous. Show that if, for example, μ is Lebesgue measure in $X = [0, 1]$, then $L_\infty(X, \mu)$ does not have order continuous norm. Show that $C([0, 1])$ does not have order continuous norm.

Hint: Use Theorem 17.9.

Exercise 17.16. If E is a normed Riesz space (but not necessarily a Banach lattice), then every order bounded increasing sequence in E is a norm Cauchy sequence if and only if every order bounded disjoint sequence in E converges in norm to zero. Prove this.

Some conditions necessary and sufficient for a normed Riesz space or a Banach lattice to have an order continuous norm were mentioned already. There are several further conditions of this kind. One of these, due to T. Ando (1965), is contained in the following theorem.

Theorem 17.17. *The normed Riesz space E has order continuous norm if and only if every norm closed ideal in E is a band. Furthermore, if E is a Banach lattice having order continuous norm, then every norm closed ideal in E is a projection band.*

Proof. Let E be a normed Riesz space with order continuous norm. Furthermore, let A be a norm closed ideal in E. For the proof that A is a band, assume that D is an upwards directed subset of A^+ with supremum u_0, so $D \uparrow u_0$. We have to show that $u_0 \in A$. Since the norm in E is order continuous, it follows from $D \uparrow u_0$ that $(\| u_0 - u \| : u \in D) \downarrow 0$. Hence, similarly as in the first part of the proof of Lemma 17.1, there exists an increasing sequence $0 \leq u_1 \leq u_2 \leq \cdots$ in D such that $\| u_0 - u_n \| \downarrow 0$ as $n \to \infty$. Since $u_n \in A$ holds for all n and A is norm closed, it follows that $u_0 \in A$.

Conversely, assume that every norm closed ideal in E is a band. It is sufficient now to prove that $D \uparrow u_0$ in E^+ implies that $(\| u_0 - u \| : u \in D) \downarrow 0$. For this purpose, choose $\epsilon > 0$ and then choose α such that $0 < \alpha < 1$ and $\| (1 - \alpha)u_0 \| < \epsilon$. It follows then from

$$0 \leq u_0 - u = (1 - \alpha)u_0 + \alpha u_0 - u \leq (1 - \alpha)u_0 + (\alpha u_0 - u)^+ \qquad (2)$$

for every $u \in D$ that

$$\| u_0 - u \| \leq \epsilon + \| (\alpha u_0 - u)^+ \| .$$

Since the set $\{(\alpha u_0 - u)^+ : u \in D\}$ is downwards directed, it is sufficient to prove now that $\| (\alpha u_0 - u^*)^+ \| \leq \epsilon$ for some $u^* \in D$ since then $\| u_0 - u \| \leq 2\epsilon$ for all $u \in D$ satisfying $u \geq u^*$. Observe that

$$(u - \alpha u_0 : u \in D) \uparrow (1 - \alpha)u_0, \text{ so } \{(u - \alpha u_0)^+ : u \in D\} \uparrow (1 - \alpha)u_0.$$

Therefore, denoting by A the ideal generated by the set of all $\{(u - \alpha u_0)^+ : u \in D\}$, it is evident that u_0 is an element of the band B generated by A. The norm closure A^- of A is a band by hypothesis, so B is a subset of A^- (since B is the smallest band containing A). Therefore, it follows from $u_0 \in B$ that $u_0 \in A^-$. This implies that for the given $\epsilon > 0$ there exists an element $v \in A$ such that $\| u_0 - v \| < \epsilon$. Then $v^+ \in A$ and

$$| u_0 - v^+ | = | u_0^+ - v^+ | \leq | u_0 - v | , \text{ so } \| u_0 - v^+ \| < \epsilon.$$

We may assume, therefore, just as well that $0 \leq v \in A$. Similarly, v may be replaced by $v \wedge u_0$, so we may assume that $0 \leq v \leq u_0$ holds. Now observe that by the definition of the ideal generated by a given set of elements (see section 7) any element in the ideal is already contained in the ideal generated by a finite number of elements of the given set. This implies in our case (since the set $\{(u - \alpha u_0)^+ : u \in D\}$ is upwards directed) that there exists an element $u^* \in D$ such that v is an element of the principal ideal generated by $(u^* - \alpha u_0)^+$. Hence

$$v \perp (u^* - \alpha u_0)^- = (\alpha u_0 - u^*)^+.$$

Each of the disjoint elements v and $(\alpha u_0 - u^*)^+$ is majorized by u_0, so

$$v + (\alpha u_0 - u^*)^+ \leq u_0.$$

It follows that $\| (\alpha u_0 - u^*)^+ \| \leq \| u_0 - v \| \leq \epsilon$, which is the desired result.

It remains to show that in a Banach lattice E with order continuous norm every norm closed ideal is not only a band but even a projection band. This follows immediately by observing that in this case E is super Dedekind complete (see Theorem 17.8), which implies that E has the projection property (in view of the main inclusion theorem; Theorem 12.3). Every band in E is therefore a projection band. ∎

The inequality in formula (2) is known as *Ando's inequality*. The inequality will play a further important role in the discussion of order continuous operators (in section 22).

Exercise 17.18. Let $E = L_1([0,1], \mu) \cap L_\infty([0,1], \mu)$, where μ is Lebesgue measure. Equipped with the L_1-norm, E is a normed Riesz space but not a Banach lattice. Show that the norm in E is order continuous and every norm closed ideal in E is a projection band. Hence, the property that every norm closed ideal is a projection band does not necessarily imply that the space is a Banach lattice.

Exercise 17.19. In section 11 we have defined a σ-ideal in the Riesz space E as an ideal A with the property that if $f_n \in A$ for $n = 1, 2, \ldots$ and $\sup f_n = f_0$ exists in E, then $f_0 \in A$. Show that the following conditions for the ideal A in E are equivalent.

(i) A is a σ-ideal in E.
(ii) If $u_n \in A^+$ for $n = 1, 2, \ldots$ and $u_n \uparrow u_0$ in E, then $u_0 \in A$ (compare this with Theorem 11.7).
(iii) If $f_n \in A$ for $n = 1, 2, \ldots$ and $f_n \to f_0$ (in order) in E, then $f_0 \in A$ (i.e., A is an order closed ideal).

Assume now that E is a normed Riesz space. Show that E has σ-order continuous norm if and only if every norm closed ideal is a σ-ideal, i.e., every norm closed ideal is order closed.

Hint: It is easy to see that (i) and (ii) are equivalent and (iii) implies (ii). For the proof that (ii) implies (iii), assume that (ii) holds and let $f_n \in A$ for $n = 1, 2, \ldots$ with $f_n \to f_0$ (in order). Then $f_n^+ \in A$ for all n and $f_n^+ \to f_0^+$. Let $g_n = f_n^+ \wedge f_0^+$ for all n. Show that

$$0 \leq f_0^+ - g_n \leq | f_0^+ - f_n^+ | \, .$$

Write $u_n = \sup(g_1, \ldots, g_n)$ for $n = 1, 2, \ldots$ and show that $0 \leq u_n$ in A with

$$0 \leq f_0^+ - u_n \leq f_0^+ - g_n \leq | f_0^+ - f_n^+ | \to 0.$$

Hence $f_0^+ \in A$ by hypothesis. Similarly $f_0^- \in A$, and so $f_0 \in A$.

The proof for the last part is very similar to the proof that E has order continuous norm if and only if every norm closed ideal is a band.

Exercise 17.20. (i) Recall (see section 11) that a linear operator (linear mapping) P from the vector space V into itself is called a projection if $P^2 = P$, i.e., $P(Pf) = Pf$ for every $f \in V$. In the case that $V = E$ is a Riesz space, the projection P in E is said to be a positive projection if $Pu \geq 0$ for every $u \geq 0$ in E. As shown in Theorem 11.4, every band projection P in E is a positive projection having the further property that $0 \leq Pu \leq u$ holds for every $u \geq 0$ in E. The range B of the band projection P (i.e., the set of all $g \in E$ satisfying $g = Pf$ for some $f \in E$) is a projection band, i.e., $B \oplus B^d = E$, and $Pf = 0$ for every $f \in B^d$. Not every positive projection is a band projection, as simple examples in $\mathbb{R}^2$ already show. We shall indicate, however, how to show that if P is a positive projection having an ideal A in E as its range, then A is a projection band and the corresponding band projection P_A on A satisfies $P_A \leq P$ (i.e., $P_A u \leq Pu$ for every $u \in E^+$). To prove that A is a projection band, show that for any $u \in E^+$ we have the decomposition

$$u = (u \wedge Pu) + \{u - (u \wedge Pu)\} \tag{3}$$

with $0 \leq u \wedge Pu \in A$ and $0 \leq u - (u \wedge Pu) \in A^d$. This will imply that $E = A \oplus A^d$, and hence A is a projection band (see Theorem 11.3). For the proof that (3) holds, observe first that any $g \in A$ is of the form $g = Pf$ for some f, so $g = Pf = P^2 f = Pg$. Now, let $u \in E^+$ be given. Then $u \wedge Pu \in A$, so it is sufficient to prove that $u - (u \wedge Pu) \in A^d$, i.e., $p = v \wedge \{u - (u \wedge Pu)\}$ satisfies $p = 0$ for every $v \in A^+$. We have $p \in A$, so $Pp = p$. Furthermore $p \leq u - (u \wedge Pu)$, which implies that

$$p = Pp \leq P\{u - (u \wedge Pu)\} = Pu - P(u \wedge Pu) = Pu - (u \wedge Pu).$$

It follows that

$$p \leq \{u - (u \wedge Pu)\} \wedge \{Pu - (u \wedge Pu)\} = (u \wedge Pu) - (u \wedge Pu) = 0.$$

It is obvious, by the uniqueness of the decomposition, that the component of $u \in E^+$ in the projection band A is $u \wedge Pu$, so $P_A u = u \wedge Pu \leq Pu$.

It is an immediate corollary that the set of ideals that are the range of some positive projection is the same as the set of all projection bands. This result is due to T. Ando (1969).

(ii) Let E be a normed Riesz space having the projection property. Show that the norm in E is order continuous if and only if every norm closed ideal in E is the range of a positive projection.

(iii) Let E be a Banach lattice. Show that the norm in E is order continuous if and only if every norm closed ideal in E is the range of a positive projection.

CHAPTER 9
Linear Operators

18. Linear Operators in Normed Spaces and in Riesz Spaces

If T is a mapping from the vector space V into the vector space W, i.e., if T has *domain* V and *range* W (notation $T : V \to W$), then (as well-known) T is called a *linear mapping* (also called a *linear transformation* or a *linear operator*) if

$$T(\alpha f + \beta g) = \alpha T f + \beta T g$$

for all f and g in V and all (real or complex) numbers α and β. For brevity we shall usually say operator instead of linear operator. It is evident that the set $\mathcal{L}(V, W)$ of all operators from V into W is a vector space if, for T_1, T_2 in $\mathcal{L}(V, W)$ and α, β real or complex, we define $\alpha T_1 + \beta T_2$ by

$$(\alpha T_1 + \beta T_2)f = \alpha T_1 f + \beta T_2 f \text{ for all } f \in V$$

If V, W, Z are vector spaces and T_1, T_2 are operators such that $T_1 : W \to Z$ and $T_2 : V \to W$, then the product operator $T_1 T_2 : V \to Z$ is defined by

$$(T_1 T_2)f = T_1(T_2 f) \text{ for all } f \in V.$$

For any vector space V the space $\mathcal{L}(V, V)$ of all operators from V into itself is usually denoted by $\mathcal{L}(V)$. The space $\mathcal{L}(V)$ is an algebra because, for T_1 and T_2 in $\mathcal{L}(V)$, the product $T_1 T_2$ is likewise an element of $\mathcal{L}(V)$. Note that the familiar distributive laws hold in $\mathcal{L}(V)$, such as for example $(T_1 + T_2)T_3 = T_1 T_3 + T_2 T_3$. In general the multiplication is not commutative. The operators $TT, TTT, \ldots$ in $\mathcal{L}(V)$ are denoted by $T^2, T^3, \ldots$. If $P \in \mathcal{L}(V)$ satisfies the equality $P^2 = P$, then P is called a *projection*. Note that if P is a projection, then $P = P^n$ for $n = 1, 2, \ldots$. The identity operator in $\mathcal{L}(V)$ is usually denoted by I, i.e., $If = f$ for all $f \in V$.

Any linear operator from the vector space V into the space $\mathbb{R}$ of real numbers or the space $\mathbb{C}$ of complex numbers (depending on whether V is a real or complex vector space) is called a *linear functional* on V. Hence, the set of all linear functionals on V is exactly the vector space $\mathcal{L}(V, \mathbb{R})$ or $\mathcal{L}(V, \mathbb{C})$.

This space is called the *algebraic dual space* of V, and it is sometimes denoted by $V^{\#}$ instead of $\mathcal{L}(V, \mathbb{R})$ or $\mathcal{L}(V, \mathbb{C})$.

Let $T : V \to W$ be an operator from the normed space V into the normed space W. Norms will be denoted by $\| \cdot \|$ as before; it will be clear from the context whether this denotes the norm in V or in W. The operator T is said to be *continuous at the point* $f_0 \in V$ if $\| f_n - f_0 \| \to 0$ as $n \to \infty$ implies $\| Tf_n - Tf_0 \| \to 0$, and T is said to be *continuous on* V if T is continuous at each point of V. Furthermore, T is called a *bounded operator* (more precisely, a *norm bounded operator*) if there exists a constant $A > 0$ such that $\| Tf \| \le A \| f \|$ for all $f \in V$. In the particular case that $W = \mathbb{R}$ or $W = \mathbb{C}$, the operator T is then a linear functional on V, continuous at the point f_0, continuous on V or bounded respectively. Since T is linear, these three properties coincide. We present the proof. Note first that T is continuous at f_0 if and only if for any $\epsilon > 0$ there exists $\delta > 0$ such that $\| f - f_0 \| \le \delta$ implies $\| Tf - Tf_0 \| \le \epsilon$.

Theorem 18.1. *If V and W are (real or complex) normed spaces and $T : V \to W$ is a linear operator, then the following conditions for T are equivalent.*

(i) T is continuous at one point $f_0 \in V$.
(ii) T is continuous on V.
(iii) T is norm bounded.

In particular, therefore, a linear functional on V is norm bounded if and only if it is continuous.

Proof. (i)$\to$ (ii) We assume that T is continuous at f_0. Choose $f^{\sim} \in V$ arbitrarily. We prove that T is continuous at $f^{\sim}$. For this purpose, assume that $\| f_n - f^{\sim} \| \to 0$ as $n \to \infty$, i.e., $f_n \to f^{\sim}$ in norm. Then, for $v = f^{\sim} - f_0$, we have $f_n - v \to f^{\sim} - v = f_0$ in norm, so $Tf_n - Tv \to Tf_0$ by hypothesis, i.e., $Tf_n \to Tf_0 + Tv = Tf^{\sim}$ in norm. This shows that T is continuous at $f^{\sim}$.

(ii)$\to$ (iii) T is continuous at the origin by hypothesis, so there exists a number $\delta > 0$ such that $\| f \| \le \delta$ implies $\| Tf \| \le 1$. In other words, $\| f \| \le 1$ implies $\| Tf \| \le \delta^{-1}$. If $f \neq 0$ is now arbitrary in V, then $\| f \|^{-1} f$ has norm one, so $T(\| f \|^{-1} f)$ has norm at most δ^{-1}, i.e., $\| Tf \| \le \delta^{-1} \| f \|$. Hence, for $A = \delta^{-1}$, we have $\| Tf \| \le A \| f \|$ for every $f \in V$.

(iii)$\to$ (i) Let $\| Tf \| \le A \| f \|$ for every $f \in V$. It is sufficient to prove that T is continuous at the origin (i.e., at the null element of V). This is clear since $\| f_n \| \to 0$ implies $\| Tf_n \| \le A \| f_n \| \to 0$. ∎

It is evident that the set $\mathcal{B}(V, W)$ of all norm bounded operators from V into W is a (real or complex) vector space, a linear subspace of the vector space $\mathcal{L}(V, W)$ of all linear operators from V into W. If $W = \mathbb{R}$ or $W = \mathbb{C}$, the corresponding $\mathcal{B}(V, \mathbb{R})$ or $\mathcal{B}(V, \mathbb{C})$ of all norm bounded linear functionals on V is called the *Banach dual space* of V (or *norm dual space* of V) and is denoted by V^* (or sometimes by V'). Linear functionals are often denoted by $\varphi, \psi, \ldots$. For any $T \in \mathcal{B}(V, W)$, let, by definition, $\| T \|$ be the infimum of all numbers $A > 0$ such that $\| Tf \| \leq A \| f \|$ holds for all $f \in V$. Then $\| Tf \| \leq \| T \| \cdot \| f \|$ for all $f \in V$, but $\| T \|$ is now the smallest number having this property. The number $\| T \|$ is called the *norm* of the operator T. This is an appropriate name because $\| T \|$ is a norm in $\mathcal{B}(V, W)$. Indeed, it is obvious that $\| T \| \geq 0$ for all T and $\| T \| = 0$ if and only if T is the null operator. Furthermore, $\| \alpha T \| = | \alpha | \cdot \| T \|$ for every $\alpha \in \mathbb{R}$ or $\alpha \in \mathbb{C}$. Finally,

$$\| (T_1 + T_2)f \| \leq \| T_1 f \| + \| T_2 f \| \leq (\| T_1 \| + \| T_2 \|) \cdot \| f \|,$$

so that, in view of the definition of the norm, we have

$$\| T_1 + T_2 \| \leq \| T_1 \| + \| T_2 \| .$$

It is evident from the definition that

$$\| T \| = \sup(\| Tf \| / \| f \| : f \neq 0) = \sup(\| Tf \| : \| f \| \leq 1).$$

In particular, if φ is a bounded linear functional on V, i.e., if $\varphi \in V^*$, then $| \varphi(f) | \leq \| \varphi \| \cdot \| f \|$ for every $f \in V$ and

$$\| \varphi \| = \sup(| \varphi(f) | / \| f \| : f \neq 0) = \sup(| \varphi(f) | : \| f \| \leq 1).$$

It is a natural question to ask for conditions under which the normed space $\mathcal{B}(V, W)$ is a Banach space. We prove that if W is a Banach space, then so is $\mathcal{B}(V, W)$. Before giving the proof, it is of use to observe that any Cauchy sequence in a normed space is norm bounded. Indeed, if $(f_n : n = 1, 2, \ldots)$ is a Cauchy sequence, then, for some natural number n_1, we have $\| f_m - f_n \| \leq 1$ for all $m \geq n \geq n_1$, so $\| f_m \| \leq \| f_{n_1} \| + 1$ for all $m \geq n_1$. Furthermore, note that if $f_n \to f_0$ in norm, then $\| f_n \| \to \| f_0 \|$. This follows from

$$| \| f_n \| - \| f_0 \| | \leq \| f_n - f_0 \| .$$

Theorem 18.2. *If V is a normed space and W is a Banach space, then $\mathcal{B}(V, W)$ is a Banach space. In particular, since $\mathbb{R}$ and $\mathbb{C}$ are Banach spaces with respect to the familiar absolute value as norm, the Banach dual V^* of any normed space V is a Banach space.*

Proof. Let $(T_n : n = 1, 2, \ldots)$ be a Cauchy sequence in $\mathcal{B}(V, W)$. Then, as observed above, the sequence $(\| T_n \| : n = 1, 2, \ldots)$ is bounded, say $\| T_n \| \leq M$ for all n. For any $f \in V$ it follows from

$$\| T_n f - T_m f \| \leq \| T_n - T_m \| \cdot \| f \| \to 0$$

as $n \geq m \to \infty$ that the sequence $(T_n f : n = 1, 2, \ldots)$ is a Cauchy sequence in W. Since W is a Banach space, the norm limit of $T_n f$ (as $n \to \infty$) exists in W. Denoting this limit by Tf, we thus define a mapping $T : V \to W$. Obviously the mapping T is linear, i.e., T is a linear operator. Since $T_n f \to Tf$, we have $\| T_n f \| \to \| Tf \|$ for every f. From

$$\| T_n f \| \leq \| T_n \| \cdot \| f \| \leq M \| f \|$$

for every n and every f it follows, therefore, that $\| Tf \| \leq M \| f \|$ for every f. This shows that T is norm bounded, i.e., $T \in \mathcal{B}(V, W)$.

Finally, let $\epsilon > 0$ be given. Since

$$\| T_n f - T_m f \| \to \| T_n f - Tf \|$$

as $m \to \infty$ and $\| T_n f - T_m f \| \leq \epsilon \| f \|$ for all $m \geq n \geq n(\epsilon)$, we have $\| T_n f - Tf \| \leq \epsilon \| f \|$ for all $n \geq n(\epsilon)$ and all f, and so $\| T_n - T \| \leq \epsilon$ for all $n \geq n(\epsilon)$. This shows that $T_n \to T$ in $\mathcal{B}(V, W)$. ■

Let now V and W be ordered real vector spaces, not necessarily Riesz spaces and not necessarily normed (see Definition 4.1 for the definition of an ordered vector space and Exercise 7.5(ii) for an example of an ordered vector space which is not a Riesz space). Recall that an element $f \in V$ is a positive element if $f \geq 0$ and that the set $(f : f \in V, f \geq 0)$ is called the positive cone of V. Similarly for W. The linear operator $T : V \to W$ is called a *positive operator* if T maps the positive cone of V into the positive cone of W, i.e., if $f \geq 0$ in V implies $Tf \geq 0$ in W. It is now evident that the space $\mathcal{L}(V, W)$ of all linear operators from V into W becomes an ordered vector space by defining that $T_1 \geq T_2$ in $\mathcal{L}(V, W)$ whenever $T_1 - T_2$ is positive. The operator $T \in \mathcal{L}(V, W)$ is said to be a *regular operator* if T can be written as $T = T_1 - T_2$ with T_1 and T_2 positive. The set $\mathcal{L}_r(V, W)$ of all regular operators from V into W is a linear subspace of $\mathcal{L}(V, W)$. It is clear from the definition that $T \in \mathcal{L}(V, W)$ is regular if and only if T is majorized by a positive operator, i.e., if there exists a positive operator T_1 such that $T \leq T_1$. The null operator 0 is the null element of $\mathcal{L}(V, W)$ as well as of $\mathcal{L}_r(V, W)$ and T is positive if and only if $T \geq 0$.

There is still another important class of linear operators from V into W. For the definition we recall that an order interval in V is a subset of V of the form $(f : f_1 \leq f \leq f_2)$, where f_1 and f_2 are elements of V satisfying $f_1 \leq f_2$ (see the definition in section 3). We shall denote the order interval $(f : f_1 \leq f \leq f_2)$ by $[f_1, f_2]$, in accordance with the notation for a closed interval in $\mathbb{R}$. The operator $T : V \to W$ is said to be an *order bounded operator* whenever T maps any order interval in V into (not necessarily onto) an order interval in W. It is easy to see that T is order bounded if and only if T maps any order interval of the form $[0, f]$ in V into an order interval in W. It follows immediately that every positive operator T is order bounded, because T maps any order interval $[0, f]$ into the order interval $[0, Tf]$. We have, therefore, the following simple theorem.

Theorem 18.3. *If V and W are ordered vector spaces and $T : V \to W$ is a regular operator, then T is order bounded.*

Finally, note that the set $\mathcal{L}_b(V, W)$ of all order bounded operators from V into W is a linear subspace of $\mathcal{L}(V, W)$. This follows by observing that the algebraic sum of two order intervals is contained in an order interval. Precisely, the algebraic sum of $[g_1, g_2]$ and $[g_3, g_4]$ is contained in, but not necessarily equal to, the interval $[g_1 + g_3, g_2 + g_4]$. Hence,

$$\mathcal{L}_r(V, W) \subseteq \mathcal{L}_b(V, W) \subseteq \mathcal{L}(V, W).$$

In one of the next sections (section 20) we shall deal with an important situation (V and W Riesz spaces with W Dedekind complete) in which $\mathcal{L}_r(V, W)$ and $\mathcal{L}_b(V, W)$ coincide.

In the case that we have to deal with normed Riesz spaces we can say more. If E is a (real) Banach lattice and F is a (real) normed Riesz space, then every order bounded operator from E into F is norm bounded. The proof will show that this is true even if E has the weak Riesz-Fischer property instead of being a Banach lattice. Since regular operators are order bounded, it follows that under these conditions for E and F every regular operator from E into F is norm bounded.

Theorem 18.4. *If E is a (real) Banach lattice and F is a (real) normed Riesz space, and $T : E \to F$ is an order bounded operator, then T is norm bounded. It follows that every regular operator from E into F is norm bounded.*

Proof. Note first that any order interval $[f_1, f_2]$ is norm bounded. To see this, note that any f in $[0, f_2 - f_1]$ satisfies $\| f \| \leq \| f_2 - f_1 \|$, so any f in $[f_1, f_2]$ satisfies

$$\| f \| \leq \| f_2 - f_1 \| + \| f_1 \| .$$

Assume now that $T : E \to F$ is order bounded but not norm bounded. Then there exists a sequence $(f_n : n = 1, 2, \ldots)$ in E such that $\| Tf_n \| > 2n^3 \| f_n \|$ for all n. Note that $f_n \neq 0$ for all n, so (replacing f_n by $f_n / \| f_n \|$) we may assume that $\| f_n \| = 1$ and $\| Tf_n \| > 2n^3$ for all n. Since $\| Tf_n \|$ is less than or equal to $\| Tf_n^+ \| + \| Tf_n^- \|$, it follows that at least one of $\| Tf_n^+ \|$ and $\| Tf_n^- \|$ exceeds n^3. Hence, we may assume that

$$f_n > 0, \| f_n \| \leq 1 \text{ and } \| Tf_n \| > n^3 \text{ for } n = 1, 2, \ldots.$$

Writing $u_n = n^{-2} f_n$, we get

$$u_n \in E^+, \| u_n \| \leq n^{-2} \text{ and } \| Tu_n \| > n \text{ for } n = 1, 2, \ldots.$$

It follows that $\Sigma \| u_n \|$ converges, so (since E is a Banach lattice and, therefore, has the Riesz-Fischer property) the element $s = \Sigma u_n$ exists in E. This shows that all u_n are contained in the order interval $[0, s]$. Since T is order bounded, all Tu_n are then contained in some order interval in F. But then the sequence of all Tu_n is norm bounded, which contradicts $\| Tu_n \| > n$ for all n. Hence, the operator T is norm bounded. $\blacksquare$

Remark. (i) If it is only given that E has the weak Riesz-Fischer property, then it follows just as well from the convergence of $\Sigma \| u_n \|$ that all u_n are contained in some order interval $[0, s]$, so the proof proceeds as above.

(ii) The space of all order bounded operators from a Banach lattice into itself may be properly larger than the space of all regular operators. As an example we mention the case that $E = F$ is the space $C[-1, 1]$ and $T : E \to F$ is defined by

$$(Tf)(x) = f\{\sin(x^{-1})\} - f\{\sin(x + x^{-1})\} \text{ for } 0 <| x |< 1,$$

$$(Tf)(0) = 0.$$

It is easy to see that $Tf \in C[-1, 1] = F$ for every $f \in E$ and T is order bounded. It is more difficult, however, to show that T is not regular.

19. Riesz Homomorphisms and Quotient Spaces

In algebraic ring theory ring homomorphisms play a particular role among the additive mappings from one ring into another, because they preserve the multiplicative structure as well as the additive structure. Similarly, Riesz homomorphisms (or lattice homomorphisms) play a particular role in the space of all linear operators from one Riesz space into another one, because they preserve the lattice structure as well as the vector space structure. We present the definition.

Definition 19.1. The linear operator T from the Riesz space E into the Riesz space F is called a *Riesz homomorphism* (or *lattice homomorphism*) if

$$T(f \vee g) = (Tf) \vee (Tg)$$

holds for all f and g in E.

Before mentioning some characterizations of Riesz homomorphisms we observe already that a Riesz homomorphism is necessarily a positive operator. To see this, note that if $f \geq 0$, then

$$Tf = T(f \vee 0) = (Tf) \vee (T0) = (Tf) \vee 0 = (Tf)^+ \geq 0.$$

Theorem 19.2. *Let $T : E \to F$ be a linear operator from the Riesz space E into the Riesz space F. Then the following conditions for T are equivalent.*

(i) T is a Riesz homomorphism, i.e.,

$$T(f \vee g) = (Tf) \vee (Tg) \text{ for all } f \text{ and } g \text{ in } E.$$

(ii) $T(f \wedge g) = (Tf) \wedge (Tg)$ for all f and g in E.
(iii) $f \wedge g = 0$ implies $(Tf) \wedge (Tg) = 0$.
(iv) $\mid Tf \mid = T(\mid f \mid)$ for all $f \in E$.
(v) $Tf^+ = (Tf)^+$ for all $f \in E$.

Proof. (i)$\to$ (ii) This follows immediately from

$$f + g = (f \vee g) + (f \wedge g).$$

(ii)$\to$ (iii) Evident.

(iii)$\to$ (iv) Since $f^+ \wedge f^- = 0$, we have $(Tf^+) \wedge (Tf^-) = 0$ by (iii). Recall now that, generally, $u \wedge v = 0$ implies $\mid u - v \mid = \mid u + v \mid = u + v$. Hence, in the present situation,

$$\mid Tf \mid = \mid Tf^+ - Tf^- \mid = Tf^+ + Tf^- = T(\mid f \mid).$$

(iv)$\to$ (i) We have

$$2T(f \vee g) = T(f + g + \mid f - g \mid) = Tf + Tg + T(\mid f - g \mid) =$$

$$Tf + Tg + \mid Tf - Tg \mid = 2\{(Tf) \vee (Tg)\}.$$

Finally, it is evident that (i) implies (v). Conversely, (v) implies (iv) in view of $f^- = (-f)^+$, which implies by (v) that $Tf^- = (Tf)^-$, and so

$$\mid Tf \mid = (Tf)^+ + (Tf)^- = Tf^+ + Tf^- = T(\mid f \mid). \qquad \blacksquare$$

Note that if T is a Riesz homomorphism, then $f \perp g$ implies $Tf \perp Tg$. This follows immediately from

$$\mid Tf \mid \wedge \mid Tg \mid = T(\mid f \mid) \wedge T(\mid g \mid) = T(\mid f \mid \wedge \mid g \mid) = T0 = 0.$$

The converse does not hold. If, for example, $-T$ is a Riesz homomorphism, then $f \perp g$ implies $Tf \perp Tg$, but T is not a Riesz homomorphism. If, however, T is a positive operator with the property that $f \perp g$ implies $Tf \perp Tg$, then $f \wedge g = 0$ implies $Tf \wedge Tg = 0$, and so T is now a Riesz homomorphism.

Note also that any order projection P on a projection band B in the Riesz space E is a Riesz homomorphism. This is clear since it follows from $0 \leq Pf \leq f$ for every $f \geq 0$ in E that $f \wedge g = 0$ implies $Pf \wedge Pg = 0$.

A Riesz homomorphism T mapping the Riesz space E onto the Riesz space F in a one-one way is called a *Riesz isomorphism*. The inverse operator T^{-1}

from F onto E is now also a Riesz isomorphism and the spaces E and F are said to be *Riesz isomorphic.*

Every Riesz isomorphism $T : E \to F$ is one-one, onto and positive. Is it true that, conversely, any linear operator $T : E \to F$ which is one-one, onto and positive is a Riesz isomorphism ?. The answer is negative, as shown by the following example. Let $E = \mathbb{R}^2$ with the familiar coordinatewise ordering and $F = \mathbb{R}^2$ with the lexicographical ordering (i.e., $(x_1, x_2) \leq (y_1, y_2)$ whenever $x_1 < y_1$ or $x_1 = y_1, x_2 \leq y_2$) and let $T : E \to F$ be the identity operator. Then T is one-one, onto and positive, but T is not a Riesz isomorphism because T^{-1} is not positive. The situation improves if T^{-1} is positive.

Theorem 19.3. *Let $T : E \to F$ be a linear operator which is one-one and onto. Then T is a Riesz isomorphism if and only if T and T^{-1} are positive.*

Proof. It is evident that if T is a Riesz isomorphism, then T and T^{-1} are positive. Assume now, conversely, that T and T^{-1} are positive. Then, for f and g in E, it follows from $f \leq f \vee g$ and $g \leq f \vee g$ that $Tf \leq T(f \vee g)$ and $Tg \leq T(f \vee g)$, so

$$(Tf) \vee (Tg) \leq T(f \vee g). \tag{1}$$

Similarly, we have $(T^{-1}h) \vee (T^{-1}k) \leq T^{-1}(h \vee k)$ for all h and k in F. There exist unique f and g in E such that $h = Tf$ and $k = Tg$, so we get that $f \vee g \leq T^{-1}\{(Tf) \vee (Tg)\}$. Applying T to both sides, we now obtain

$$T(f \vee g) \leq (Tf) \vee (Tg). \tag{2}$$

Combination of (1) and (2) shows that $T(f \vee g) = (Tf) \vee (Tg)$, so T is a Riesz homomorphism. Since T is one-one and onto, T is a Riesz isomorphism. ∎

The *kernel* or *null space* (i.e., the set $(f : Tf = 0)$) of a linear operator $T : V \to W$ between vector spaces is a linear subspace of V and the range of T (i.e., the set $(g : g = Tf$ for some $f \in V)$) is a linear subspace of W. If E and F are Riesz spaces and $T : E \to F$ is a Riesz homomorphism, then it is clear that the range of T is a Riesz subspace of F and it is not difficult to see that the kernel $k(T)$ of T is an ideal in E. For the proof, observe first that it follows from $T(|\,f\,|) = |\,Tf\,|$ that $f \in k(T)$ if and only if $|\,f\,| \in k(T)$. Hence, if $f \in k(T)$ and $|\,g\,| \leq |\,f\,|$, then

$$0 \leq T(|\,g\,|) \leq T(|\,f\,|) = 0,$$

so $|\,g\,| \in k(T)$, which implies $g \in k(T)$. Thus, the kernel $k(T)$ is an ideal.

If A is a linear subspace of the (real or complex) vector space V, we can divide V into equivalence classes modulo A. This is done by introducing in V the equivalence relation defined by saying that f_1 and f_2 in V are equivalent whenever $f_1 - f_2 \in A$. The set of all elements in V equivalent to a given $f \in V$

is called the equivalence class of f and denoted by $[f]$. Hence $[f_1] = [f_2]$ if and only if f_1 and f_2 are equivalent, i.e., if and only if $f_1 - f_2 \in A$. The linear subspace A itself is one of the equivalence classes; it is the equivalence class containing the null element of V, so $A = [0]$. In other words, $f \in A$ if and only if $[f] = [0]$. The set of all equivalence classes is called the *quotient space of V modulo A* and it is denoted by V/A. As well-known, the set V/A is a vector space if we define the vector space operations by

$$[f] + [g] = [f + g] \text{ for all } [f] \text{ and } [g],$$

$$\alpha[f] = [\alpha f] \text{ for all } [f] \text{ and all } \alpha.$$

It is easy to see that these definitions are not contradictory. For example, note that if f_1, f_2 are in $[f]$ and g_1, g_2 are in $[g]$, then $f_1 + g_1$ is in the same equivalence class as $f_2 + g_2$, i.e.,

$$(f_1 + g_1) - (f_2 + g_2) \in A.$$

Assume now that E is a Riesz space and A is an ideal in E. We shall prove that the vector space E/A can be partially ordered in such a manner that E/A is a Riesz space with respect to the partial ordering.

Definition 19.4. Let A be an ideal in the Riesz space E. Given $[f]$ and $[g]$ in E/A, we shall write $[f] \le [g]$ whenever there exist elements $f_1 \in [f]$ and $g_1 \in [g]$ satisfying $f_1 \le g_1$.

Before showing that this defines a partial ordering in E/A, we observe that the following properties are immediately evident from the definition.

(i) $[f] \le [g]$ if and only if for every $f_1 \in [f]$ there exists an element $g_1 \in [g]$ satisfying $f_1 \le g_1$.

(ii) $[f] \le [g]$ if and only if for every $f_1 \in [f]$ and every $g_1 \in [g]$ there exists an element $q \in A$ such that $g_1 - f_1 \ge q$.

Theorem 19.5. *If A is an ideal in the Riesz space E, the quotient space E/A is a Riesz space with respect to the ordering defined in Definition 19.4.*

Proof. We show first that E/A is an ordered vector space. To do this we have to prove first of all that what was defined in Definition 19.4 is indeed a partial ordering.

(i) It is clear that $[f] \le [f]$ for every $[f]$.

(ii) If $[f] \le [g]$ and $[g] \le [h]$, let f_1 in $[f], g_1$ and g_2 in $[g]$ and h_1 in $[h]$ satisfy $f_1 \le g_1$ and $g_2 \le h_1$. Then

$$f_1 \le g_1 = g_2 + (g_1 - g_2) \le h_1 + (g_1 - g_2).$$

Since $g_1 - g_2 \in A$, we have $h_1 + (g_1 - g_2) \in [h]$, which shows that $[f] \leq [h]$.

(iii) Let $[f] \leq [g]$ as well as $[g] \leq [f]$. Then there exist f_1, f_2 in $[f]$ and g_1, g_2 in $[g]$ such that $f_1 \leq g_1$ and $g_2 \leq f_2$. It follows that

$$0 \leq g_1 - f_1 \leq (g_1 - f_1) + (f_2 - g_2) = (f_2 - f_1) + (g_1 - g_2) \in A,$$

so $g_1 - f_1 \in A$, i.e., $[f] = [g]$.

Having established that we have a partial ordering in E/A, we prove now that the ordering is compatible with the vector space structure. It is obvious that $[f] \leq [g]$ implies $\alpha[f] \leq \alpha[g]$ for every $\alpha \geq 0$. Finally, to show that $[f] \leq [g]$ implies $[f] + [h] \leq [g] + [h]$, choose $f \in [f]$ and $g \in [g]$ satisfying $f \leq g$ and choose $h \in [h]$ arbitrarily. Then $f + h \leq g + h$, so $[f + h] \leq [g + h]$, i.e., $[f] + [h] \leq [g] + [h]$.

It remains to show that E/A is a Riesz space with respect to the ordering. Precisely, we shall prove that $[f] \vee [g]$ exists for all $[f], [g]$ and is equal to $[f \vee g]$. It is evident that $[f \vee g] \geq [f]$ as well as $\geq [g]$, so it will be sufficient to show that any upper bound $[h]$ of $[f]$ and $[g]$ satisfies $[h] \geq [f \vee g]$. Given that $[h]$ is such an upper bound, choose elements f, g, h in $[f], [g], [h]$ respectively. Then there exist q_1, q_2 in A such that $h \geq f + q_1$ and $h \geq g + q_2$, so for $q = q_1 \wedge q_2 \in A$ we have $h \geq f + q$ and $h \geq g + q$. Therefore,

$$h \geq (f + q) \vee (g + q) = (f \vee g) + q.$$

Since $q \in A$, this shows that $[h] \geq [f \vee g]$, as desired. Note that it follows from $f + g = (f \vee g) + (f \wedge g)$ that $[f] \wedge [g]$ exists and satisfies

$$[f] \wedge [g] = [f \wedge g]. \qquad \blacksquare$$

The Riesz space E/A is called the *quotient Riesz space of E modulo the ideal A*. The mapping $f \to [f]$ is a Riesz homomorphism of E onto E/A and A is the kernel of the homomorphism. This shows that every quotient space of E modulo some ideal is a Riesz homomorphic image of E. Conversely, let T be a Riesz homomorphism of the Riesz space E into the Riesz space F. The image $T(E)$ of E is a Riesz subspace of F, so T is a Riesz homomorphism onto the Riesz space $T(E)$. Let A be the kernel of T. Then A is an ideal in E, and the mapping $[f] \to Tf$ is now a Riesz isomorphism of E/A onto $T(E)$. Hence E/A and $T(E)$ are Riesz isomorphic. This shows that every Riesz homomorphic image of a Riesz space E is Riesz isomorphic to the quotient space of E modulo the kernel of the homomorphism.

The situation is, therefore, analogous to the situation in algebraic ring theory, where we have ring homomorphisms and quotient spaces modulo ring ideals.

One of the problems when dealing with Riesz homomorphisms is to determine under which conditions the Riesz homomorphic image of an Archimedean Riesz space is again Archimedean. In other words, the problem is to determine which properties an ideal A in the space E should possess in order that E/A will be Archimedean. In general E/A is not Archimedean even in the case that E is super Dedekind complete. Conversely, a non-Archimedean space may possess an Archimedean quotient space.

Example 19.6. (i) Let E be the space ℓ_∞ of all (real) bounded sequences and A the principal ideal generated by the element $u = (u(1), u(2), \ldots)$ with $u(n) = n^{-2}$ for all $n = 1, 2, \ldots$. Note already that any element in E with only finitely many non-zero coordinates is contained in A, so elements differing only in a finite number of coordinates are contained in the same equivalence class modulo A. Now, let $e = (1, 1, \ldots)$ and $v = (v(1), v(2), \ldots)$ with $v(n) = n^{-1}$ for all n. Given the natural number k, we have $v(n) \leq k^{-1}e(n)$ for all $n \geq k$ and so, taking into account the above remark, we see that $[v] \leq k^{-1}[e]$ with $[v] \neq [0]$ holds in E/A. This holds for all $k = 1, 2, \ldots$, and so E/A is not Archimedean.

(ii) If $E = \mathbb{R}^2$ with lexicographical ordering and A is the set of all points $(0, x_2)$ in $\mathbb{R}^2$, then E is not Archimedean, A is a band in E and E/A is Riesz isomorphic to the space $\mathbb{R}$ of real numbers (with the familiar ordering), so E/A is even super Dedekind complete.

Before presenting necessary and sufficient conditions for a quotient space to be Archimedean we recall that if E is a Riesz space and $u \in E^+$, then $f_n(n = 1, 2, \ldots)$ is said to converge u-uniformly to f whenever, for any $\epsilon > 0$, there exists a natural number $n(\epsilon)$ such that $\mid f - f_n \mid \leq \epsilon u$ holds for all $n \geq n(\epsilon)$ and f_n is said to converge (relatively) uniformly to f (notation $f_n \to f$ (un)) whenever f_n converges u-uniformly to f for some $u \in E^+$. Furthermore the subset D of E is said to be uniformly closed (un-closed) if it follows from $f_n \in D$ for all n and $f_n \to f$ (un) that $f \in D$ (see section 15 for this definition). The main result in the next theorem is now that if A is an ideal in the Riesz space E, then E/A is Archimedean if and only if A is un-closed.

Theorem 19.7. *If E is an arbitrary Riesz space and A is an ideal in E, then the following conditions for A are equivalent.*

(i) E/A is Archimedean.
(ii) A is uniformly closed.
(iii)   If $(u_n : n = 1, 2, \ldots)$ is an increasing sequence in A^+ which is uniformly convergent to u, then $u \in A$.
(iv) . If $u, v \in E^+$ and $(nu - v)^+ \in A$ for $n = 1, 2, \ldots$, then $u \in A$.

Proof. (i)$\to$ (ii) Let A be an ideal in E such that E/A is Archimedean and let $f_n \to f(e\text{-un})$ for some $e \in E^+$ with $f_n \in A$ for all n. We have to prove that $f \in A$. Given $\epsilon > 0$, it follows from

$$\mid f \mid \leq \mid f - f_n \mid + \mid f_n \mid \leq \epsilon e + \mid f_n \mid$$

for $n \geq n(\epsilon)$ that $[f] \leq \epsilon[e]$. The last inequality holds for every $\epsilon > 0$, so $[f] = [0]$ since E/A is Archimedean. This shows that $f \in A$.

(ii)$\rightarrow$ (iii) This is evident.

(iii)$\rightarrow$ (iv) Let $u, v \in E^+$ and $(nu - v)^+ \in A$ for $n = 1, 2, \ldots$. We assume that A satisfies the condition in (iii) and we have to prove that u is a member of A. On account of

$$0 \leq u - (u - n^{-1}v)^+ = \mid u^+ - (u - n^{-1}v)^+ \mid \leq \mid u - (u - n^{-1}v) \mid = n^{-1}v$$

it is evident that the increasing sequence

$$w_n = (u - n^{-1}v)^+; n = 1, 2, \ldots$$

converges v-uniformly to u. Since $w_n \in A$ for all n, it follows from condition (iii) that $u \in A$.

(iv)$\rightarrow$(i) We assume that (iv) is satisfied and we prove that E/A is Archimedean. To this end, assume that $u, v \in E^+$ and $n[u] \leq [v]$ for $n = 1, 2, \ldots$. Then

$$[0] = (n[u] - [v])^+ = [(nu - v)^+]$$

for all n, and so $(nu - v)^+ \in A$ for all n. It follows from (iv) that $u \in A$, i.e., $[u] = [0]$. This shows that E/A is Archimedean.   ∎

Corollary 19.8. *If E is Archimedean and A is a σ-ideal in E, then E/A is Archimedean.*

Proof. Let A be a σ-ideal in the Archimedean Riesz space E and assume that the increasing sequence $(u_n : n = 1, 2, \ldots)$ in A^+ converges uniformly to u. Then, by Theorem 10.3(i), u_n converges to u in order, i.e., $0 \leq u_n \uparrow u$. Here we use that E is Archimedean. Since $u_n \in A$ for all n and A is a σ-ideal, it follows that $u \in A$. This shows that condition (iii) in the last theorem is satisfied, and so E/A is Archimedean.   ∎

Exercise 19.9. Let T be a Riesz homomorphism from the Riesz space E into the Riesz space F.

(i) Show that for any Riesz subspace V of E the image $T(V)$ is a Riesz subspace of F. In particular, the image $T(E)$ of E itself is a Riesz subspace of F. Show that for any Riesz subspace W of F the inverse image $T^{-1}(W) = (f : f \in V, Tf \in W)$ is a Riesz subspace of E.

(ii) Show that for any ideal A in E the image $T(A)$ is an ideal in $T(E)$. Show that if A_1 and A_2 are ideals in E, then

$$T(A_1 \cap A_2) = T(A_1) \cap T(A_2).$$

(iii) Show that for any ideal B in $T(E)$ the inverse image $T^{-1}(B)$ is an ideal in E and for any ideal B_1 in F the inverse image $T^{-1}(B_1)$ is an ideal in E.

(iv) Show that for any principal ideal A_u in E, generated by $u \in E^+$, the image $T(A_u)$ is the principal ideal generated in $T(E)$ by Tu.

(v) Show that for any subset D of E the set $T(D^d)$ is a subset of $\{T(D)\}^d$.

(vi) Show that the image of a projection band in E is a projection band in $T(E)$.

Hint: (i) Let W be a Riesz subspace of F. It is evident that $T^{-1}(W)$ is a linear subspace of E. If $f, g \in T^{-1}(W)$, then Tf and Tg are members of W, so $T(f \vee g) = (Tf) \vee (Tg) \in W$, i.e., $f \vee g \in T^{-1}(W)$.

(ii) To show that $T(A)$ is an ideal in $T(E)$, it is sufficient to show that if $0 \leq u \in A$ and $0 \leq m \leq Tu$ with $m \in T(E)$, then $m \in T(A)$. Since $m = Tf$ for some $f \in E$ and $m = m^+$, it is easy to see that $m = Tf^+$ and so $m = Tw$ for $w = u \wedge f^+ \in A$.

For ideals A_1, A_2 in E it is sufficient to prove that $T(A_1) \cap T(A_2)$ is a subset of $T(A_1 \cap A_2)$. If f is a member of $T(A_1) \cap T(A_2)$, then so if $|\, f\, |$, so let $|\, f\, |= Tu$ with $0 \leq u \in A_1$ and $|\, f\, |= Tv$ with $0 \leq v \in A_2$. Then $w = u \wedge v$ satisfies $0 \leq w \in A_1 \cap A_2$ and $Tw =|\, f\, |$. Hence $|\, f\, |\in T(A_1 \cap A_2)$.

(iii) Let B be an ideal in $T(E)$. To show that $T^{-1}(B)$ is an ideal in E, it is sufficient to show that $0 \leq u \leq v \in T^{-1}(B)$ implies $u \in T^{-1}(B)$. This is easy. It follows from $0 \leq Tu \leq Tv \in B$ that $Tu \in B$ and so $u \in T^{-1}(B)$. If B_1 is an ideal in F, then $B_1 \cap T(E)$ is an ideal in $T(E)$ and so $T^{-1}(B_1) = T^{-1}\{B_1 \cap T(E)\}$ is an ideal in E.

(iv) The image $T(A_u)$ is an ideal in $T(E)$. Since $Tu \in T(A_u)$, the ideal generated in $T(E)$ by Tu is a subset of $T(A_u)$. For the proof of the converse, assume $0 \leq m \in T(A_u)$, so $m = Tv$ for some $v \in A_u$. Then $v \leq \alpha u$ for some real $\alpha > 0$, so $m = Tv \leq \alpha Tu$, i.e., m is a member of the ideal generated in $T(E)$ by Tu.

(v) If $m \in T(D^d)$, then $m = Tf$ for some $f \in D^d$, so $f \perp g$ for all $g \in D$. Then $m = Tf \perp Tg$ for all $g \in D$, so $m \in \{T(D)\}^d$.

(vi) Let B be a projection band in E and let C be the disjoint complement of B, so $B \oplus C = E$. The images $T(B)$ and $T(C)$ are ideals in $T(E)$ and $T(B) \perp T(C)$. Given $m \in T(E)$, we have $m = Tf$ for some $f \in E$; let $f = f_1 + f_2$ with $f_1 \in B$ and $f_2 \in C$. It follows that $m = Tf = Tf_1 + Tf_2$, so $T(E) = T(B) + T(C)$. Then, by Theorem 11.1, $T(B)$ and $T(C)$ are bands in $T(E)$.

CHAPTER 10
Order Bounded Operators

20. Order Bounded Operators

In the present section we shall prove an important result which lies at the foundations of operator theory in Riesz spaces. Briefly stated, it says that if E and F are Riesz spaces with F Dedekind complete and $T : E \to F$ is a linear operator, then T is regular if and only if T is order bounded. In other words, using the notations introduced in section 18, we have $\mathcal{L}_r(E, F) = \mathcal{L}_b(E, F)$. Moreover (and this is the really interesting point), the vector space $\mathcal{L}_b(E, F)$ is now a Dedekind complete Riesz space with the set of positive operators (from E into F) as positive cone. This implies that every $T \in \mathcal{L}_b(E, F)$ can be written as $T = T^+ - T^-$, where T^+ and T^- are positive operators such that $T^+ = T \vee 0, T^- = (-T) \vee 0$ and $T^+ \wedge T^- = 0$. The case that $F = \mathbb{R}$ is of special interest. Since $\mathbb{R}$ is Dedekind complete, it follows that regular linear functionals on E are the same as order bounded linear functionals and the space $\mathcal{L}_r(E, \mathbb{R}) = \mathcal{L}_b(E, \mathbb{R})$ is a Dedekind complete Riesz space. This space, denoted by $E^\sim$ for convenience, is called the *order dual* of E. The theorem stating that $\mathcal{L}_r(E, F) = \mathcal{L}_b(E, F)$ is a Dedekind complete Riesz space is due to L.V. Kantorovitch (1936) in the Soviet Union and to H. Freudenthal (1936) in the Netherlands. The theorem on $E^\sim$, with further results on order continuous linear functionals (see the next section), is due to F. Riesz (1937) in Hungary who, already in 1928 at the Bologna International Mathematical Congress, presented his ideas about linear functionals on certain ordered vector spaces, which is one of the reasons why lattice ordered vector spaces are now known as Riesz spaces. We proceed with the details, but first we recall the definition of an additive mapping. Let D be a subset of the vector space V having the property that $f, g \in D$ implies $f + g \in D$. Let W be another vector space. The mapping $\tau : D \to W$ is now said to be an *additive mapping* whenever $\tau(f + g) = \tau(f) + \tau(g)$ holds for all f and g in D.

Lemma 20.1. *(Extension lemma). Let E and F be Riesz spaces with F Archimedean and let the mapping $\tau : E^+ \to F^+$ be additive. Then there exists a uniquely determined positive linear operator $T : E \to F$ such that T extends τ, i.e., $Tu = \tau(u)$ for every $u \in E^+$.*

Proof. It is immediately evident by induction that $\tau(nu) = n\tau(u)$ for any $u \in E^+$ and $n = 0, 1, 2, \ldots$. Then, for $n = 1, 2, \ldots$, we have $\tau(u) = \tau(n \cdot n^{-1}u) = n\tau(n^{-1}u)$, so $\tau(n^{-1}u) = n^{-1}\tau(u)$. Hence $\tau(ru) = r\tau(u)$ for any rational $r \geq 0$. If $\alpha > 0$ is irrational, choose rational r and r' such that $0 < r < \alpha < r'$. Then $ru \leq \alpha u \leq r'u$, so

$$r\tau(u) = \tau(ru) \leq \tau(\alpha u) \leq \tau(r'u) = r'\tau(u).$$

Letting $r \uparrow \alpha$ and $r' \downarrow \alpha$, we have $r\tau(u) \uparrow \alpha\tau(u)$ and $r'\tau(u) \downarrow \alpha\tau(u)$; note that we use here that F is Archimedean. It follows that $\alpha\tau(u) \leq \tau(\alpha u) \leq \alpha\tau(u)$, so $\tau(\alpha u) = \alpha\tau(u)$. Now observe that if $u, v, u_1, v_1 \in E^+$ satisfy $u - v = u_1 - v_1$ and so $u + v_1 = u_1 + v$, then

$$\tau(u) + \tau(v_1) = \tau(u + v_1) = \tau(u_1 + v) = \tau(u_1) + \tau(v),$$

so $\tau(u) - \tau(v) = \tau(u_1) - \tau(v_1)$. Hence, since every $f \in E$ has at least one decomposition $f = u - v$ with u and v in E^+, it follows that if we define T on E by $Tf = \tau(u) - \tau(v)$, then Tf depends only on f and not on the particular decomposition of f. Note already that T is now defined on E and $Tu = \tau(u)$ holds for every $u \in E^+$.

The thus defined mapping $T : E \to F$ is linear. For the proof that T is additive, let f and g in E be given. Then there exist u, v, u_1, v_1 in E^+ such that $f = u - v$ and $g = u_1 - v_1$, and so

$$T(f + g) = T\{u + u_1 - (v + v_1)\} = \tau(u + u_1) - \tau(v + v_1) =$$

$$\tau(u) + \tau(u_1) - \tau(v) - \tau(v_1) = T(f) + T(g).$$

From $\tau(\alpha u) = \alpha\tau(u)$, holding for all $u \in E^+$ and all $\alpha \geq 0$, it follows easily that $T(\alpha f) = \alpha T(f)$ holds for all $f \in E$ and all real α. It has thus been proved that $T : E \to F$ is a positive linear operator which extends τ.

Finally, for the proof that T is unique, assume that $T_1 : E \to F$ is another operator extending τ. Given $f = u - v$ with u and v in E^+, we have

$$T_1 f = T_1 u - T_1 v = \tau(u) - \tau(v) = Tf,$$

so $T_1 = T$. ∎

Theorem 20.2. *Let E and F be Riesz spaces with F Dedekind complete. Then the linear operator $T : E \to F$ is order bounded if and only if T is regular. In other words, $\mathcal{L}_b(E, F) = \mathcal{L}_r(E, F)$. Furthermore the space $\mathcal{L}_b(E, F)$ is a Dedekind complete Riesz space with the set of positive operators from E into F as positive cone.*

Proof. It was proved already in Theorem 18.3 that every regular operator is order bounded. For the proof in the converse direction, assume now that $T : E \to F$ is order bounded. Then, for any given $u \in E^+$, the set $(Tv : 0 \leq v \leq u)$ is contained in an order interval in F, and so the set is bounded above. It follows, since F is Dedekind complete, that the supremum

$$\tau(u) = \sup(Tv : 0 \leq v \leq u)$$

exists in F. Note that $\tau(u) \geq 0$ and $\tau(u) \geq Tu$. We prove that τ is additive on E^+. Given u_1 and u_2 in E^+, we have

$$\tau(u_1) + \tau(u_2) = \sup\{T(v_1 + v_2) : 0 \leq v_1 \leq u_1, 0 \leq v_2 \leq u_2\} \leq \tau(u_1 + u_2).$$

In the converse direction, if $v \in E^+$ satisfies $0 \leq v \leq u_1 + u_2$, there exist v_1 and v_2 in E^+ satisfying $v_1 + v_2 = v$ such that $0 \leq v_1 \leq u_1$ and $0 \leq v_2 \leq u_2$ (by the Riesz decomposition property). Hence

$$Tv = Tv_1 + Tv_2 \leq \tau(u_1) + \tau(u_2).$$

Taking the supremum on the left for all v satisfying $0 \leq v \leq u_1 + u_2$, we get $\tau(u_1 + u_2) \leq \tau(u_1) + \tau(u_2)$. It has been shown thus that τ is an additive mapping from E^+ into F^+, so by the extension lemma above there exists a positive operator $T_1 : E \to F$ which extends τ. Now define the operator T_2 by $T_2 = T_1 - T$, so $T = T_1 - T_2$. In view of the definition of τ we have $T_1 u = \tau(u) \geq Tu$ for every $u \in E^+$, so $T_2 u \geq 0$, i.e., T_2 is a positive operator. It follows that $T = T_1 - T_2$ with T_1 and T_2 positive, and so T is regular.

It remains to prove that the vector space $\mathcal{L}_b(E, F)$ is a Dedekind complete Riesz space. Let $T \in \mathcal{L}_b(E, F)$ and the positive operator T_1 be as above, so

$$T_1 u = \sup(Tv : 0 \leq v \leq u) \text{ for every } u \in E^+.$$

Then $T_1 u \geq 0$ and $T_1 u \geq Tu$, so T_1 is an upper bound of T as well as of the null operator 0. If T' is another upper bound, then $T'u \geq T'v \geq Tv$ for $0 \leq v \leq u$, so

$$T'u \geq \sup(Tv : 0 \leq v \leq u) = T_1 u.$$

It follows that T_1 is the least upper bound of T and 0, i.e., $T_1 = T \vee 0 = T^+$. For arbitrary S and T in $\mathcal{L}_b(E, F)$ the operator

$$T + (S - T)^+ = T + \{(S - T) \vee 0\} = S \vee T$$

is then the least upper bound of S and T and $-\{(-S) \vee (-T)\}$ is the greatest lower bound of S and T. It has thus been proved that $\mathcal{L}_b(E, F)$ is a Riesz space having the set of positive operators from E into F as positive cone.

To show that $\mathcal{L}_b(E, F)$ is Dedekind complete, assume that D is an upwards directed set in $\{\mathcal{L}_b(E, F)\}^+$ such that D is bounded above. We have to prove that D has a supremum in $\mathcal{L}_b(E, F)$. To this end, let $\tau(u) = \sup(Tu : T \in D)$

for every $u \in E^+$. Note that $\tau(u)$ exists in F since D is bounded and F is Dedekind complete. For T_1 and T_2 in D there exists $T_3 \in D$ such that $T_3 \geq T_1$ and $T_3 \geq T_2$, so for $u_1, u_2 \in E^+$ we have

$$T_1 u_1 + T_2 u_2 \leq T_3(u_1 + u_2) \leq \tau(u_1 + u_2),$$

which implies, by taking suprema on the left, that

$$\tau(u_1) + \tau(u_2) \leq \tau(u_1 + u_2).$$

In the converse direction, it follows from

$$T(u_1 + u_2) = Tu_1 + Tu_2 \leq \tau(u_1) + \tau(u_2)$$

for every $T \in D$ that $\tau(u_1 + u_2) \leq \tau(u_1) + \tau(u_2)$. Hence, the mapping $\tau : E^+ \to F^+$ is additive, so (by the extension lemma) there exists a positive operator $T_0 : E \to F$ which extends τ. Since $T_0 u = \sup(Tu : T \in D)$ for every $u \in E^+$, the operator T_0 clearly satisfies $T_0 = \sup D$.   ∎

The space $\mathbb{R}$ is Dedekind complete. Therefore, as observed already, we may specialize to the case that $F = \mathbb{R}$, which results then in the following corollary.

Corollary 20.3. *For a Riesz space E, a linear functional on E is order bounded if and only if it is regular. The vector space $E^\sim$ of all order bounded linear functionals on E, called the order dual of E, is a Dedekind complete Riesz space.*

We proceed with some formulas concerning the operators T, T^+, T^- and $|T| = T^+ + T^- = T^+ \vee T^-$.

Theorem 20.4. *As before, let E and F be Riesz spaces with F Dedekind complete and let T, T_1, T_2 be order bounded operators from E into F. Then, for any $u \in E^+$, we have*

(1) $T^+ u = \sup(Tv : 0 \leq v \leq u)$,
(2) $-T^- u = \inf(Tv : 0 \leq v \leq u)$,
(3) $(T_1 \vee T_2)(u) = \sup(T_1 v + T_2 w : v \geq 0, w \geq 0, v + w = u)$,
(4) $(T_1 \wedge T_2)(u) = \inf(T_1 v + T_2 w : v \geq 0, w \geq 0, v + w = u)$,
(5) $|T|(u) = \sup(Tf : |f| \leq u) = \sup(|Tf| : |f| \leq u)$,

and for any $f \in E$ we have

(6) $|Tf| \leq |T|(|f|)$.

Proof. T^+ is the operator called T_1 in the proof of Theorem 20.2 and T_1 is the extension of τ, defined in that proof by

$$\tau(u) = \sup(Tv : 0 \leq v \leq u) \text{ for } u \in E^+.$$

Hence, (1) follows. For (2), note that in view of (1) we have

$$-T^- u = Tu - T^+ u = \inf\{T(u - w) : 0 \leq w \leq u\} = \inf(Tv : 0 \leq v \leq u).$$

For the proof of (3), write $T_3 = T_1 \vee T_2$, so $T_3 = (T_1 - T_2)^+ + T_2$. It follows that

$$(T_1 \vee T_2)(u) = T_3 u = \sup(T_1 v - T_2 v : 0 \leq v \leq u) + T_2 u =$$

$$\sup\{T_1 v + T_2(u - v) : 0 \leq v \leq u),$$

which is the desired result. The proof of (4) is similar.

Next, we prove the inequality in (6). For $u \in E^+$ we have $Tu \leq T^+ u \leq \mid T \mid (u)$ and $-Tu \leq T^- u \leq \mid T \mid (u)$, so

$$\mid Tu \mid = (Tu) \vee (-Tu) \leq \mid T \mid (u).$$

If $f \in E$ is arbitrary, then

$$\mid Tf \mid = \mid Tf^+ - Tf^- \mid \leq \mid Tf^+ \mid + \mid Tf^- \mid \leq$$

$$\mid T \mid (f^+) + \mid T \mid (f^-) = \mid T \mid (\mid f \mid).$$

It remains to prove the formulas (5) for $\mid T \mid (u)$. We have

$$\mid T \mid (u) = T^+ u + T^- u =$$

$$\sup(Tv : 0 \leq v \leq u) + \sup(-Tw : 0 \leq w \leq u) =$$

$$\sup\{T(v - w) : 0 \leq v \leq u, 0 \leq w \leq u\}.$$

From $0 \leq v \leq u, 0 \leq w \leq u$ it follows that $v - w \leq u$ and $w - v \leq u$, so $\mid v - w \mid \leq u$. Hence

$$\mid T \mid (u) = \sup\{T(v - w) : 0 \leq v \leq u, 0 \leq w \leq u\} \leq \sup(Tf : \mid f \mid \leq u).$$

In the converse direction, $\mid f \mid \leq u$ implies $Tf \leq \mid Tf \mid \leq \mid T \mid (\mid f \mid) \leq \mid T \mid (u)$, so

$$\sup(Tf : \mid f \mid \leq u) \leq \sup(\mid Tf \mid : \mid f \mid \leq u) \leq \mid T \mid (u).$$

Therefore, (5) follows. ∎

Before presenting another application of the extension lemma, we recall some definitions and notations. If σ is a mapping from the set X into the set Y (with X and Y non-empty) and X_1 is a subset of X, then the *restriction* of σ to X_1 is denoted by $\sigma \mid X_1$. Hence, $\sigma \mid X_1$ maps X_1 into Y and $(\sigma \mid X_1)(x) = \sigma(x)$ for every $x \in X_1$. Furthermore, if E and F are Riesz spaces and ρ is a mapping from E^+ into F^+, then ρ is said to be *monotone* if $0 \leq v \leq u$ in E^+ implies that $0 \leq \rho(v) \leq \rho(u)$ in F^+. Examples: $(i)\rho$ is the restriction $T \mid E^+$ of a positive linear operator $T : E \rightarrow F, (ii)F = \mathbb{R}$ and ρ is a Riesz norm in E^+ (or, rather, ρ is the restriction to E^+ of a Riesz norm in E).

Theorem 20.5. *As before, let E and F be Riesz spaces with F Dedekind complete and let A be an ideal in E. Furthermore, let $T : A \rightarrow F$ be a positive linear operator and let $\rho : E^+ \rightarrow F^+$ be a monotone mapping majorizing T on A^+, i.e., $\rho(u) \geq Tu$ for every $u \in A^+$. Then there exists a positive operator $T_A : E \rightarrow F$ which extends T and which vanishes on A^d, i.e., $T_A f = Tf$ for all $f \in A$ and $T_A f = 0$ for all $f \in A^d$. The operator T_A is the minimal positive extension of T, that is to say, if T_1 is another positive extension of T, then $T_1 u \geq Tu$ for all $u \in E^+$.*

Proof. For any $u \in E^+$, let

$$\tau(u) = \sup(Tv : v \in A, 0 \leq v \leq u).$$

Note that the supremum exists in F^+ since $Tv \leq \rho(v) \leq \rho(u)$ for every $v \in A$ satisfying $0 \leq v \leq u$ and F is Dedekind complete. The mapping $\tau : E^+ \rightarrow F^+$ is additive; the additivity proof is almost the same as for the mapping τ in Theorem 20.2. Hence, by the extension lemma, there exists a positive operator $T_A : E \rightarrow F$ extending τ. Since $\tau(u) = Tu$ for every $u \in A^+$ and $\tau(u) = 0$ for $0 \leq u \in A^d$, it is evident that T_A extends T and T_A vanishes on A^d. If T_1 is another positive extension of T, then $T_1 u \geq T_1 v = Tv$ for $v \in A^+$ and $0 \leq v \leq u \in E^+$, so

$$T_1 u \geq \sup(Tv : v \in A, 0 \leq v \leq u) = T_A u.$$

This shows that T_A is the minimal positive extension of T.   ∎

We mention some special cases.

(i) If $S : E \rightarrow F$ is a positive operator and T is the restriction $S \mid A$, then T_A is, therefore, the minimal positive extension to E of $S \mid A$, i.e., $T_A = (S \mid A)_A$. Since S itself is a positive extension of $S \mid A$, it is evident now that $(S \mid A)_A \leq S$. Usually $(S \mid A)_A$ is simply denoted by S_A. Observe that if A and B are ideals in E such that $A \subseteq B$, then $S_A \leq S_B \leq S$.

(ii) If $F = \mathbb{R}$ and $\sigma : E^+ \rightarrow \mathbb{R}^+$ is a monotone mapping majorizing on A^+ the positive linear functional φ defined on A, then φ can be extended to a positive linear functional φ_A on E in such a way that $\varphi_A \leq \psi$ for any other linear extension ψ of φ. This holds in particular if there exists a Riesz norm in E majorizing φ on A^+.

Exercise 20.6. Show that if in the last theorem the ideal A is a principal ideal A_w generated by an element $w \in E^+$, then

$$\tau(u) = \sup(Tv : v \in A_w, 0 \leq v \leq u) = \sup\{T(u \wedge nw) : n = 1, 2, \ldots\}$$

for every $u \in E^+$. Note that it is sufficient for the existence of $\tau(u)$ to assume that F is Dedekind σ-complete.

Hint: If $v \in A_w$ satisfies $0 \leq v \leq u$, then $v \leq u \wedge nw$ for some n (n depending on v). Conversely, if $v = u \wedge nw$, then $0 \leq v \leq u$ and $v \in A_w$.

Example 20.7. In this example, let $E = \mathbb{R}^n$ and $F = \mathbb{R}^m$, where n and m are natural numbers greater than or equal to one and where the ordering in both spaces is coordinatewise, which makes them Dedekind complete Riesz spaces. For $k = 1, \ldots, n$, let a_k be the element of $\mathbb{R}^n$ having its k-th coordinate one and all other coordinates zero. Similarly, for $j = 1, \ldots, m$, let b_j be the element of $\mathbb{R}^m$ having its j-th coordinate one and all other coordinates zero. The positive cone of $\mathbb{R}^n$ is, therefore, the set of all $u = \alpha_1 a_1 + \cdots + \alpha_n a_n$ with $\alpha_k \geq 0$ for all $k = 1, \ldots, n$. Similarly for $\mathbb{R}^m$. Any linear operator $T : \mathbb{R}^n \to \mathbb{R}^m$ is completely determined if one knows the images $Ta_k(k = 1, \ldots, n)$. Hence, if $Ta_k = \Sigma_{j=1}^m t_{jk} b_j$ for $k = 1, \ldots, n$, then T is completely determined by the coefficient matrix (t_{jk}). Evidently, T is a positive operator if and only if all $Ta_k(k = 1, \ldots, n)$ are elements of the positive cone in $\mathbb{R}^m$, i.e., if and only if $t_{jk} \geq 0$ holds for all $j = 1, \ldots, m$ and $k = 1, \ldots, n$. It follows that if S and T are linear operators from $\mathbb{R}^n$ into $\mathbb{R}^m$ with matrices (s_{jk}) and (t_{jk}) respectively, then $S \leq T$ if and only if $s_{jk} \leq t_{jk}$ for all j, k. For an arbitrary $T : \mathbb{R}^n \to \mathbb{R}^m$ with matrix (t_{jk}), let us write $t_{jk}^+ = \max(t_{jk}, 0)$ and $t_{jk}^- = \max(-t_{jk}, 0)$, so t_{jk}^+ and t_{jk}^- are non-negative numbers (one of them or both zero) satisfying $t_{jk}^+ - t_{jk}^- = t_{jk}$ and $t_{jk}^+ + t_{jk}^- =\mid t_{jk} \mid$. Let T_1 and T_2 be the operators with matrices (t_{jk}^+) and (t_{jk}^-) respectively. Then T_1 and T_2 are positive operators such that $T_1 - T_2 = T$. This shows that every linear operator from $\mathbb{R}^n$ into $\mathbb{R}^m$ is regular (and, therefore, order bounded). Since $t_{jk}^+ \geq t_{jk}$ as well as $t_{jk}^+ \geq 0$, the operator T_1 majorizes T as well as the null operator, i.e., T_1 is an upper bound of T and of 0. If S, with matrix (s_{jk}), is another upper bound, then $s_{jk} \geq t_{jk}$ and $s_{jk} \geq 0$, so $s_{jk} \geq t_{jk}^+$. This shows that T_1 is the least upper bound of T and 0, i.e., $T_1 = T \vee 0 = T^+$. Similarly, $T_2 = (-T) \vee 0 = T^-$. Hence, $T = T^+ - T^-$, where T^+ and T^- have matrices (t_{jk}^+) and (t_{jk}^-) respectively. It follows that $\mid T \mid = T^+ \vee T^- = T^+ + T^-$ has the matrix $(\mid t_{jk} \mid)$.

This matrix result can be generalized to certain operators T from E into F, where E and F are Riesz spaces of (real) measurable functions on point sets Y and X respectively and where the operator T is of the form

$$(Tf)(x) = \int_Y T(x, y) f(y) dy$$

for all $f \in E$. The function $T(x, y)$, called the kernel of T, takes over the role of the matrix (t_{jk}). One of the results is that the operator $\mid T \mid = T^+ \vee T^-$ in the space $\mathcal{L}_b(E, F)$ has the kernel $\mid T(x, y) \mid$.

In Theorem 20.4 it has been shown that if E and F are Riesz spaces with F Dedekind complete and the operator $T : E \to F$ is order bounded, then

$$T^+ u = \sup(Tv : 0 \le v \le u) \text{ and } \mid T \mid (u) = \sup(Tf : -u \le f \le u)$$

for every $u \in E^+$. There exist dual formulas, due to I.Namioka (1957), as follows. It is sufficient for the validity of these formulas to assume that F is Dedekind σ-complete.

Theorem 20.8. *Let E and F be Riesz spaces with F Dedekind σ-complete. Furthermore, let $T : E \to F$ be a positive operator. Then*

$$Tf^+ = \max(Sf : 0 \le S \le T) \quad and \quad T(\mid f \mid) = \max(Sf : -T \le S \le T)$$

for every $f \in E$. In particular, if φ is a positive linear functional on the Riesz space E, then

$$\varphi(f^+) = \max(\psi(f) : 0 \le \psi \le \varphi) \quad and \quad \varphi(\mid f \mid) = \max(\psi(f) : -\varphi \le \psi \le \varphi)$$

for every $f \in E$.

Proof. For $0 \le S \le T$ and $f \in E$ fixed, we have $Sf \le Sf^+ \le Tf^+$, so Tf^+ is an upper bound of the set $(Sf : 0 \le S \le T)$. For the proof that Tf^+ is the maximum of the set, let T_{f+} be the minimal positive extension of the restriction of T to the principal ideal generated by f^+, i.e., (see Exercise 20.6 above)

$$T_{f+}(u) = \sup\{T(u \wedge nf^+) : n = 1, 2, \ldots\}$$

for $u \in E^+$. Here it is used that Dedekind σ-completeness of F is sufficient. The operator T_{f+} satisfies $0 \le T_{f+} \le T$, so T_{f+} is an admissible S. Note that $T_{f+}(f^-) = 0$ since f^- is disjoint to the principal ideal generated by f^+. For the operator T_{f+} the desired maximum is now obtained because

$$T_{f+}(f) = T_{f+}(f^+) - T_{f+}(f^-) = Tf^+ - 0 = Tf^+.$$

To prove the second formula, observe first that if $-T \le S \le T$, then

$$Sf = Sf^+ - Sf^- \le Tf^+ + Tf^- = T(\mid f \mid),$$

so $T(\mid f \mid)$ is an upper bound of the set $(Sf : -T \le S \le T)$. For the proof that $T(\mid f \mid)$ is the maximum, consider the operator $S = (T_{f+}) - (T_{f-})$. Then $-T \le S \le T$ and

$$Sf = T_{f+}(f) - T_{f-}(f) = Tf^+ + Tf^- = T(\mid f \mid),$$

so this S yields the desired maximum.   ∎

In Exercise 6.8 it was indicated how to prove in a Riesz space E that if $0 = z_0 \leq z_1 \leq \cdots \leq z_n = z$ and the positive elements u and v satisfy $u + v = z$, then there exist sequences

$$0 = u_0 \leq u_1 \leq \cdots \leq u_n = u \text{ and } 0 = v_0 \leq v_1 \leq \cdots \leq v_n = v$$

such that $u_k + v_k = z_k$ for $k = 1, \ldots n$. By induction it follows easily that if $0 = z_0 \leq z_1 \leq \cdots \leq z_n = z$ and the positive elements $u_j(j = 1, \ldots, m)$ satisfy $\Sigma_{j=1}^m u_j = z$, then there exist sequences

$$0 = u_{j0} \leq u_{j1} \leq \cdots \leq u_{jn} = u_j (j = 1, \ldots, m)$$

such that $u_{1k} + u_{2k} + \cdots + u_{mk} = z_k$ for $k = 1, \ldots, n$. This result can be reformulated. Write $v_k = z_k - z_{k-1}$ for $k = 1, \ldots, n$, so $v_k \geq 0$ for all k and $\Sigma_{k=1}^n v_k = z = \Sigma_{j=1}^m u_j$. Also write $w_{jk} = u_{jk} - u_{j,k-1}$ for $k = 1, \ldots, n$, so $w_{jk} \geq 0$ for all j, k and $\Sigma_{k=1}^n w_{jk} = u_{jn} = u_j$ for all j as well as

$$\Sigma_{j=1}^m w_{jk} = \Sigma_j u_{jk} - \Sigma_j u_{j,k-1} = z_k - z_{k-1} = v_k$$

for all k. Hence, the following theorem (known as the *general Riesz decomposition theorem*) holds.

Theorem 20.9. *If in the Riesz space E the positive elements $u_j(j = 1, \ldots, m)$ and $v_k(k = 1, \ldots, n)$ satisfy $\Sigma_{j=1}^m u_j = \Sigma_{k=1}^n v_k$, then there exist positive elements w_{jk} for $j = 1, \ldots, m$ and $k = 1, \ldots, n$ such that*

$$\Sigma_{k=1}^n w_{jk} = u_j \text{ and } \Sigma_{j=1}^m w_{jk} = v_k \text{ for all } j, k \text{ involved.}$$

The general Riesz decomposition theorem plays an important role in one more theorem concerning $S \vee T$ and $S \wedge T$, where S and T are order bounded operators.

Theorem 20.10. *Let E and F be Riesz spaces with F Dedekind complete and let S and T be order bounded operators from E into F. Then the following holds.*

(i) For any fixed $u \in E^+$, the set

$$D_u = (\Sigma_{j=1}^n \mid Tu_j \mid : \text{ all } u_j \in E^+ \text{ and } \Sigma_j u_j = u),$$

where n is also variable, is upwards directed and has $\mid T \mid (u)$ as its supremum, i.e., $D_u \uparrow \mid T \mid (u)$.

(ii) Similarly,

$$P_u = \{\Sigma_{j=1}^n (Su_j) \vee (Tu_j) : \text{ all } u_j \in E^+ \text{ and } \Sigma_j u_j = u\}$$

satisfies $P_u \uparrow (S \vee T)(u)$.

(iii) Also,

$$\{\Sigma_{j=1}^n (Su_j) \wedge (Tu_j): \text{ all } u_j \in E^+ \text{ and } \Sigma_j u_j = u\} \downarrow (S \wedge T)(u).$$

Proof. If $u = \Sigma_1^n u_j = \Sigma_1^m v_k$ (all $u_j, v_k \in E^+$), there exist in view of the last theorem elements $w_{jk}(j = 1, \ldots, n; k = 1, \ldots, m)$ such that $u_j = \Sigma_k w_{jk}$ for all j and $v_k = \Sigma_j w_{jk}$ for all k. Then $\Sigma_j \mid Tu_j \mid$ and $\Sigma_k \mid Tv_k \mid$ are both less than or equal to $\Sigma_j \Sigma_k \mid Tw_{jk} \mid$. This shows that the set D_u is upwards directed. Since $\Sigma \mid Tu_j \mid \leq \Sigma \mid T \mid (u_j) = \mid T \mid (u)$, the set D_u has $\mid T \mid (u)$ as an upper bound, so $\sup D_u$ exists and satisfies $\sup D_u \leq \mid T \mid (u)$. For an inequality in the converse direction, note that if $f \in E$ satisfies $\mid f \mid \leq u$, then

$$Tf \leq \mid Tf \mid = \mid T(f^+) - T(f^-) \mid \leq$$

$$\mid T(f^+) \mid + \mid T(f^-) \mid + \mid T(u- \mid f \mid) \mid \in D_u,$$

so $Tf \leq \sup D_u$. It follows that

$$\mid T \mid (u) = \sup(Tf : \mid f \mid \leq u) \leq \sup D_u.$$

This, together with $\sup D_u \leq \mid T \mid (u)$, shows that $\mid T \mid (u) = \sup D_u$.

(ii) In Theorem 5.2 it was proved that $2(f \vee g)$ equals $f + g+ \mid f - g \mid$ for all f and g in a Riesz space. Hence

$$2(S \vee T)(u) = Su + Tu+ \mid S - T \mid (u) =$$

$$Su + Tu + \sup(\Sigma_1^n \mid Su_j - Tu_j \mid : \text{ all } u_j \in E^+, \Sigma u_j = u) =$$

$$\sup\{\Sigma_1^n (Su_j + Tu_j+ \mid Su_j - Tu_j \mid): \text{ all } u_j \in E^+, \Sigma u_j = u\} =$$

$$2\sup\{\Sigma_1^n (Su_j) \vee (Tu_j): \text{ all } u_j \in E^+, \Sigma u_j = u\} = 2 \sup P_u.$$

It is evident in view of what was proved in part (i) that the set

$$(\Sigma_1^n \mid Su_j - Tu_j \mid : \text{ all } u_j \in E^+, \Sigma u_j = u)$$

is upwards directed, and so the set P_u is upwards directed.

(iii)   Similarly, using now that $2(f \wedge g) = f + g- \mid f - g \mid$.   ∎

Exercise 20.11. Once more, let E and F be Riesz spaces with F Dedekind complete and let T_1 and T_2 be order bounded operators from E into F. It has been shown in Theorem 20.4 in the present section that, for any $u \in E^+$, we have

$$(T_1 \wedge T_2)(u) = \inf(T_1 v + T_2 w : v \geq 0, w \geq 0, v + w = u)$$

and similarly for $(T_1 \vee T_2)(u)$, with sup instead of inf. It may be asked if the same holds if we require in addition that v and w in $v + w = u$ are disjoint, so $v \wedge w = 0$. In order that such suitable components v and w of u may exist it is reasonable to assume additionally that E has the principal projection property. Hence, assume now that E has the principal projection property and F is Dedekind complete. Prove that

$$(T_1 \wedge T_2)(u) = \inf\{T_1 z + T_2(u - z) : z \wedge (u - z) = 0\},$$
$$(T_1 \vee T_2)(u) = \sup\{T_1 z + T_2(u - z) : z \wedge (u - z) = 0\}.$$

Show first that the second formula follows from the first by observing that

$$T_1 \vee T_2 = -\{(-T_1) \wedge (-T_2)\}.$$

Next, show that the formula for $T_1 \wedge T_2$ holds if it holds for the special case that $T_1 \wedge T_2 = 0$. To see this, assume that it holds in the special case and prove that it holds generally by observing that

$$\{T_1 - (T_1 \wedge T_2)\} \wedge \{T_2 - (T_1 \wedge T_2)\} = 0.$$

For the proof in the special case, assume that $T_1 \wedge T_2 = 0$ and let $u \in E^+$ be given. We know that

$$(T_1 \wedge T_2)(u) = \inf\{T_1 v + T_2(u - v) : 0 \le v \le u\} = 0$$

and we wish to show that

$$\inf\{T_1 z + T_2(u - z) : z \wedge (u - z) = 0\} = 0.$$

The idea is now to indicate for any v satisfying $0 \le v \le u$ an appropriate z satisfying $z \wedge (u - z) = 0$ and such that $z \le 2v$ and $u - z \le 2(u - v)$, because then

$$T_1 z + T_2(u - z) \le 2\{T_1 v + T_2(u - v)\},$$

which will immediately imply the desired result.

At this point it will help to assume for a moment that the elements of E are real functions on a point set X with u the unit function, so if we look for functions z satisfying $z \wedge (u - z) = 0$ we have to find them among the characteristic functions of subsets of X. Now, if v is given such that $0 \le v(x) \le 1$ for all x and we wish to find a characteristic function z such that

$$z(x) \le 2v(x) \text{ and } 1 - z(x) \le 2\{1 - v(x)\},$$

it is obvious that z must be the characteristic function of the set on which $2v(x) \ge 1$, i.e., z is the order projection of u on the band generated by $(2v - u)^+$. This is what we also do in the general case. Let P be the order projection on the band generated by $(2v - u)^+$ and set $z = Pu$, so $z \wedge (u - z) = 0$. Observe that $u \le 2v + (u - 2v)^+$ and $P\{(u - 2v)^+\} = 0$, so $z = Pu \le 2Pv \le 2v$. It remains to show that $u - z \le 2(u - v)$, i.e., $2v - u \le z = Pu$. Now,

$$2v - u \le (2v - u)^+ = P\{(2v - u)^+\},$$

so the proof will be complete if we can prove that $(2v - u)^+ \le u$. Show that this holds indeed. The result in this exercise is due to Yu.A. Abramovitch (1971).

CHAPTER 11
Order Continuous Operators

21. Order Continuous Operators

We shall now turn our attention to a special kind of linear operators mapping a Riesz space E into another Riesz space F, called order continuous operators. As long as nothing more is assumed about F the interest is focused mainly on positive operators, but if F is Dedekind complete any regular operator $T : E \to F$ has an absolute value $|\,T\,|$, and then we can say more. We begin by presenting the definition of an order continuous operator.

Definition 21.1. The linear operator T mapping the Riesz space E into the Riesz space F is said to be *order continuous* if, for any downwards directed set D in E having infimum the null element (i.e., $D \downarrow 0$ in E), we have $\inf(|\,Tf\,|: f \in D) = 0$ in F.

Some immediate consequences are listed below.

Theorem 21.2. *(i) If T is order continuous, then so is αT for any real number α.*

(ii) If $T \geq 0$ (i.e., if T is positive), then T is order continuous if and only if it follows from $D \downarrow 0$ in E that $(Tf : f \in D) \downarrow 0$ in F (equivalently, $D \uparrow u$ in E implies $(Tf : f \in D) \uparrow Tu$ in F).

(iii) If $T \geq 0$ and T is order continuous, and if $0 \leq S \leq T$, then S is order continuous.

For a discussion of further properties we recall a lemma about subsets of E.

Lemma 21.3. *If D and G are subsets of the ordered vector space E such that $d_0 = \inf D$ and $g_0 = \inf G$, then*

$$d_0 + g_0 = \inf(d + g : d \in D, g \in G).$$

Similarly for suprema.

Proof. It is obvious that $d_0 + g_0$ is a lower bound of the set of all $d + g$. If h is another lower bound we have $h - g \leq d$ for every $g \in G$ and every $d \in D$, so

$$h - g \leq \inf(d : d \in D) = d_0.$$

Hence $h - d_0 \leq g$ for every $g \in G$, which implies

$$h - d_0 \leq \inf(g : g \in G) = g_0.$$

This shows that $h \leq d_0 + g_0$, and so $d_0 + g_0$ is the greatest lower bound of the set of all $d + g$ for $d \in D, g \in G$. ∎

Theorem 21.4. *(i) If the positive operators T and S are order continuous, then $\alpha T + \beta S$ is order continuous for all real numbers α, β. In particular, if $T = T_1 - T_2$ with T_1 and T_2 positive and order continuous, then T is order continuous.*

(ii) If $-T_1 \leq T \leq T_2$ with T_1, T_2 positive and order continuous, then T is order continuous.

Proof. We prove first that if T and S are positive and order continuous, then so is $T + S$. To this end, we have to show that $D \downarrow 0$ implies $(Tf + Sf : f \in D) \downarrow 0$. It is clear that the set of all $Tf + Sf$ is downwards directed and the null element is a lower bound. If h is another lower bound and f_1, f_2 are given in D, there exists $f_3 \in D$ such that $f_3 \leq f_1$ and $f_3 \leq f_2$, so

$$h \leq Tf_3 + Sf_3 \leq Tf_1 + Sf_2.$$

Hence $h \leq \inf(Tf_1 + Sf_2 : f_1, f_2 \in D) = 0$, where the last equality (in view of Lemma 21.3 above) follows from

$$\inf(Tf : f \in D) = \inf(Sf : f \in D) = 0.$$

This shows that $\inf(Tf + Sf : f \in D) = 0$.

For α, β real numbers and $f \in D$ we have

$$\mid (\alpha T + \beta S)f \mid \leq \mid \alpha Tf \mid + \mid \beta Sf \mid =$$

$$\mid \alpha \mid \cdot Tf + \mid \beta \mid \cdot Sf \leq (\mid \alpha \mid + \mid \beta \mid)(Tf + Sf) \downarrow 0,$$

so $\inf\{\mid (\alpha T + \beta S)f \mid : f \in D\} = 0$.

(ii) If $-T_1 \leq T \leq T_2$ with T_1, T_2 positive and order continuous, then

$$0 \leq T + T_1 \leq T_2 + T_1$$

with $T_2 + T_1$ positive and order continuous. It follows that $T + T_1$ is positive and order continuous, and so $T = (T + T_1) - T_1$ is order continuous. ∎

As the reader may already have expected the notion of a σ-order continuous operator (countably order continuous operator) may be defined as well.

Definition 21.5. The linear operator $T : E \to F$ is said to be *σ-order continuous* if, for any monotone sequence $f_n \downarrow 0 (n = 1, 2, \ldots)$, we have $\inf(\mid Tf_n \mid: n = 1, 2, \ldots) = 0$. For T positive this means, therefore, that $f_n \downarrow 0$ implies $Tf_n \downarrow 0$.

All properties proved so far for order continuous operators hold (with appropriate modifications) for σ-order continuous operators. Observe that if $T : E \to F$ is positive, then T is σ-order continuous if and only if T has the property that if $f_n (n = 1, 2, \ldots)$ converges in order to f in E, then Tf_n converges in order to Tf in F (equivalently, $f_n \to 0$ in E implies $Tf_n \to 0$ in F). Since $T \geq 0$, it is trivial to see that if $f_n \to 0$ implies $Tf_n \to 0$, then $f_n \downarrow 0$ implies $Tf_n \downarrow 0$, so in this case T is σ-order continuous. Conversely, if T is σ-order continuous and $f_n \to 0$, then there exists a sequence $(p_n : n = 1, 2, \ldots)$ such that $\mid f_n \mid \leq p_n \downarrow 0$, so

$$\mid Tf_n \mid \leq \mid Tf_n^+ \mid + \mid Tf_n^- \mid = Tf_n^+ + Tf_n^- = T(\mid f_n \mid) \leq Tp_n \downarrow 0,$$

which shows that $Tf_n \to 0$.

As a further remark observe that if E is order separable, then order continuous and σ-order continuous operators from E into F are the same. This is clear because any set $D \downarrow 0$ in E contains a decreasing sequence $f_n \downarrow 0$ (by Theorem 17.6), so if T is σ-order continuous, then $\inf(\mid Tf_n \mid) = 0$ and hence $\inf(\mid Tf \mid: f \in D) = 0$, which shows that T is order continuous.

It will be proved now that there exist E and F for which every positive linear operator from E into F is order continuous and there also exist E and F for which the only order continuous (or even σ-order continuous) positive linear operator from E into F is the null operator.

Example 21.6. (i) Let E be a normed Riesz space of type $L_p = L_p(X, \mu)$ for some p satisfying $1 \leq p < \infty$ and let T be a positive operator from $E = L_p$ into some normed Riesz space F (for example into L_p itself). The norm in L_p is order continuous, i.e., $D \downarrow 0$ in L_p implies $(\parallel f \parallel: f \in D) \downarrow 0$ in $\mathbb{R}$ (see Exercise 17.15). Furthermore, T is norm bounded (since by Theorem 18.4 every positive operator from the Banach lattice L_p into a normed Riesz space is norm bounded). Hence $\parallel Tf \parallel \leq \parallel T \parallel \cdot \parallel f \parallel$ for every $f \in L_p$. Therefore, $D \downarrow 0$ in L_p implies $(\parallel f \parallel: f \in D) \downarrow 0$, and so $(\parallel Tf \parallel: f \in D) \downarrow 0$. Since $(Tf : f \in D)$ is downwards directed in F, it follows that $(Tf : f \in D) \downarrow 0$. It has been shown thus that $D \downarrow 0$ implies $(Tf : f \in D) \downarrow 0$, so T is order continuous. Therefore, every positive operator from E into F is order continuous. In particular, choosing $F = \mathbb{R}$, it follows that every positive linear functional on L_p is order continuous.

(ii) As on earlier occasions, let $C([0, 1])$ denote the Riesz space of all real continuous functions on $[0,1]$ with the familiar pointwise ordering. We shall prove that the only positive σ-order continuous operator from $C([0, 1])$ into

$\mathbb{R}$ is the null operator. In other words, the only positive σ-order continuous linear functional on $C([0,1])$ is the null functional. For the proof, let $(r_n : n = 1, 2, \ldots)$ be the set of all rational numbers in $[0,1]$ and, for any pair (m, n) of rational numbers, let u_{mn} be a continuous function on $[0,1]$ such that

(a) $0 \le u_{mn}(x) \le 1$ for all x and $u_{mn}(r_n) = 1$,
(b) $u_{mn}(x) = 0$ for $\mid x - r_n \mid > (2^n m)^{-1}$.

Furthermore, let $v_{mn} = \sup(u_{m1}, \ldots, u_{mn})$. Then $v_{m1} \le v_{m2} \le \cdots$, the pointwise limit f_m of this increasing sequence exists on $[0,1]$ and f_m satisfies $0 \le f_m(x) \le 1$ for all x and $f_m(r_n) = 1$ for all r_n. In view of condition (b) the set $F_{mn} = (x : v_{mn}(x) > 0)$ is of Lebesgue measure at most $2/m$ for every n. Therefore, since F_{mn} is ascending as n increases, the set $(x : f_m(x) > 0)$ is of measure at most $2/m$. Keeping in mind that $f_m(r_n) = 1$ for all n, it is easy to see now that (for $m > 2$) the function f_m is not continuous on the whole of $[0,1]$, i.e., f_m is not a member of $C([0,1])$. However, once more since $f_m(r_n) = 1$ for all r_n, it is evident that the supremum of the sequence $(v_{m1}, v_{m2}, \ldots)$ exists in $C([0,1])$; the supremum is in fact the function e, identically one on $[0,1]$. Hence $v_{mn} \uparrow e$.

Assume now that there exists a σ-order continuous positive linear functional φ on $C([0,1])$, not the null functional. Without loss of generality we may assume that $\varphi(e) = 1$. In view of $v_{mn} \uparrow e$ we have $\varphi(v_{mn}) \uparrow \varphi(e) = 1$, so there exists a natural number n_m such that

$$\varphi(e - v_{mn_m}) < 2^{-m-1}.$$

As observed, the set $(x : v_{mn}(x) > 0)$ is of measure at most $2/m$ for every n, in particular for $n = n_m$. Finally, let

$$w_m = \inf(v_{1n_1}, v_{2n_2}, \ldots, v_{mn_m})$$

for every m. Then $w_m \downarrow$ and $(x : w_m(x) > 0)$ is of measure at most $2/m$. Therefore, $w_m \downarrow 0$ in $C([0,1])$; this holds even pointwise almost everywhere. It follows that $\varphi(w_m) \downarrow 0$. But

$$e - w_m = \sup(e - v_{1n_1}, \ldots, e - v_{mn_m}) \le (e - v_{1n_1}) + \cdots + (e - v_{mn_m})$$

for every m, so

$$\varphi(e - w_m) < 2^{-2} + \cdots + 2^{-m-1} < 2^{-1},$$

which implies $\varphi(w_m) > 2^{-1}$ on account of $\varphi(e) = 1$. This contradicts $\varphi(w_m) \downarrow 0$. It follows that the only σ-order continuous linear functional on $C([0,1])$ is the null functional. We present an extension of this result in the exercise which follows.

Exercise 21.7. Let μ be a σ-finite measure in the point set X and let p and q be real numbers satisfying $1 \leq p \leq \infty, p^{-1}+q^{-1} = 1$ (so $q = \infty$ if $p = 1$ and $q = 1$ if $p = \infty$). Recall that if $f \in L_p = L_p(X,\mu)$ and $g \in L_q = L_q(X,\mu)$, then

$$\mid \int_X fgd\mu \mid \leq \int_X \mid fg \mid d\mu < \infty$$

and observe that if $0 \leq g \in L_q$ is fixed, then $\varphi(f) = \int_X fgd\mu$ is a positive linear functional on L_p. In view of what was proved in part (i) of the last example, the positive linear functional φ is order continuous for $1 \leq p < \infty$.

(i) Show that if $p = \infty$, then φ is order continuous as well.

(ii) Let T be a positive σ-order continuous operator from the Riesz space E into $L_p(X,\mu)$ and let $0 \leq g \in L_q(X,\mu)$ be given. Show that ψ, defined by

$$\psi(f) = \int_X \{(Tf)(x)\}g(x)d\mu$$

for all $f \in E$, is a positive σ-order continuous linear functional on E. If T is order continuous, then so is ψ.

(iii) Show that if $0 \leq f \in L_p$ is given and f has the property that $\int_X fgd\mu = 0$ for all $0 \leq g \in L_q$, then $f = 0$ (almost everywhere).

(iv) As before, $C([0,1])$ denotes the Riesz space of all real continuous functions on [0,1]. Show that the only σ-order continuous operator from $C([0,1])$ into $L_p(X,\mu)$ is the null operator.

Hint: For (iii), observe that $\int_E fd\mu = 0$ for every subset E of X of finite measure. If f is not almost everywhere zero, there exists a set E of finite measure such that $f(x) > \epsilon$ for some $\epsilon > 0$ and all $x \in E$. Contradiction. For (iv), choose $E = C([0,1])$ in (ii) and use part (ii) of the last example.

The result that the only σ-order continuous linear functional on $C([0,1])$ is the null functional may seem surprising at first because it follows for example that the familiar Riemann integral on $C([0,1])$ is not σ-order continuous, although the Riemann integral has the property that if $(u_n : n = 1, 2, \ldots)$ is a sequence in $C([0,1])$ satisfying $u_n(x) \downarrow 0$ for every $x \in [0,1]$, then $\int u_n(x)dx \downarrow 0$. This follows from Lebesgue's theorem on integration of monotone sequences or, alternatively, from a theorem known as Dini's theorem asserting that u_n converges uniformly to zero. This seems to contradict the result mentioned above. Observe, however, that $u_n \downarrow 0$ in $C([0,1])$ is not the same as $u_n(x) \downarrow 0$ for every $x \in [0,1]$. It is evident that $u_n(x) \downarrow 0$ pointwise implies $u_n \downarrow 0$ in $C([0,1])$, but the converse does not hold (remember that we had $v_{mn} \uparrow e$ in the last example, but the pointwise limit of the sequence is not the function e). This explains why a positive linear functional φ on $C([0,1])$ satisfying $\varphi(u_n) \downarrow 0$ if $u_n(x) \downarrow 0$ pointwise need not yet be σ-order continuous.

22. The Band of Order Continuous Operators

We return to the situation that E and F are Riesz spaces with F Dedekind complete. As proved in Theorem 20.2, the set $\mathcal{L}_b(E, F)$ of all order bounded operators from E into F is a Dedekind complete Riesz space. We recall that this is at the same time the set of all regular operators from E into F. The set of all order continuous operators in $\mathcal{L}_b(E, F)$ will be denoted by $\mathcal{L}_n(E, F)$ and, similarly, the set of all σ-order continuous operators in $\mathcal{L}_b(E, F)$ will be denoted by $\mathcal{L}_c(E, F)$. As long as E and F are fixed, we write $\mathcal{L}_b, \mathcal{L}_n$ and $\mathcal{L}_c$. Evidently we have $\mathcal{L}_n \subseteq \mathcal{L}_c \subseteq \mathcal{L}_b$. We prove first that $\mathcal{L}_n$ and $\mathcal{L}_c$ are bands in $\mathcal{L}_b$.

Lemma 22.1. *Let* $T \in \mathcal{L}_b$. *Then* $T \in \mathcal{L}_n$ *if and only if* T^+ *and* T^- *are members of* $\mathcal{L}_n$, *i.e., if and only if* $|T| \in \mathcal{L}_n$. *Similarly,* $T \in \mathcal{L}_c$ *if and only if* T^+ *and* T^- *are members of* $\mathcal{L}_c$, *i.e., if and only if* $|T| \in \mathcal{L}_c$.

Proof. The proof will be given for $\mathcal{L}_n$; the proof for $\mathcal{L}_c$ is quite similar. We show first that $T \in \mathcal{L}_n$ implies that $T^+ \in \mathcal{L}_n$. Hence, assume that $T \in \mathcal{L}_n$ and $D \downarrow 0$ in E. We have to prove that $\inf(T^+ u : u \in D) = 0$. For this purpose we may assume that there exists an element $u_0 \in D$ such that $0 \le u \le u_0$ for all $u \in D$ (choose $u_0 \in D$ arbitrarily and observe that for any $v \in D$ there exists $u \in D$ satisfying $u \le v \wedge u_0$). Now choose $v \in E^+$ such that $0 \le v \le u_0$. Then, for any $u \in D$, we have

$$0 \le v - (v \wedge u) = (v \wedge u_0) - (v \wedge u) \le u_0 - u,$$

so (since $0 \le p \le q$ implies $Tp \le T^+ p \le T^+ q$) we get

$$T\{v - (v \wedge u)\} \le T^+(u_0 - u).$$

It follows that

$$0 \le T^+ u \le |T(v \wedge u)| + T^+ u_0 - Tv.$$

Since $T \in \mathcal{L}_n$ and $(v \wedge u : u \in D) \downarrow 0$, we have

$$\inf\{|T(v \wedge u)| : u \in D\} = 0,$$

so $0 \le \inf(T^+ u : u \in D) \le T^+ u_0 - Tv$. This holds for all $v \in E$ satisfying $0 \le v \le u_0$. Hence

$$0 \le \inf(T^+ u : u \in D) \le$$

$$T^+ u_0 - \sup(Tv : 0 \le v \le u_0) = T^+ u_0 - T^+ u_0 = 0,$$

and so $\inf(T^+ u : u \in D) = 0$, which is the desired result. It follows that $T^+ \in \mathcal{L}_n$. For the proof that $T \in \mathcal{L}_n$ implies $T^- \in \mathcal{L}_n$, observe that $T^- = (-T)^+$. Therefore, $T \in \mathcal{L}_n$ implies $|T| = T^+ + T^- \in \mathcal{L}_n$ (since by Theorem 21.4(i) the sum of positive order continuous operators is order continuous). Conversely, it follows from $0 \leq T^+ \leq |\,T\,|$ and $0 \leq T^- \leq |\,T\,|$ that order continuity of $|\,T\,|$ implies order continuity of T^+ and T^-, so that (by Theorem 21.4(i) once more) $T = T^+ - T^-$ is order continuous as well. $\qquad\blacksquare$

Theorem 22.2. *$\mathcal{L}_n$ and $\mathcal{L}_c$ are bands in $\mathcal{L}_b$.*

Proof. We present the proof for $\mathcal{L}_n$. Recall Theorem 21.4(i) in which it was shown that if T and S are positive and order continuous, then $\alpha T + \beta S$ is order continuous for all real α, β. Now, if T_1 and T_2 are members of $\mathcal{L}_n$, then

$$T_1 + T_2 = (T_1^+ + T_2^+) - (T_1^- + T_2^-)$$

with $T_1^+ + T_2^+$ and $T_1^- + T_2^-$ positive and order continuous. Hence, $T_1 + T_2$ is order continuous. It follows immediately that $\mathcal{L}_n$ is a linear subspace of $\mathcal{L}_b$. The last lemma shows that if $0 \leq |\,T_2\,| \leq |\,T_1\,|$ and $T_1 \in \mathcal{L}_n$, then $T_2 \in \mathcal{L}_n$. Hence, $\mathcal{L}_n$ is an ideal in $\mathcal{L}_b$. For the proof that $\mathcal{L}_n$ is a band, assume that G is an upwards directed set of positive operators in $\mathcal{L}_n$ having supremum T_0 in $\mathcal{L}_b$, so $\sup(T : T \in G) = T_0$. This implies that

$$T_0 u = \sup(Tu : T \in G) \text{ in } F \text{ for every } u \in E^+$$

(see the last part of the proof of Theorem 20.2). We have to show that $T_0 \in \mathcal{L}_n$, i.e., we have to show that if $D \downarrow 0$ in E, then $\inf(T_0 u : u \in D) = 0$ in F. It may be assumed that there exists an element $u_0 \in D$ such that $0 \leq u \leq u_0$ for all $u \in D$. Then $T(u_0 - u) \leq T_0(u_0 - u)$ for all $T \in G$ and all $u \in D$, so

$$T_0 u \leq Tu + T_0 u_0 - Tu_0.$$

Keep $T \in G$ fixed. Because T_0 is positive and D is directed downwards, the set $(T_0 u : u \in D)$ is directed downwards as well and so $\inf(T_0 u : u \in D)$ exists in F^+ since F is Dedekind complete. Thus, we get

$$\inf(T_0 u : u \in D) \leq Tu + T_0 u_0 - Tu_0$$

for every $u \in D$. It follows that

$$\inf(T_0 u : u \in D) \leq \inf(Tu : u \in D) + T_0 u_0 - Tu_0.$$

Since $\inf(Tu : u \in D) = 0$, we thus obtain

$$\inf(T_0 u : u \in D) \leq T_0 u_0 - Tu_0.$$

From now on we do not keep T fixed any more. Instead, observing that the last inequality holds for every $T \in G$, we derive from this inequality that

$$\inf(T_0 u : u \in D) \leq T_0 u_0 - \sup(Tu_0 : T \in G) = T_0 u_0 - T_0 u_0 = 0,$$

which is the desired result. $\qquad\blacksquare$

The disjoint complements of $\mathcal{L}_c$ and $\mathcal{L}_n$ in $\mathcal{L}_b$ will be denoted by $\mathcal{L}_s$ and $\mathcal{L}_t$ respectively. Elements of $\mathcal{L}_s$ are called *singular operators*. Since $\mathcal{L}_b$ is Dedekind complete, the space $\mathcal{L}_b$ is the direct sum of $\mathcal{L}_c$ and $\mathcal{L}_s$ as well as the direct sum of $\mathcal{L}_n$ and $\mathcal{L}_t$, so

$$\mathcal{L}_b = \mathcal{L}_c \oplus \mathcal{L}_s = \mathcal{L}_n \oplus \mathcal{L}_t.$$

In view of $\mathcal{L}_n \subseteq \mathcal{L}_c$ we have $\mathcal{L}_s \subseteq \mathcal{L}_t$ and it follows that each $T \in \mathcal{L}_b$ has a unique decomposition

$$T = T_n + T_{c,t} + T_s \text{ with } T_n \in \mathcal{L}_n, T_s \in \mathcal{L}_s \text{ and } T_{c,t} \in \mathcal{L}_c \cap \mathcal{L}_t.$$

Observe here that the intersection $\mathcal{L}_c \cap \mathcal{L}_t$ is a band in $\mathcal{L}_c$ as well as in $\mathcal{L}_t$, and so it is a band in $\mathcal{L}_b$. The operator T_n is called the *order continuous component* of T and $T_c = T_n + T_{c,t}$ is called the *σ-order continuous component* of T. Finally, T_s is the *singular component* of T.

Note that $\mathcal{L}_c \cap \mathcal{L}_t = \{0\}$ if and only if $\mathcal{L}_c = \mathcal{L}_n$, i.e., if and only if every σ-order continuous operator (from E into F) is order continuous. This occurs for example if E is order separable because then, if $D \downarrow 0$ in E, there exists a sequence $u_n \downarrow 0$ contained in D (see Theorem 17.6(i)). As we have seen in the example in the preceding section, it may happen that $\mathcal{L}_n = \mathcal{L}_b$ (so $\mathcal{L}_c = \mathcal{L}_b$ and $\mathcal{L}_t = \mathcal{L}_s = \{0\}$) and it may also happen that $\mathcal{L}_c = \{0\}$ (so $\mathcal{L}_n = \{0\}$ and $\mathcal{L}_s = \mathcal{L}_t = \mathcal{L}_b$).

Let $T : E \to F$. As well-known from linear algebra, the linear subspace of E consisting of all $f \in E$ satisfying $Tf = 0$ is called the *null space* or *kernel* of T. In the present case in which we have available the notion of absolute values, the set N_T of all $f \in E$ satisfying $|T|(|f|) = 0$ is called the *null ideal* or *absolute kernel* of T. It is immediately evident that N_T is an ideal in E. Also, if T is order continuous, then $|T|$ is order continuous, so N_T is now a band. The converse does not always hold, as the following simple example shows. Let the positive linear functionals φ_1 and φ_2 on $E = C([0,1])$ be defined by $\varphi_1(f) = \int_0^{1/2} f(x)dx$ and $\varphi_2(f) = f(0)$ for all $f \in E$. It is easily seen that N_{φ_1} is a band but N_{φ_2} is not. Note that both φ_1 and φ_2 are singular (since every positive linear functional on $E = C([0,1])$ is singular). Returning to the general case that we have an order bounded operator $T : E \to F$, the condition that N_T is a band is necessary, but not sufficient, for T to be order continuous. Strengthening the condition a little, we can make it necessary and sufficient. It was observed by W.A.J.Luxemburg (1979) that if not only N_T is a band but N_S is a band for every S in the ideal generated by T in $\mathcal{L}_b$, then T is order continuous. This was known earlier for linear functionals, i.e., if φ is a linear functional on E and N_ψ is a band in E for every ψ in the ideal generated by φ in $E^\sim$, then φ is order continuous. We present the proof (due to C.D.Aliprantis and O.Burkinshaw, 1983, and simpler than earlier proofs), but first we recall (see Theorem 20.5 and the remarks belonging to that theorem) that if $0 \leq T : E \to F$ and A is an ideal in E, then T_A denotes the minimal positive extension of the restriction $T \mid A$, so $T_A = T$ on A and $T_A = 0$ on A^d. Furthermore, as already observed there, if A and B are ideals such that $A \subseteq B$, then $0 \leq T_A \leq T_B \leq T$.

Theorem 22.3. *Let $T : E \to F$ be an order bounded operator and let A_T be the ideal generated by T in $\mathcal{L}_b = \mathcal{L}_b(E, F)$. Then T is order continuous (σ-order continuous) if and only if the null ideal N_S is a band (σ-ideal) for every $S \in A_T$.*

Proof. If T is order continuous, then so is every $S \in A_T$, and hence N_S is a band. For the proof in the converse direction we may assume that T is positive. Let $0 \leq D \uparrow u_0$ in E. Then $0 \leq (Tu : u \in D) \uparrow \leq Tu_0$ in F. Hence, since F is Dedekind complete, there exists an element p in F^+ such that $0 \leq (Tu : u \in D) \uparrow p$. We have to prove that $p = Tu_0$. For this purpose, choose ϵ such that $0 < \epsilon < 1$, let A_u denote the ideal generated by $(\epsilon u_0 - u)^+$ for $u \in D$ and, for brevity, write T_u instead of T_{A_u}, where $T_u = T_{A_u}$ is the minimal positive extension of the restriction $T \mid A_u$, so $T_u = T$ on A_u and $T_u = 0$ on A_u^d. In particular,

$$T_u(\epsilon u_0 - u)^+ = T(\epsilon u_0 - u)^+ \text{ and } T_u(\epsilon u_0 - u)^- = 0,$$

the first part of which implies $T(\epsilon u_0 - u)^+ \leq T_u u_0$. The main idea is now to use again, for any $u \in D$, Ando's inequality

$$0 \leq u_0 - u = (1 - \epsilon)u_0 + (\epsilon u_0 - u) \leq (1 - \epsilon)u_0 + (\epsilon u_0 - u)^+,$$

introduced already in Theorem 17.17. This implies, for any $u \in D$, that

$$0 \leq Tu_0 - p \leq T(u_0 - u) \leq$$

$$(1 - \epsilon)Tu_0 + T(\epsilon u_0 - u)^+ \leq (1 - \epsilon)Tu_0 + T_u u_0.$$

Since $D \uparrow$, the set $\{(\epsilon u_0 - u)^+ : u \epsilon D\}$ is downwards directed, and so the set of operators $(T_u : u \epsilon D)$ is downwards directed, i.e., $T \geq T_u \downarrow \geq 0$. Hence, because $\mathcal{L}_b(E, F)$ is Dedekind complete, there exists an operator $S \in \mathcal{L}_b(E, F)$ such that $T_u \downarrow S \geq 0$, so $T_u u_0 \downarrow Su_0$. If we can prove that $Su_0 = 0$, it will follow from $T_u u_0 \downarrow Su_0 = 0$ that

$$0 \leq Tu_0 - p \leq (1 - \epsilon)Tu_0.$$

This holds then for every ϵ satisfying $0 < \epsilon < 1$. It follows that $Tu_0 - p = 0$, as desired (note that we use here that $\epsilon \uparrow 1$ implies $(1 - \epsilon)Tu_0 \downarrow 0$ since F is Archimedean).

For the proof that $Su_0 = 0$, observe first that $S \in A_T$ and $S(\epsilon u_0 - u)^- = 0$ for every $u \in D$ (since $0 \leq S \leq T_u$ for every $u \in D$), so $(\epsilon u_0 - u)^- \in N_S$ for every $u \in D$. Hence, since

$$(\epsilon u_0 - u)^- = (u - \epsilon u_0)^+ \uparrow (u_0 - \epsilon u_0)^+ = u_0 - \epsilon u_0$$

and N_S is a band, we get $(1 - \epsilon)u_0 \in N_S$, i.e., $u_0 \in N_S$. In other words, $Su_0 = 0$, which is what we wish to prove. It is only at the last moment that we use the hypothesis that N_S is a band.

The proof for the case of σ-order continuity is similar.   ∎

Corollary 22.4. *Every order bounded operator from E into F is order continuous (σ-order continuous) if and only if the null ideal of every order bounded operator from E into F is a band (σ-ideal).*

Example 22.5. This example serves to illustrate the last theorem. Let X be an infinite point set and let E be the set of all real functions f on X having the property that there exists a real number, call it $f(\infty)$, such that for any $\epsilon > 0$ the inequality $\mid f(x) - f(\infty) \mid > \epsilon$ holds for only finitely many $x \in X$. The set E is a Riesz space with respect to pointwise addition, scalar multiplication and ordering. Clearly, every $f \in E$ is a bounded function. The space E is not Dedekind complete or even Dedekind σ-complete; it is easy to indicate an increasing sequence $0 \leq u_n \leq e$ (where e is the function identically one) such that the sequence does not possess a supremum. If, however, D is an upwards directed set in E^+, then $u_0 = \sup D$ exists in E if and only if $u_0 \in E$ and $u_0(x) = \sup(u(x) : x \in D)$ holds for every $x \in X$. To see this observe that if $u_0 \in E$ is an upper bound of D such that $u_0(x_0) > \sup_D u(x_0)$ at some point x_0 we can change the value of u_0 at x_0 to $\sup_D u(x_0)$ and we still have an upper bound of D. Fix a countable subset $(x_1, x_2, \ldots)$ of X and define the positive linear functional φ on E by

$$\varphi(f) = f(\infty) + \Sigma_1^\infty 2^{-n} f(x_n) \text{ for every } f \in E.$$

The null ideal N_φ is the set of all $f \in E$ satisfying $f(x_n) = 0$ for $n = 1, 2, \ldots$. Note that $f(\infty) = 0$ for every $f \in N_\varphi$. Since $0 \leq D \uparrow u_0$ in E implies that $u_0(x) = \sup(u(x) : u \in D)$ for every $x \in X$ (as observed already), it follows that N_φ is a band in E. The functional φ, however, is not order continuous. To see this, let D consist of all characteristic functions of finite subsets of X. Then $D \uparrow e$ (where e is the function identically one), but $\sup\{\varphi(u) : u \in D\}$ is not $\varphi(e)$, simply because the supremum is $\Sigma_1^\infty 2^{-n}$ and $\varphi(e) = 1 + \Sigma_1^\infty 2^{-n}$. It follows now from the last theorem that there must exist a linear functional ψ on E such that $0 \leq \psi \leq \varphi$ and N_ψ is not a band. It is easy to find a ψ having these properties. Let $\psi(f) = f(\infty)$ for all $f \in E$. Then $0 \leq \psi \leq \varphi$ and $N_\psi = (f \in E : f(\infty) = 0)$. Let D be the same as above, so $0 \leq D \uparrow e$. Then $u \in N_\psi$ for every $u \in D$, but e is not a member of N_ψ. Hence, N_ψ is not a band.

Exercise 22.6. Let X, E and φ be as in the last example. Show that if X is uncountable, then φ is σ-order continuous, whereas for X countable φ is not σ-order continuous.

There exist some formulas for the order continuous and σ-order continuous components T_n and T_c of a positive operator $T : E \to F$. To understand these formulas it is convenient to prove first some further results about directed sets. We know already that if D_1 and D_2 are subsets of the ordered vector space E such that $\sup D_1 = u_0$ and $\sup D_2 = v_0$, then the algebraic sum

$$D_1 + D_2 = (u + v : u \in D_1, v \in D_2)$$

has supremum $u_0 + v_0$ (see Lemma 21.3). Hence, if $0 \le D_1 \uparrow u_0$ and $0 \le D_2 \uparrow v_0$, then $0 \le D_1 + D_2 \uparrow u_0 + v_0$. We shall need the following variant.

Lemma 22.7. *Assume that D_1 and D_2 are subsets of the ordered vector space E such that $\sup D_1 = u_0$ and $\sup(D_1 + D_2) = u_0 + v_0$. Then $\sup D_2$ exists and $\sup D_2 = v_0$. In particular, if $0 \le D_1 \uparrow u_0, 0 \le D_2 \uparrow$ and $(D_1 + D_2) \uparrow (u_0 + v_0)$, then $0 \le D_2 \uparrow v_0$.*

Proof. For every fixed $v \in D_2$ we have $u + v \le u_0 + v_0$ for every $u \in D_1$, so

$$\sup(u : u \in D_1) + v \le u_0 + v_0,$$

i.e., $u_0 + v \le u_0 + v_0$, which shows that v_0 is an upper bound of D_2. If w is any other upper bound of D_2, then $u + v \le u_0 + w$ for all $u \in D_1$ and all $v \in D_2$, so

$$\sup(u + v : u \in D_1, v \in D_2) \le u_0 + w,$$

i.e., $u_0 + v_0 \le u_0 + w$, which implies that $v_0 \le w$. This shows that $\sup D_2 = v_0$. ∎

We specialize to indexed equidirected sets in a Riesz space (see Definition 10.7 and Theorem 10.8). Recall that the upwards directed sets (u_τ) and (v_τ) are called equidirected if for any τ_1 and τ_2 in the index set (τ) there exists $\tau_3 \in (\tau)$ such that

$$u_{\tau_3} \ge u_{\tau_1} \vee u_{\tau_2} \text{ and } v_{\tau_3} \ge v_{\tau_1} \vee v_{\tau_2}$$

hold simultaneously. In Theorem 10.8 it was indicated also that if $0 \le u_\tau \uparrow u_0$ and $0 \le v_\tau \uparrow v_0$ with equidirected sets, then $0 \le (u_\tau + v_\tau) \uparrow (u_0 + v_0)$.

Lemma 22.8. *If (u_τ) and (v_τ) are equidirected with $0 \le u_\tau \uparrow u_0, 0 \le v_\tau \uparrow$ and $0 \le (u_\tau + v_\tau) \uparrow (u_0 + v_0)$, then $0 \le v_\tau \uparrow v_0$.*

Proof. For any τ_1 and τ_2 in (τ) there exists $\tau_3 \in (\tau)$ with the property mentioned above, so

$$u_{\tau_1} + v_{\tau_2} \le u_{\tau_3} + v_{\tau_3} \le u_0 + v_0.$$

This shows that $u_0 + v_0$ is an upper bound of the algebraic sum $\{(u_\tau) + (v_\tau)\}$. Recall that this algebraic sum is the set of all $u_{\tau_1} + v_{\tau_2}$. Since $u_0 + v_0$ is already the supremum of the set of all $\{u_\tau + v_\tau : \tau \in (\tau)\}$, it follows that $u_0 + v_0$

is the supremum of the algebraic sum $\{(u_\tau) + (v_\tau)\}$. But then, by the last lemma, $v_0 = \sup_\tau v_\tau$, so that, since it is given that $0 \le v_\tau$, we may conclude that $0 \le v_\tau \uparrow v_0$. ∎

We now turn to the problem of expressing the order continuous component T_n of a positive operator T (from E into F with F Dedekind complete) in terms of T itself. Let $u_0 \in E^+$ be given and choose an arbitrary upwards directed set D in E^+ having supremum u_0, so $0 \le D \uparrow u_0$. Then the element $\sup(Tu : u \in D)$ exists in F^+. Now do this for all such D in E^+ and take the infimum $T^\# u_0$ of the thus obtained elements, so

$$T^\# u_0 = \inf_{u \in D \uparrow u_0} \{\sup(Tu : u \in D)\}.$$

Evidently, the set consisting of u_0 only is one of the possible sets D, so $T^\# u_0 \le T u_0$. We show first that $T^\#(u_0 + v_0) \le T^\# u_0 + T^\# v_0$ for all u_0 and v_0 in E^+. For this purpose, assume that $0 \le D_1 \uparrow u_0$ and $0 \le D_2 \uparrow v_0$, so in view of Lemma 21.3 we have $0 \le (D_1 + D_2) \uparrow (u_0 + v_0)$, where $D_1 + D_2$ is the algebraic sum of D_1 and D_2. Once more by Lemma 21.3 we have

$$T^\#(u_0 + v_0) \le \sup(Tu + Tv : u \in D_1, v \in D_2) =$$

$$\sup(Tu : u \in D_1) + \sup(Tv : v \in D_2).$$

Taking infima on the right we get $T^\#(u_0 + v_0) \le T^\# u_0 + T^\# v_0$. For the inequality in the converse direction, assume that $0 \le D \uparrow (u_0 + v_0)$. Then, writing $u_w = w \wedge u_0$ and $v_w = w - u_w$ for any $w \in D$, we have

$$0 \le u_w \uparrow u_0, u_w + v_w = w \text{ and } 0 \le v_w \uparrow v_0.$$

The last fact follows by observing that $w_1 \le w_2$ in D implies

$$0 \le (u_w)_2 - (u_w)_1 = (w_2 \wedge u_0) - (w_1 \wedge u_0) \le w_2 - w_1$$

(by one of the Birkhoff inequalities), and so

$$(v_w)_1 = w_1 - (u_w)_1 \le w_2 - (u_w)_2 = (v_w)_2.$$

This shows already that the sets (u_w) and (v_w), indexed by means of the index set (w), are upwards directed and equidirected. It follows then from the last lemma that $0 \le v_w \uparrow v_0$. Applying now $T^\#$ to u_0 and v_0, we obtain

$$T^\# u_0 + T^\# v_0 \le \sup_w T u_w + \sup_w T v_w =$$

$$\sup_w T(u_w + v_w) = \sup_w(Tw : w \in D).$$

This holds for all upwards directed sets $0 \le D \uparrow (u_0 + v_0)$. Hence, taking the infimum on the right for all such sets, we get $T^\# u_0 + T^\# v_0 \le T^\#(u_0 + v_0)$.

Combining the two inequalities, we find the result that

$$T^{\#}(u_0 + v_0) = T^{\#}u_0 + T^{\#}v_0 \text{ for all } u_0, v_0 \text{ in } E^{+}.$$

Thus, $T^{\#}$ is additive on E^{+} and can be extended, therefore, to a positive operator (still denoted by $T^{\#}$) from E into F. Observe already that $T^{\#} = T$ if T is order continuous. This is immediately clear from the definition of $T^{\#}$.

In a similar manner as in the linearity proof for $T^{\#}$ it can be proved also that $(S + T)^{\#} = S^{\#} + T^{\#}$ holds for all positive operators S and T (from E into F). We indicate how to show that $S^{\#}u_0 + T^{\#}u_0 \leq (S + T)^{\#}u_0$. If $0 \leq D \uparrow u_0$, the upwards directed sets $(Su : u \in D)$ and $(Tu : u \in D)$ are equidirected, so

$$S^{\#}u_0 + T^{\#}u_0 \leq$$

$$\sup(Su : u \in D) + \sup(Tu : u \in D) = \sup\{(S + T)u : u \in D\}.$$

This holds for all sets $0 \leq D \uparrow u_0$, so that, by taking the infimum on the right for all such sets, we obtain

$$S^{\#}u_0 + T^{\#}u_0 \leq (S + T)^{\#}u_0.$$

Similarly as we have assigned $T^{\#}$ to T by means of upwards directed sets we can assign an operator T^{o} to T by means of sequences. This is done by defining, for any $u_0 \in E^{+}$, the element $T^{o}u_0$ in F^{+} by

$$T^{o}u_0 = \inf_{0 \leq u_n \uparrow u_0} \{\sup T u_n : n = 1, 2, \ldots\},$$

where, as indicated, the infimum is taken over all possible sequences $0 \leq u_n \uparrow u_0$ in E^{+}. The proof that T^{o} is a positive linear operator is similar to that for $T^{\#}$. The same holds for the proof that $(S + T)^{o} = S^{o} + T^{o}$.

The necessary preparations have been made now to prove that the thus defined operators $T^{\#}$ and T^{o} are exactly the components T_n and T_c of T. The main difficulty in showing that $T_n = T^{\#}$ (or $T_c = T^{o}$ respectively) holds is the proof that $T^{\#}$ is order continuous (or T^{o} is σ-order continuous respectively).

Theorem 22.9. *Let T be a positive operator from the Riesz space E into the Dedekind complete Riesz space F with order continuous component T_n and σ-order continuous component T_c. Then, for any $u_0 \in E^{+}$, we have*

$$T_n u_0 = \inf_{0 \leq D \uparrow u_0} \{\sup(Tu : u \in D)\},$$

$$T_c u_0 = \inf_{0 \leq u_n \uparrow u_0} \{\sup(Tu_n : n = 1, 2, \ldots)\}.$$

Proof. We present the proof for T_n. Defining $T^{\#}$ as above, we have to prove that $T_n = T^{\#}$. Since T_n is order continuous, the equality $T_n^{\#} = T_n$ holds and hence it follows from $0 \leq T_n \leq T$ that $0 \leq T_n = T_n^{\#} \leq T^{\#}$. Assuming for a moment that it has been proved already that $T^{\#}$ is order continuous, we have $(T^{\#})_n = T^{\#}$, and so it follows from $0 \leq T^{\#} \leq T$ that $T^{\#} = (T^{\#})_n \leq T_n$. Combining the results, we get $T_n = T^{\#}$.

All that remains to be proved is the order continuity of $T^{\#}$. The proof is very similar to the proof of Theorem 22.3, where in the last part we have a positive operator S such that its null ideal N_S is a band and we also have a directed set $0 \leq G \uparrow (u_0 - \epsilon u_0)$ having the property that $Sv = 0$ for all $v \in G$. Since N_S is a band we may then conclude that $S(u_0 - \epsilon u_0) = 0$. In the present situation it will not be known that N_S is a band so that we can only conclude from $G \uparrow (u_0 - \epsilon u_0)$ and $Sv = 0$ for all $v \in G$ that $S^{\#}(u_0 - \epsilon u_0) = 0$. This, however, will be sufficient for what we wish to prove. The fully written out proof follows now.

Let $0 \leq D \uparrow u_0$ in E. Then $0 \leq (T^{\#}u : u \in D) \uparrow \leq T^{\#}u_0$ in F. Hence, since F is Dedekind complete, there exists an element p in F^{+} such that

$$0 \leq (T^{\#}u : u \in D) \uparrow p \leq T^{\#}u_0.$$

We have to prove that $p = T^{\#}u_0$. For this purpose, choose ϵ such that $0 < \epsilon < 1$, let A_u denote the ideal generated in E by $(\epsilon u_0 - u)^{+}$ for $u \in D$ and, for brevity, write T_u instead of T_{A_u}, where $T_u = T_{A_u}$ is the minimal positive extension of the restriction $T \mid A_u$, so $T_u = T$ on A_u and $T_u = 0$ on $(A_u)^d$. In particular,

$$T_u w = T w \text{ for } 0 \leq w \leq (\epsilon u_0 - u)^{+} \quad \text{and} \quad T_u(\epsilon u_0 - u)^{-} = 0,$$

the first part of which implies

$$T^{\#}(\epsilon u_0 - u)^{+} = \inf_{0 \leq G \uparrow (\epsilon u_0 - u)^{+}} \{\sup Tw : w \in G\} =$$

$$\inf_{0 \leq G \uparrow (\epsilon u_0 - u)^{+}} \{\sup T_u w : w \in G\} = T_u^{\#}(\epsilon u_0 - u)^{+} \leq T_u^{\#}u_0.$$

The main idea is now to use Ando's inequality (as in the proof of Theorem 22.3)

$$0 \leq u_0 - u = (1 - \epsilon)u_0 + (\epsilon u_0 - u) \leq (1 - \epsilon)u_0 + (\epsilon u_0 - u)^{+}.$$

It follows, for any $u \in D$, that

$$0 \leq T^{\#}u_0 - p \leq T^{\#}(u_0 - u) \leq$$

$$(1 - \epsilon)T^{\#}u_0 + T^{\#}(\epsilon u_0 - u)^{+} \leq (1 - \epsilon)T^{\#}u_0 + T_u^{\#}u_0.$$

Since $D \uparrow$, the set $\{(\epsilon u_0 - u)^{+} : u \in D\}$ is downwards directed, and so the set of operators $(T_u : u \in D)$ is downwards directed, i.e., $T \geq T_u \downarrow \geq 0$. Hence, because $\mathcal{L}_b(E, F)$ is Dedekind complete, there exists an operator $S \in \mathcal{L}_b(E, F)$ such that $T_u \downarrow S \geq 0$. In view of

$$0 \leq T_u^{\#} - S^{\#} = (T_u - S)^{\#} \leq T_u - S \downarrow 0$$

it follows that $T_u^{\#} \downarrow S^{\#}$. In particular, $T_u^{\#} u_0 \downarrow S^{\#} u_0$. If we can prove that $S^{\#} u_0 = 0$, it will follow from $T_u^{\#} u_0 \downarrow S^{\#} u_0 = 0$ that

$$0 \leq T^{\#} u_0 - p \leq (1 - \epsilon) T^{\#} u_0.$$

This holds then for every ϵ satisfying $0 < \epsilon < 1$. It follows that $T^{\#} u_0 - p = 0$, as desired. Note that we use here that $\epsilon \uparrow 1$ implies $(1 - \epsilon) T^{\#} u_0 \downarrow 0$ since F is Dedekind complete.

For the proof that $S^{\#} u_0 = 0$, note that it follows from $0 \leq S \leq T_u$ and $T_u(\epsilon u_0 - u)^- = 0$ that

$$S(u - \epsilon u_0)^+ = S(\epsilon u_0 - u)^- = 0 \text{ for every } u \in D.$$

Hence, since $(u - \epsilon u_0)^+ \uparrow (u_0 - \epsilon u_0)^+ = u_0 - \epsilon u_0$, we see now from the definition of $S^{\#}$ that $S^{\#}(u_0 - \epsilon u_0) = 0$, i.e., $S^{\#} u_0 = 0$, which is what remained to be proved.

The proof that $T_c = T^o$ holds is similar.   ∎

The theorem thus proved is due to A.R. Schep (1978) and the present proof is due to C.D. Aliprantis and O. Burkinshaw (1983). The case that $F = \mathbb{R}$, i.e., the case that we have a positive linear functional φ with components φ_n and φ_c was known earlier; the formula for φ_n was established by W.A.J.Luxemburg (1965) and the one for φ_c by W.A.J.Luxemburg and A.C.Zaanen (1963).

Exercise 22.10. Let E and F be Riesz spaces with F Dedekind complete and let T be a positive operator from E into F.

(i) Let A be an ideal in E. As before, the minimal positive extension of the restriction $T \mid A$ is denoted by T_A. Show that if $T \mid A$ is order continuous (σ-order continuous) on A, then T_A is order continuous (σ-order continuous) on E.

(ii) The order continuous and σ-order continuous components of T are denoted by T_n and T_c respectively. Recall that $T_s = T - T_c$ is the singular component of T. Furthermore, let us write $T_t = T - T_n$. Show that for any $u_0 \in E^+$ we have

$$T_t u_0 = \sup_{u_0 \geq D \downarrow 0} \{\inf(Tu : u \in D)\},$$

$$T_s u_0 = \sup_{u_0 \geq u_n \downarrow 0} \{\inf(Tu_n : n = 1, 2, \ldots)\}.$$

Hint: Assuming $T \mid A$ to be order continuous on A, we have to prove that $0 \leq D \uparrow u_0$ in E implies $(T_A u : u \in D) \uparrow T_A u_0$. It is sufficient to prove that

$T_A u_0 \leq \sup(T_A u : u \in D)$, the inequality in the other direction being evident. Choose $v \in A$ such that $0 \leq v \leq u_0$. Then $(v \wedge u : u \in D) \uparrow v$ in A, so $\{T(v \wedge u) : u \in D\} \uparrow Tv$. Since $u \geq v \wedge u$ implies $T_A u \geq T_A(v \wedge u) = T(v \wedge u)$, it follows that

$$\sup(T_A u : u \in D) \geq \sup\{T(v \wedge u) : u \in D\} = Tv,$$

so

$$\sup(T_A u : u \in D) \geq \sup(Tv : v \in A, 0 \leq v \leq u_0) = T_A u_0.$$

CHAPTER 12
Carriers of Operators

23. Order Denseness

In this brief section we discuss the notion of order denseness. This could have happened earlier, but there was no need for that.

Definition 23.1. The Riesz subspace V of the Riesz space E is said to be *order dense* in E if for each $u > 0$ in E (i.e., $u \geq 0, u \neq 0$) there exists an element $v \in V$ such that $0 < v \leq u$.

It is evident that if V is order dense in E, then $V^d = \{0\}$. To see this, observe that if $u \in V^d$ and u would satisfy $u > 0$, then there would exist an element $v \in V$ such that $0 < v \leq u$. But then $v \in V \cap V^d$, so $v = 0$, which would yield a contradiction.

Also, if V is an order dense Riesz subspace of E and V_1 is an order dense Riesz subspace of V, then V_1 is an order dense Riesz subspace of E. This follows immediately from the definition.

It is not true that a Riesz subspace V satisfying $V^d = \{0\}$ is always order dense in E. Example: If $E = C([0,1])$ and V is the Riesz subspace of all constant real functions on $[0,1]$, then $V^d = \{0\}$, but V is not order dense in E since if $u(x) = x$ on $[0,1]$, then $0 < u \in E$ but there is no $v \in V$ satisfying $0 < v \leq u$. Another example: If E is the lexicographically ordered plane $\mathbb{R}^2$ and V is the horizontal axis (i.e., V is the set of all points $(x,0), x \in \mathbb{R}$), then V is a Riesz subspace such that $V^d = \{0\}$, but V is not order dense since if $u = (0,1)$, there is no $v \in V$ satisfying $0 < v \leq u$.

Theorem 23.2. *Let V be an order dense Riesz subspace of the Riesz space E and let D be an upwards directed subset of V. Furthermore, let $f_0 \in V$. Then $D \uparrow f_0$ holds in V if and only if $D \uparrow f_0$ holds in E. Similarly if D is downwards directed.*

Proof. It is sufficient to show that if D is a downwards directed subset of V, then $D \downarrow 0$ holds in V if and only if $D \downarrow 0$ holds in E. It is evident that $D \downarrow 0$ in E implies $D \downarrow 0$ in V. Assume now that $D \downarrow 0$ in V, but D has a lower bound $u > 0$ in E. Then there exists an element $v \in V$ satisfying $0 < v \leq u$, so $v > 0$ is a lower bound of D in V. This contradicts $D \downarrow 0$ in V. Hence, $D \downarrow 0 \in E$. ∎

If V is a Riesz subspace of E and V is order dense, then $V^d = \{0\}$, but as we have seen it does not always follow conversely from $V^d = \{0\}$ that V is order dense. For ideals the situation is different; we shall prove now that if A is an ideal in E, then A is order dense if and only if $A^d = \{0\}$.

Theorem 23.3. *For an ideal A in the Riesz space E the following holds.*

(i) A is order dense in E if and only if $A^d = \{0\}$.
(ii)   The ideal $A \oplus A^d$ is order dense in E.
(iii) A is order dense in A^{dd} (and, therefore, A is order dense in the band $[A]$ generated by A).

Proof. (i) Assume first that A is order dense in E and let $0 \leq u \in A^d$. We have to prove that $u = 0$. If $u > 0$, there exists some $v \in A$ such that $0 < v \leq u$. Then $v \in A$ as well as $v \subset A^d$, so $v = 0$. Contradiction. For the converse, assume that $A^d = \{0\}$. Let $0 < u \in E$ be given. Then $w \wedge u > 0$ for at least one $w \in A^+$ (because if not, we should have $u \in A^d = \{0\}$, which contradicts $u > 0$). Writing $v = w \wedge u$, we have now that $0 < v \leq u$ and $v \in A$ (because $0 < v \leq w \in A$). Thus, for any $u > 0$ in E there exists some $v \in A$ such that $0 < v \leq u$. This shows that A is order dense.

(ii) In view of part (i) it is sufficient to show that $(A \oplus A^d)^d = \{0\}$. If $f \in (A \oplus A^d)^d$, then $f \perp A$ and $f \perp A^d$, so $f \in A^d \cap A^{dd} = \{0\}$.

(iii) The disjoint complement of A in A^{dd} is $A^d \cap A^{dd} = \{0\}$. Hence, by part (i), A is order dense in A^{dd}. ∎

If E is Archimedean, more can be said about order dense Riesz subspaces.

Theorem 23.4. *The Riesz subspace V of the Archimedean Riesz space E is order dense in E if and only if, for any $u \in E^+$, we have*

$$(v \in V : 0 \leq v \leq u) \uparrow u. \tag{1}$$

Hence, if A is an ideal in E, the ideal A is order dense in E if and only if the band generated by A is the entire space E.

Proof. If (1) holds for every $u \in E^+$, then V is obviously order dense in E. For the converse, assume that V is order dense, but for some $u \in E^+$ the element u is not the supremum of the set $D = (v \in V : 0 \leq v \leq u)$. Then there exists an upper bound z of D not satisfying $z \geq u$, and so $w = z \wedge u$ is an upper bound of D satisfying $w < u$, i.e., $v \leq w < u$ for all $v \in V$. By the order denseness of V there exists an element $v_1 \in V$ satisfying $0 < v_1 \leq u - w$. Since it is evident now that $v_1 \in D$, we have $v_1 \leq w$, so

$$2v_1 = v_1 + v_1 \le w + (u - w) = u,$$

which shows that $2v_1 \in D$. By induction $0 < nv_1 \le u$ holds for $n = 1, 2, \ldots$. This contradicts the hypothesis that E is Archimedean. Hence, $u = \sup D$, i.e., (1) holds.

The statement for the ideal A follows by combining the formula (1) above and formula (4) in Theorem 7.8, in which it is stated that the band generated by A consists of all $f \in E$ satisfying

$$\mid f \mid = \sup(u \in A : 0 \le u \le \mid f \mid). \qquad \blacksquare$$

If E is not Archimedean, the result in the last theorem does not always hold. Example: If $E = \mathbb{R}^2$ with lexicographical ordering and V is the vertical axis, i.e., V is the set of all points (0,y) with $y \in \mathbb{R}$, then V is an order dense band in E, but the point $u = (1, 1)$, for example, is not the supremum of the set of all $v \in V$ satisfying $0 \le v \le u$.

In section 10 the notion of a strong unit in a Riesz space was introduced. The element $u > 0$ in the Riesz space E is called a strong unit if the principal ideal A_u generated by u in E satisfies $A_u = E$. In a similar manner, we can define weak units.

Definition 23.5. The element $e > 0$ in the Riesz space E is called a *weak unit* if the principal band B_e generated by e in E satisfies $B_e = E$.

If E is Archimedean, we see from the last theorem that $e \in E^+$ is a weak unit in E if and only if the principal ideal A_e is order dense in E, i.e., if and only if $A_e^d = \{0\}$ (equivalently, $\{e\}^d = \{0\}$). Observe that if E is not Archimedean and e is a weak unit in E, then A_e is order dense, but in the converse direction there may exist elements $e > 0$ for which A_e is order dense without e being a weak unit. Example: $E = \mathbb{R}^2$ with lexicographical ordering and $e = (0, 1)$.

Exercise 23.6. Show that if A and B are order dense ideals in the Riesz space E, then $A \cap B$ is an order dense ideal in E.

Hint: For any given $0 < u \in E$ there exists $v \in A$ such that $0 < v \le u$, and so there exists $w \in B$ such that $0 < w \le v$.

Exercise 23.7. If V is a Riesz subspace of the Riesz space E such that the ideal A_V generated by V is order dense (equivalently, $(A_V)^d = \{0\}$), then V is not necessarily order dense, but it remains true that $V^d = \{0\}$. Show this.

24. The Carrier of an Operator

In the present section we assume that E and F are Riesz spaces with F Dedekind complete. As we have seen in earlier sections (Theorem 20.2 and Theorem 22.2), the vector space $\mathcal{L}_b(E, F)$ of all order bounded linear operators from E into F is a Dedekind complete Riesz space and the subset $\mathcal{L}_n = \mathcal{L}_n(E, F)$ of all order continuous operators in $\mathcal{L}_b$ is a band in $\mathcal{L}_b$. Recall that for $T \in \mathcal{L}_b$ the set

$$N_T = (f \in E : |T|(|f|) = 0)$$

is an ideal in E, called the null ideal of T. For T order continuous, the null ideal N_T is a band in E, but the converse does not always hold. As proved in Theorem 22.3, T is order continuous if and only if N_S is a band in E for every operator S in the ideal generated by T in $\mathcal{L}_b$. By definition, the disjoint complement $(N_T)^d$ of N_T is called the *carrier* of T and denoted by C_T. The carrier C_T is, therefore, a band in E. Note that T and $|T|$ have the same null ideal and hence have the same carrier. Also, $|S| \leq |T|$ implies $N_T \subseteq N_S$, and so $C_S \subseteq C_T$. In particular, if T_n is the order continuous component of T, then $|T_n| \leq |T|$, so $C_{T_n} \subseteq C_T$. Similarly $C_{T_c} \subseteq C_T$, where T_c is the σ-order continuous component of T. We shall first prove a simple lemma.

Lemma 24.1. *For S and T positive operators from E into F, we have*

(i) $N_{S \vee T} = N_{S+T} = N_S \cap N_T$, $C_{S \vee T} = C_{S+T} = (C_S + C_T)^{dd}$.

(ii) $C_S \subseteq (N_T)^{dd}$ *if and only if* $C_S \perp C_T$ *and also if and only if* $C_T \subseteq (N_S)^{dd}$.

Proof. (i) From $\{(S + T)/2\} \leq (S \vee T) \leq (S + T)$ it follows immediately that $N_{S \vee T} = N_{S+T}$, and so $C_{S \vee T} = C_{S+T}$. It is evident from the definition of a null ideal that $N_{S+T} = N_S \cap N_T$. Furthermore,

$$(C_S + C_T)^d = (C_S)^d \cap (C_T)^d = (N_S)^{dd} \cap (N_T)^{dd} =$$

$$(N_S \cap N_T)^{dd} = (N_{S \vee T})^{dd},$$

where we have used that $(A \cap B)^{dd} = A^{dd} \cap B^{dd}$ for ideals A and B (see Exercise 8.6). It follows that

$$(C_S + C_T)^{dd} = (N_{S \vee T})^{ddd} = (N_{S \vee T})^d = C_{S \vee T}.$$

Note that it follows from $C_{S+T} = (C_S + C_T)^{dd}$ that the order bounded operators T satisfying $C_T = \{0\}$ form an ideal in $\mathcal{L}_b$.

(ii) $C_S \subseteq (N_T)^{dd}$ implies $C_S \perp (N_T)^d = C_T$. Conversely, $C_S \perp C_T$ implies $C_S \subseteq (C_T)^d = (N_T)^{dd}$. ∎

More can be proved if, besides assuming that F is Dedekind complete, we also assume that E is Archimedean. In this case it is not only so that the carrier C_T of an order continuous operator T is the disjoint complement of N_T, but conversely N_T is now also the disjoint complement of C_T. To see this, observe that N_T is now a band, so $N_T = (N_T)^{dd}$ since E is Archimedean. Hence

$$N_T = (N_T)^{dd} = \{(N_T)^d\}^d = (C_T)^d.$$

Theorem 24.2. *Assume that E is Archimedean and F is Dedekind complete. Then the following holds.*

(i) If $T \in \mathcal{L}_b(E, F)$ is order continuous and $C_T = \{0\}$, then $T = 0$. Hence, if T is order continuous and $T \neq 0$, then $C_T \neq \{0\}$. It follows that $C_S \neq \{0\}$ for every operator $S \neq 0$ in the ideal generated by an order continuous operator T. In particular, if $T \in \mathcal{L}_b$ has an order continuous component $T_n \neq 0$, then $C_T \neq \{0\}$. Hence, if $T \in \mathcal{L}_b$ and $C_T = \{0\}$, then $T \perp \mathcal{L}_n$, i.e., T is a member of the disjoint complement $\mathcal{L}_t$ of $\mathcal{L}_n$.

(ii) If $T \in \mathcal{L}_b$ and $C_S \neq \{0\}$ for every $S \neq 0$ in the ideal generated by T, then T is order continuous.

(iii) Every order bounded operator (from E into F) is order continuous, i.e., $\mathcal{L}_b = \mathcal{L}_n$, if and only if $C_T \neq \{0\}$ for every $T \neq 0$ in $\mathcal{L}_b$.

Proof. (i) Let T be order continuous and $C_T = \{0\}$. The null ideal is now a band. Since E is Archimedean, the band N_T satisfies $N_T = (N_T)^{dd}$. Hence

$$N_T = (N_T)^{dd} = \{(N_T)^d\}^d = (C_T)^d = \{0\}^d = E.$$

This shows that $|T|(|f|) = 0$ for every $f \in E$. It follows that $|T| = 0$, i.e., $T = 0$.

(ii) Assume first that T is a positive operator (from E into F) such that T is not order continuous. Then, in view of Theorem 22.3, there exists an operator R satisfying $0 < R \leq T$ such that N_R is not a band. Hence, if B is the band generated in E by N_R, the operator R is not identically zero on B. Let S be the minimal positive extension of the restriction $R \mid B$ to E, so $S = R$ on B and $S = 0$ on $B^d = (N_R)^d$. Then $0 < S \leq R \leq T$ and $S = 0$ on $N_R \oplus (N_R)^d$, so $N_R \oplus (N_R)^d \subseteq N_S$. Since $N_R \oplus (N_R)^d$ is order dense in E, it follows that the ideal N_S is order dense in E. Hence (in view of Theorem 23.3 in the preceding section) the disjoint complement $C_S = (N_S)^d$ satisfies $C_S = \{0\}$. It has been proved thus that if the positive operator T is not order continuous, there exists an operator S such that $0 < S \leq T$ and $C_S = \{0\}$.

Assume now that $T \in \mathcal{L}_b$ and $C_S \neq \{0\}$ for every $S \neq 0$ in the ideal generated by T. Then T is order continuous because, if not, there would exist an operator S such that $0 < |S| \leq |T|$ and $C_S = C_{|S|} = \{0\}$, which would contradict our hypothesis.

(iii) Follows immediately from parts (i) and (ii).   ∎

We have seen in part (i) of the last theorem that if $T \in \mathcal{L}_b$ and $C_T = \{0\}$, then T is disjoint to every order continuous operator in $\mathcal{L}_b$, i.e., $T \perp \mathcal{L}_n$. In other words, T is a member of the disjoint complement $\mathcal{L}_t$ of $\mathcal{L}_n$. Furthermore, in the proof of Lemma 24.1, it was observed that the set of operators T satisfying $C_T = \{0\}$ is an ideal in $\mathcal{L}_b$. Combining these results, we see that the set of operators $T \in \mathcal{L}_b$ satisfying $C_T = \{0\}$ is an ideal in $\mathcal{L}_t$. As shown in part (ii) of the last theorem there exists for any $0 < T \in \mathcal{L}_t$ an operator S satisfying $0 < S \leq T$ such that $C_S = \{0\}$. The ideal of all operators with carrier equal to $\{0\}$ is, therefore, order dense in $\mathcal{L}_t$. We state this as a theorem.

Theorem 24.3. *Let E and F be Riesz spaces such that E is Archimedean and F is Dedekind complete. The set of all operators $T \in \mathcal{L}_b$ such that $C_T = \{0\}$ is now an order dense ideal in $\mathcal{L}_t$.*

It may be asked whether perhaps the set of all T with carrier $\{0\}$ is not always equal to the entire space $\mathcal{L}_t$. The answer is negative, as indicated in the following exercise.

Exercise 24.4. In Example 21.6(ii) it was shown that the only order continuous operator from $E = C([0,1])$ into $\mathbb{R}$ is the null operator. In other words, the only order continuous linear functional on E is the null functional, i.e., every order bounded linear functional on E is a member of $\mathcal{L}_t = \mathcal{L}_t(E, \mathbb{R})$. Fix $x_0 \in [0,1]$ and let the positive linear functional φ on E be defined by $\varphi(f) = f(x_0)$ for all $f \in E$. Show that N_φ consists of all $f \in E$ satisfying $f(x_0) = 0$, and so $C_\varphi = \{0\}$. Let the positive linear functional ψ on E be defined by $\psi(f) = \int_0^1 f(x)dx$ for all $f \in E$. Show that $N_\psi = \{0\}$, and so $C_\psi = E$.

Under certain conditions, as stated in the next theorem, it follows from $C_S \perp C_T$ that $S \perp T$. For convenience, we shall write "iff" instead of "if and only if" at some places.

Theorem 24.5. *Let E and F be Riesz spaces such that E is Archimedean and F is Dedekind complete. Furthermore, let S and T be order bounded operators from E into F such that one of these, say S, is order continuous. Now,*

$$C_S \subseteq (N_T)^{dd} \quad \text{iff} \quad C_S \perp C_T \quad \text{iff} \quad C_T \subseteq N_S,$$

and if these conditions are satisfied, then $S \perp T$.

Proof. We may assume that S and T are positive operators. Since E is Archimedean and S is order continuous, we have $N_S = (N_S)^{dd}$, and so it follows immediately from Lemma 24.1 above that

$$C_S \subseteq (N_T)^{dd} \quad \text{iff} \quad C_S \perp C_T \quad \text{iff} \quad C_T \subseteq N_S.$$

Assume now that these conditions are satisfied. Then, for any $0 \le u \in N_T \oplus C_T$, we have $0 \le u = v + w$ with $v \in N_T$ and $w \in C_T$. Hence

$$0 \le (S \wedge T)u = (S \wedge T)v + (S \wedge T)w \le Tv + Sw = 0 + 0 = 0,$$

where $Sw = 0$ in view of $C_T \subseteq N_S$. This shows that $S \wedge T$ vanishes on $N_T \oplus C_T$. Since $N_T \oplus C_T = N_T \oplus (N_T)^d$ is order dense in E (see Theorem 23.3), it follows now from Theorem 23.4 that every positive element of the Archimedean space E is the supremum of a set of elements in $N_T \oplus C_T$. Observing also that $S \wedge T$ is order continuous (because $0 \le (S \wedge T) \le S$ and S is order continuous), we see that $S \wedge T$ vanishes on E, i.e., $S \perp T$. ∎

In the last theorem it was shown that under certain conditions $C_S \perp C_T$ implies that $S \perp T$. Does the converse hold as well? The answer is that this is not so in general (see the exercise at the end), but at least the converse holds for linear functionals. We present the details.

Theorem 24.6. *Let φ and ψ be order bounded linear functionals on the Archimedean space E (i.e., φ and ψ are members of the order dual $E^\sim$) such that one of these is order continuous. Then $\varphi \perp \psi$ if and only if $C_\varphi \perp C_\psi$.*

Proof. The functionals φ and ψ are order bounded operators from the Archimedean space E into the Dedekind complete space $\mathbb{R}$ and one (at least) of φ and ψ, say ψ, is order continuous. Hence, by the preceding theorem, it follows from $C_\varphi \perp C_\psi$ that $\varphi \perp \psi$. To prove the converse, we may assume that φ and ψ are positive, $\varphi \perp \psi$ and ψ is order continuous. Let $0 \le u \in C_\varphi = (N_\varphi)^d$ and $\epsilon > 0$ be given. Since $(\varphi \wedge \psi)u = 0$, there exists a sequence $(u_n : n = 1, 2, \ldots)$ in E^+ such that $0 \le u_n \le u$ and

$$\varphi(u_n) + \psi(u - u_n) \le 2^{-n}\epsilon \text{ for } n = 1, 2, \ldots.$$

Defining $v_n = \inf(u_k : k = 1, \ldots, n)$, we have $v_n \downarrow$ and any v satisfying $0 \le v \le v_n$ for all n also satisfies

$$0 \le \varphi(v) \le \varphi(v_n) \le \varphi(u_n) \le 2^{-n}\epsilon$$

for all n, so $\varphi(v) = 0$. But then $v \in C_\varphi \cap N_\varphi = \{0\}$, i.e., $v = 0$. It follows that if $0 \le v \le v_n$ for all n, then $v = 0$. Hence $v_n \downarrow 0$, which implies (since ψ is order continuous) that $\psi(u - v_n) \uparrow \psi(u)$. Observe now that

$$u - v_n = u - \inf(u_k : k = 1, \ldots, n) =$$

$$\sup\{(u - u_k) : k = 1, \ldots, n\} \le \Sigma_{k=1}^n (u - u_k),$$

so

$$0 \le \psi(u - v_n) \le \Sigma_{k=1}^n \psi(u - u_k) < \epsilon$$

for all n. Thus, since $\psi(u - v_n) \uparrow \psi(u)$, we see that $0 \le \psi(u) \le \epsilon$. This holds for every $\epsilon > 0$, and so $\psi(u) = 0$, i.e., $u \in N_\psi$. Any u satisfying $0 \le u \in C_\varphi$ is, therefore, a member of N_ψ, i.e., $C_\varphi \subseteq N_\psi$. It follows that $C_\varphi \perp C_\psi$.

Note that for the proof that $\varphi \perp \psi$ implies $C_\varphi \subseteq N_\psi$ (and hence $C_\varphi \perp C_\psi$) it is sufficient to assume that ψ is σ-order continuous. ∎

Exercise 24.7. In this exercise we present an example showing that $S \wedge T = 0$ does not always imply $C_S \perp C_T$, even if S and T are order continuous. Let $A = [0, 1]$ and $B = [1, 2]$, and let the positive operators S and T from $L_1([0, 2])$ into $L_1([0, 2])$ be defined by

$$Sf = (\int_0^2 f\, dx)\chi_A \text{ and } Tf = (\int_0^2 f\, dx)\chi_B,$$

where χ_A and χ_B are the characteristic functions of A and B respectively. Show that S and T are order continuous (see Example 21.6) and $C_S = C_T = L_1([0, 2])$. Also, for $0 \le f \in L_1([0, 2])$, show that $(S \wedge T)f = 0$ by observing that $0 \le (S \wedge T)f \le (Sf) \wedge (Tf)$.

CHAPTER 13
Order Duals and Adjoint Operators

25. The Order Dual of a Riesz Space

According to Corollary 20.3 the linear functional φ on the Riesz space E is order bounded if and only if φ is regular, i.e., if and only if $\varphi = \varphi_1 - \varphi_2$ with φ_1 and φ_2 positive. The vector space $E^\sim$ of all order bounded linear functionals on E is a Dedekind complete Riesz space with respect to the ordering defined by saying that $\varphi_1 \leq \varphi_2$ holds whenever $\varphi_2 - \varphi_1$ is positive. It follows that for any $\varphi \in E^\sim$ the functionals $\varphi^+ = \varphi \vee 0$ and $\varphi^- = (-\varphi) \vee 0$ exist in $E^\sim$ and $\varphi = \varphi^+ - \varphi^-$. Also, $| \varphi | = \varphi \vee (-\varphi)$ exists in $E^\sim$ and $| \varphi | = \varphi^+ + \varphi^-$. The space $E^\sim$ is called the order dual of E (see Corollary 20.3).

It is evident that if E consists of the null element only, then the same is true of $E^\sim$. It may come as a disappointment, however, that also if E is a space of infinite dimension it may occur that $E^\sim = \{0\}$. The standard example is the space $E = L_0(\mathbb{R}, \mu)$ of all Lebesgue measurable (real) functions on $\mathbb{R}$. The proof that $E^\sim = \{0\}$ in this case is based on the property that if Y is a subset of $\mathbb{R}$ satisfying $0 < \mu(Y) < \infty$ and α is a number satisfying $0 < \alpha < \mu(Y)$, then there exists a subset Y_1 of Y such that $\mu(Y_1) = \alpha$. To see this, define for any $a \in \mathbb{R}$ the subset Y_a of Y by $Y_a = (x \in Y : x \leq a)$. It is evident that $\mu(Y_a) \to 0$ as $a \to -\infty$ and $\mu(Y_a) \to \mu(Y)$ as $a \to \infty$. Furthermore, $\mu(Y_a)$ is a continuously increasing function of a (prove this). Hence $\mu(Y_a) = \alpha$ for some value of a. In particular $\mu(Y_a) = \mu(Y)/2$ for some a.

Theorem 25.1. *The order dual $E^\sim$ of the Riesz space $E = L_0(\mathbb{R}, \mu)$ of all Lebesgue measurable (real) functions on $\mathbb{R}$ satisfies $E^\sim = \{0\}$.*

Proof. If $0 \leq \varphi \in E^\sim$ and $u \in E^+$ satisfies $\varphi(u\chi_X) = \alpha > 0$ for some set $X \subseteq \mathbb{R}$, and if furthermore the measurable sets $X_n (n = 1, 2, \ldots)$ satisfy $X_n \uparrow X$, then $\varphi(u\chi_{X_n}) > 0$ for some n. This follows because if $\varphi(u\chi_{X_n}) = 0$ would hold for all n, then $Y_n = X \backslash X_n$ satisfies $\varphi(u\chi_{Y_n}) = \varphi(u\chi_X) = \alpha$ for every n, and hence the function

$$u_0 = \Sigma_{n=1}^\infty u\chi_{Y_n} \in E^+$$

satisfies $\varphi(u_0) \geq k\alpha$ for $k = 1, 2, \ldots$, which leads to a contradiction. Note here that $u_0 \in E^+$ follows from the observation that for (almost) every $x \in \mathbb{R}$ only finitely many terms of $\Sigma_1^\infty u\chi_{Y_n}(x)$ differ from zero. As a consequence of what has been proved we see that if φ is a positive linear functional on E such that $\varphi(u) > 0$ for some $u \in E^+$, then $\varphi(u\chi_Y) > 0$ for some set Y of finite measure (since the sets $X_n = [-n, n]$ satisfy $X_n \uparrow \mathbb{R}$). Hence, as observed above, it is possible to decompose Y into disjoint sets Z_1 and Z_1' of equal measure $\mu(Y)/2$. For at least one of these, say Z_1, we must have $\varphi(u\chi_{Z_1}) > 0$. Similarly, Z_1 has a subset Z_2 of measure $\mu(Y)/4$ such that $\varphi(u\chi_{Z_2}) > 0$. Proceeding by induction, we obtain a sequence $(Z_n : n = 1, 2, \ldots)$ of subsets of Y, descending to a set of measure zero and such that $\alpha_n = \varphi(u\chi_{Z_n}) > 0$ for all n. Then

$$u_1 = \Sigma_{n=1}^\infty \alpha_n^{-1} u\chi_{Z_n} \in E^+ \text{ and } \varphi(u_1) \geq k \text{ for } k = 1, 2, \ldots.$$

This is impossible. There does not exist, therefore, any positive linear functional on $E = L_0(\mathbb{R}, \mu)$ differing from the null functional.   ∎

Exercise 25.2. Let μ be a non-negative σ-finite and countably additive measure in the point set X such that $\mu(X) > 0$. Assume furthermore that μ has the property that for any μ-measurable subset Y of X satisfying $0 < \mu(Y) < \infty$ and for any number α satisfying $0 < \alpha < \mu(Y)$ there exists a subset Y_1 of Y such that $\mu(Y_1) = \alpha$. Show that the order dual $E^\sim$ of the Riesz space $E = L_0(X, \mu)$ satisfies $E^\sim = \{0\}$.

How can we find out whether a given linear functional φ on a given Riesz space E is order bounded or not? One possible answer can be found by looking at the behaviour of φ towards (relatively) uniformly converging sequences. Observe that if φ is order bounded and $(u_n : n = 1, 2, \ldots)$ is a decreasing sequence in E^+ converging (relatively) uniformly to zero, then $0 \leq u_n \leq \epsilon_n u_0$ for some $u_0 \in E^+$ and an appropriate sequence of numbers $(\epsilon_n : n = 1, 2, \ldots)$ satisfying $\epsilon_n \downarrow 0$, so

$$0 \leq \varphi^+(u_n) \leq \epsilon_n \varphi^+(u_0) \downarrow 0,$$

and similarly for φ^-. We shall prove now in the converse direction that if φ is a linear functional on E such that $(\varphi(u_n) : n = 1, 2, \ldots)$ is bounded for every decreasing sequence (u_n) in E^+ converging (relatively) uniformly to zero, then φ is order bounded. Precisely, we have the following theorem.

Theorem 25.3. *The linear functional φ on the Riesz space E is order bounded if* $\sup | \varphi(u_n) |$ *is finite for every decreasing sequence $(u_n : n = 0, 1, 2, \ldots)$ in E^+ satisfying $0 \leq u_n \leq 2^{-n} u_0$ for $n = 1, 2, \ldots$. Hence, the linear functional φ on E is order bounded if and only if $\varphi(f_n) \to 0$ for every sequence $(f_n : n = 1, 2, \ldots)$ in E that converges (relatively) uniformly to zero.*

Proof. It is sufficient to give a proof in one direction. Assume that φ is a linear functional on E such that $\sup \mid \varphi(u_n) \mid < \infty$ for every decreasing sequence $(u_n : n = 0, 1, 2, \ldots)$ in E^+ satisfying $0 \le u_n \le 2^{-n}u_0$ for $n = 1, 2, \ldots$. To prove that φ is order bounded we have to show (according to the definitions in section 18) that φ maps every order interval $[0, u]$ in E into a bounded interval in $\mathbb{R}$. In other words, we have to show that the set $(\mid \varphi(v) \mid : 0 \le v \le u)$ is a bounded set in $\mathbb{R}$. If this does not hold, there exists an element $u_0 \in E^+$ such that $\sup(\mid \varphi(v) \mid : 0 \le v \le u_0) = +\infty$. Then

$$\sup(\mid \varphi(v) \mid : 0 \le v \le u_0/2) = +\infty, \tag{1}$$

so there exists an element v_1 satisfying $0 < v_1 \le u_0/2$ such that

$$\mid \varphi(v_1) \mid \ge 3 \mid \varphi(u_0) \mid +2.$$

It follows that $w_1 = (u_0/2) - v_1$ satisfies

$$\mid \varphi(w_1) \mid \ge \mid \varphi(v_1) \mid - \frac{1}{2} \mid \varphi(u_0) \mid \ge 2 \mid \varphi(u_0) \mid +2.$$

Observe now that the suprema

$$\sup(\mid \varphi(v) \mid : 0 \le v \le v_1) \text{ and } \sup(\mid \varphi(v) \mid : 0 \le v \le w_1)$$

cannot both be finite, since this would contradict the formula (1). Now, if the first supremum is $+\infty$, denote v_1 by u_1, and in the other case denote w_1 by u_1. Then

$$0 < u_1 \le u_0/2, \mid \varphi(u_1) \mid \ge 2 \mid \varphi(u_0) \mid +2, \sup(\mid \varphi(v) \mid : 0 \le v \le u_1) = +\infty.$$

Repeating the procedure with u_0 replaced by u_1, we obtain the result that there exists an element $u_2 \in E^+$ such that

$$0 < u_2 \le u_1/2 \le u_0/4, \mid \varphi(u_2) \mid \ge 2 \mid \varphi(u_1) \mid \ge 4 \mid \varphi(u_0) \mid +4,$$

$$\sup(\mid \varphi(v) \mid : 0 \le v \le u_2) = +\infty.$$

Proceeding by induction, we obtain a sequence $(u_n : n = 0, 1, 2, \ldots)$ such that

$$0 < u_n \le 2^{-n}u_0 \text{ and } \mid \varphi(u_n) \mid \ge 2^n \mid \varphi(u_0) \mid +2^n$$

for $n = 1, 2, \ldots$. By hypothesis, however, a sequence having these properties cannot exist.

The reader should note that in this proof it would not be sufficient to obtain a sequence satisfying $\mid \varphi(u_n) \mid \ge 2^n \mid \varphi(u_0) \mid$ for all n. ∎

In Lemma 24.1 it was shown that if E and F are Riesz spaces and S and T are positive operators from E into F, then the carriers C_S and C_T satisfy $C_{S \vee T} = C_{S+T} = (C_S + C_T)^{dd}$. In particular, if φ and ψ are positive linear functionals on E, then

$$C_{\varphi \vee \psi} = C_{\varphi + \psi} = (C_\varphi + C_\psi)^{dd}.$$

We shall complete this result by proving that if E is Archimedean and one at least of φ and ψ is σ-order continuous, then

$$C_{\varphi \wedge \psi} = C_\varphi \cap C_\psi.$$

The corresponding formula for more general operators does not always hold; see Exercise 24.7 of the preceding section, in which S and T are order continuous operators in $L_1([0,2])$ such that $S \wedge T = 0$ and $C_S = C_T = E$. It is even so that the formula does not necessarily hold if both φ and ψ fail to be σ-order continuous; see Exercise 25.5 which will follow soon.

Theorem 25.4. *If φ and ψ are positive linear functionals on the Archimedean Riesz space E such that one at least of φ and ψ is σ-order continuous, then the carrier of $\varphi \wedge \psi$ satisfies*

$$C_{\varphi \wedge \psi} = C_\varphi \cap C_\psi.$$

Proof. We first recall several properties of ideals. For ideals A, B, C in E we have $(A + B) \cap C = (A \cap C) + (B \cap C)$(Exercise 7.7), $(A + B)^d = A^d \cap B^d$ and $(A \cap B)^{dd} = A^{dd} \cap B^{dd}$ (Exercise 8.6). Next, let us inspect once more the proof of Theorem 24.6. It was shown that if φ and ψ are positive, $\varphi \perp \psi$ and ψ order continuous, then $C_\varphi \subseteq N_\psi$ (and hence $C_\varphi \perp C_\psi$). As observed already in the proof, it is sufficient here to assume that ψ is σ-order continuous. In other words, $\varphi \perp \psi$ with ψ σ-order continuous implies $C_\varphi \subseteq N_\psi$.

We are ready now to present the proof itself. It is assumed therefore that φ and ψ are positive and ψ is σ-order continuous. Write $\chi = \varphi \wedge \psi$ for brevity and let $\varphi_1 = \varphi - \chi$ and $\psi_1 = \psi - \chi$. Then $\varphi = \varphi_1 + \chi$ and $\psi = \psi_1 + \chi$ with all occurring functionals positive. Hence

$$N_\varphi = N_{\varphi_1} \cap N_\chi \text{ and } N_\psi = N_{\psi_1} \cap N_\chi.$$

Furthermore we have $\varphi_1 \perp \psi_1$ with $0 \leq \varphi_1 \in E^\sim$ and $0 \leq \psi_1 \in (E^\sim)_c$ (i.e., ψ_1 is σ-order continuous), so $C_{\varphi_1} \subseteq N_{\psi_1}$ in view of what was observed above. Since

$$N_{\varphi_1} \oplus C_{\varphi_1} = N_{\varphi_1} \oplus (N_{\varphi_1})^d$$

satisfies $(N_{\varphi_1} \oplus C_{\varphi_1})^d = \{0\}$, it follows that $(N_{\varphi_1} + N_{\psi_1})^d = \{0\}$, so $(N_{\varphi_1} + N_{\psi_1})^{dd} = E$. Observing now that

$$N_\varphi + N_\psi = (N_{\varphi_1} \cap N_\chi) + (N_{\psi_1} \cap N_\chi) = (N_{\varphi_1} + N_{\psi_1}) \cap N_\chi,$$

we derive from this that

$$(N_\varphi + N_\psi)^{dd} = (N_{\varphi_1} + N_{\psi_1})^{dd} \cap (N_\chi)^{dd} = E \cap (N_\chi)^{dd} = (N_\chi)^{dd},$$

so

$$C_{\varphi \wedge \psi} = C_\chi = (N_\chi)^d = (N_\chi)^{ddd} = (N_\varphi + N_\psi)^{ddd} =$$

$$(N_\varphi + N_\psi)^d = (N_\varphi)^d \cap (N_\psi)^d = C_\varphi \cap C_\psi.$$ ∎

Exercise 25.5. We present an example of positive linear functionals φ and ψ on the space $E = C([0,1])$ of all real continuous functions on $[0,1]$ such that $\varphi \perp \psi$ but neither $C_\varphi \subseteq N_\psi$ or $C_\psi \subseteq N_\varphi$ nor $C_{\varphi \wedge \psi} = C_\varphi \cap C_\psi$ holds. Let $\varphi(f) = \int_0^1 f dx$ for all $f \in E$ and $\psi(f) = \Sigma_{n=1}^\infty 2^{-n} f(r_n)$, where $(r_n : n = 1, 2, \ldots)$ is the set of all rational numbers in $[0,1]$. Show that φ and ψ are strictly positive (i.e., $f > 0$ implies $\varphi(f) > 0$ and $\psi(f) > 0$), so $N_\varphi = N_\psi = \{0\}$ and $C_\varphi = C_\psi = E$. Show that for $\psi_n = 2^{-n} f(r_n)$ we have $\varphi \perp \psi_n$. Show that $\psi = \sup_k \Sigma_{n=1}^k \psi_n$, and so $\varphi \perp \psi$, which implies that $C_{\varphi \wedge \psi} = \{0\}$.

If E is a normed Riesz space, the order dual $E^\sim$ and the norm dual E^* are related. Recall that E^* is the vector space of all norm bounded linear functionals on E. The space E^* is a Banach space with respect to the norm

$$\| \varphi \| = \sup(| \varphi(f) | / \| f \|: f \neq 0) = \sup(| \varphi(f) |: \| f \| \leq 1)$$

for $\varphi \in E^*$ (see Theorem 18.2).

Lemma 25.6. *For $0 \leq \varphi \in E^\sim$ we have*

$$\sup(\varphi(u) : u \in E^+, \| u \| \leq 1) = \sup(| \varphi(f) |: f \in E, \| f \| \leq 1), \qquad (2)$$

where both sides may be infinite. It follows that if $0 \leq \varphi \in E^\sim$ and φ is norm bounded, then

$$\| \varphi \| = \sup(\varphi(u) : u \in E^+, \| u \| \leq 1).$$

Hence, if $0 \leq \psi \leq \varphi$ and φ is norm bounded, then ψ is norm bounded and $\| \psi \| \leq \| \varphi \|$.

Proof. For convenience, denote the suprema in (2) by l and r respectively. It is evident that $l \leq r$ holds. In the converse direction we have

$$r \leq \sup(\varphi(| f |) : \| f \| \leq 1) = \sup(\varphi(| f |) : \| | f | \| \leq 1) = l.$$

Hence, $l = r$. ∎

Lemma 25.7. *If $\varphi \in E^*$, then $\varphi \in E^\sim$. In other words, E^* is a subset of $E^\sim$. Furthermore, it follows then that $| \varphi | \in E^*$ and $\| | \varphi | \| = \| \varphi \|$.*

Proof. Let $\varphi \in E^*$. To show that $\varphi \in E^\sim$ holds it is sufficient to prove that φ maps any order interval $[-u, u]$ in E into an order interval in $\mathbb{R}$, i.e., we have to show that $\sup(|\varphi(f)| : |f| \leq u)$ is finite. For $|f| \leq u$ we have $|\varphi(f)| \leq \|\varphi\| \cdot \|f\|$, so

$$\sup(|\varphi(f)| : |f| \leq u) \leq \|\varphi\| \cdot \|u\| < \infty.$$

This shows, therefore, that $\varphi \in E^\sim$. It follows that $|\varphi|$ exists in $E^\sim$. For any $u \in E^+$ we have

$$|\varphi|(u) = \sup(|\varphi(f)| : |f| \leq u) \leq \|\varphi\| \cdot \|u\|$$

(see Theorem 20.4(5)), so

$$\sup(|\varphi|(u) : u \in E^+, \|u\| \leq 1) \leq \|\varphi\|.$$

In view of the preceding lemma it follows from this inequality that $|\varphi| \in E^*$ with norm satisfying $\||\varphi\|| \leq \|\varphi\|$. Conversely, $\|\varphi\| \leq \||\varphi\||$ follows from $|\varphi(f)| \leq |\varphi|(|f|)$ for every f. ∎

We shall prove now that E^* is an ideal in $E^\sim$ and $E^* = E^\sim$ if E is a Banach lattice. Furthermore, the norm in E^* is a Riesz norm.

Theorem 25.8. *(i) The space E^* is an ideal in $E^\sim$, but E^* is not always a band in $E^\sim$.*

(ii) The space E^*, as a Riesz space with the order inherited from $E^\sim$, is Dedekind complete and the norm in E^* is a Riesz norm. Furthermore, if $0 \leq \varphi_0 \in E^*$ and G is an upwards directed set of positive elements in E^* such that $G \uparrow \varphi_0$, then $(\|\varphi\| : \varphi \in G) \uparrow \|\varphi_0\|$.

(iii) If E is a Banach lattice, then $E^* = E^\sim$.

Proof. (i) In the last lemma it was shown already that E^* is a linear subspace of $E^\sim$. For the proof that E^* is an ideal in $E^\sim$, assume that $|\psi| \leq |\varphi|$ with $\psi \in E^\sim$ and $\varphi \in E^*$. We have to prove that ψ is a member of E^*. In virtue of the last lemma we have $|\varphi| \in E^*$, and so ψ^+ and ψ^- are members of E^* by Lemma 25.6 (since $\psi^+ \leq |\varphi|$ and $\psi^- \leq |\varphi|$). It follows that ψ is a member of E^*.

For an example in which E^* is an ideal in $E^\sim$, but not a band, we refer to the next exercise.

(ii) Since E^* is an ideal in the Dedekind complete space $E^\sim$, the space E^* is Dedekind complete. It was proved in the preceding lemmas that $\||\varphi\|| = \|\varphi\|$ for every $\varphi \in E^*$ and $\|\psi\| \leq \|\varphi\|$ if $0 \leq \psi \leq \varphi \in E^*$. This shows that the norm in E^* is a Riesz norm. Let now G be an upwards directed set of positive elements in E^* such that $G \uparrow \varphi_0$, where φ_0 is also an element of E^*. Then $\varphi_0(u) = \sup(\varphi(u) : \varphi \in G)$ for every $u \in E^+$. Hence, for $\|u\| \leq 1$ we have

$$\varphi_0(u) = \sup(\varphi(u) : \varphi \in G) \leq \sup(\| \varphi \| : \varphi \in G),$$

so

$$\| \varphi_0 \| = \sup(\varphi_0(u) : u \in E^+, \| u \| \leq 1) \leq \sup(\| \varphi \| : \varphi \in G).$$

Since $0 \leq \varphi \leq \varphi_0$ for every $\varphi \in G$, it is evident that $\| \varphi_0 \| \geq \sup(\| \varphi \| : \varphi \in G)$. It follows that

$$(\| \varphi \| : \varphi \in G) \uparrow \| \varphi_0 \| .$$

(iii) Let E be a Banach lattice. To show that $E^* = E^\sim$ holds it is sufficient that any $\varphi \in E^\sim$ is norm bounded. This follows from Theorem 18.4, wherein it was shown more generally that any order bounded operator from a Banach lattice into a normed Riesz space is norm bounded. ∎

Exercise 25.9. (i) Recall that ℓ_∞ is the Riesz space of all bounded sequences of (real) numbers. Let E be the ideal in ℓ_∞ consisting of all $f = (f_1, f_2, \ldots)$ with $f_k \neq 0$ for only finitely many k. Assume that E is equipped with the uniform norm. Show that φ, defined by $\varphi(f) = \Sigma_{k=1}^\infty k f_k$ for $f \in E$, is a member of $E^\sim$ since φ is a positive linear functional on E. Show that φ is not a member of E^*. Let $\varphi_n(f) = \Sigma_{k=1}^n k f_k$ for $n = 1, 2, \ldots$. Show that $0 \leq \varphi_n \in E^*$ for every n and $\varphi_n \uparrow \varphi$ in $E^\sim$. Hence, E^* is not a band in $E^\sim$.

(ii) Show that if the normed Riesz space E has the weak Riesz-Fischer property, then $E^* = E^\sim$.

Hint: For (ii), assume the existence of a positive φ in $E^\sim$ such that φ is not a member of E^*. Then, for some sequence $(u_1, u_2, \ldots)$ in E^+, we have $\| u_n \| \leq 1$ and $\varphi(u_n) > n^3$ for $n = 1, 2, \ldots$. For $v_n = n^{-2} u_n$ we have $\Sigma \| v_n \| < \infty$, so by the weak Riesz- Fischer property there exists an element $w \in E$ such that $v_n \leq w$ for all n. Since $\varphi(w) \geq \varphi(v_n) \geq n$ for $n = 1, 2, \ldots$, we have a contradiction.

In accordance with the notations in section 22 the set of all order continuous linear functionals and the set of all σ-order continuous linear functionals on E are denoted by $E_n^\sim$ and $E_c^\sim$ respectively. These sets are projection bands in $E^\sim$. Their disjoint complements are denoted by $E_t^\sim$ and $E_s^\sim$ respectively, so

$$E^\sim = E_c^\sim \oplus E_s^\sim = E_n^\sim \oplus E_t^\sim .$$

Recall that the members of $E_s^\sim$ are called singular linear functionals. In Example 21.6(i) it was shown that every positive linear functional on a space E of type $L_p(X, \mu); 1 \leq p < \infty$, is order continuous, so $E^\sim = E_n^\sim = E_c^\sim$ and $E_t^\sim = E_s^\sim = \{0\}$. In Example 21.6(ii) it was shown that on $E = C([0, 1])$ the only σ-order continuous positive linear functional is the null functional, so $E_c^\sim = E_n^\sim = \{0\}$ and $E_s^\sim = E_t^\sim = E^\sim$. More generally, we shall prove now that if E is a Banach lattice having order continuous norm, then every positive linear functional on E is order continuous, i.e., $E^\sim = E_n^*$.

Theorem 25.10. *Let E be a Banach lattice having order continuous norm. Then $E^* = E^\sim = E_n^\sim$.*

Proof. It was shown in Theorem 25.8(iii) above that $E^* = E^\sim$ holds, so it is sufficient to show now that every positive linear functional on E is order continuous. For this purpose, let $0 \le \varphi \in E^\sim = E^*$ and $D \downarrow 0$ in E. Then $(\| u \| : u \in D) \downarrow 0$ since the norm in E is order continuous, and so

$$(\varphi(u) : u \in D) \le (\| \varphi \| \cdot \| u \| : u \in D) \downarrow 0.$$

This shows that φ is order continuous. ∎

Let, once more, E and F be Riesz spaces with F Dedekind complete and let S and T_0 be order bounded operators from E into F. It is evident that if S is a member of the ideal generated by T_0 in $\mathcal{L}_b(E, F)$, then $| T_0 | (| f |) = 0$ implies $| S | (| f |) = 0$, and so $N_{T_0} \subseteq N_S$. Now, let S be a member of the band generated by T_0. Then there exists a set G in the ideal generated by T_0 such that $0 \le G \uparrow | S |$, so

$$(T(| f |) : T \in G) \uparrow | S | (| f |) \text{ for all } f \in E.$$

Hence, if $f \in N_{T_0}$, then $T(| f |) = 0$ for all $T \in G$, and so $| S | (| f |) = 0$. This shows that $N_{T_0} \subseteq N_S$, i.e.,

$$C_S = (N_S)^d \subseteq (N_{T_0})^d = C_{T_0}.$$

It has been proved thus that if S is contained in the band generated by some operator T, then $C_S \subseteq C_T$. Does the converse hold? In general the answer is negative. In Exercise 24.7 we have strictly positive order continuous operators S and T from $E = L_1([0, 2])$ into E itself such that $C_S = C_T = E$ and $S \perp T$, so neither S nor T is contained in the band generated by the other one. Does the converse then perhaps hold if we restrict ourselves to linear functionals? Here also the answer is still negative. In Exercise 25.5 we have strictly positive linear functionals φ and ψ on $E = C([0, 1])$ such that $C_\varphi = C_\psi = E$ and $\varphi \perp \psi$, so once more neither φ nor ψ is contained in the band generated by the other one. The situation is better if one at least of φ and ψ is order continuous, as indicated in the following exercise.

Exercise 25.11. Let φ and ψ be order bounded linear functionals on the Archimedean Riesz space E such that ψ is order continuous. Show that ψ is contained in the band generated by φ in $E^\sim$ if and only if $C_\psi \subseteq C_\varphi$.

Hint: As observed above, if ψ is in the band generated by φ, then $C_\psi \subseteq C_\varphi$. Conversely, assume that $C_\psi \subseteq C_\varphi$, i.e., $(N_\psi)^d \subseteq (N_\varphi)^d$, so $(N_\varphi)^{dd} \subseteq (N_\psi)^{dd}$. Since ψ is order continuous the null ideal N_ψ is a band, and since E is Archimedean the band generated by N_ψ is $(N_\psi)^{dd}$. Hence $(N_\psi)^{dd} = N_\psi$. It follows that

$$N_\varphi \subseteq (N_\varphi)^{dd} \subseteq (N_\psi)^{dd} = N_\psi.$$

Therefore, it is given that $N_\varphi \subseteq N_\psi$ and it has to be proved that $\mid \psi \mid$ is in the band generated by φ. If not, $\mid \psi \mid$ has a non-zero component ψ_1 in the disjoint complement, i.e., $0 < \psi_1 \leq \mid \psi \mid$ and $\psi_1 \perp \varphi$ with $C_{\psi_1} \neq \{0\}$ (since ψ_1 is order continuous, see Theorem 24.2(i)). Since ψ_1 is order continuous, it follows that $C_\varphi \subseteq N_{\psi_1}$ (see the proof of Theorem 24.6), and so $C_{\psi_1} \subseteq (N_\varphi)^{dd}$ (see Lemma 24.1). Choose an arbitrary $u > 0$ in C_{ψ_1}, so $\psi_1(u) > 0$. Then $u \in (N_\varphi)^{dd}$, so there exists a subset D of $(N_\varphi)^+$ such that $D \uparrow u$, and hence

$$(\psi_1(v) : v \in D) \uparrow \psi_1(u) > 0.$$

Therefore, $\psi_1(v) > 0$ for some $v \in (N_\varphi)^+$. Then $\mid \varphi \mid (v) = 0$ and $\mid \psi \mid (v) \geq \psi_1(v) > 0$, contradicting the hypothesis that $N_\varphi \subseteq N_\psi$.

26. Adjoint Operators

As defined in section 18, the vector space of all linear functionals on the (real or complex) vector space V is called the algebraic dual of V. We denote the algebraic dual of V by $V^\#$. Now, let $T : V \to W$ be a linear operator from the vector space V into the vector space W. For any $\psi \in W^\#$, defining $\varphi(f) = \psi(Tf)$ for all $f \in V$, we obtain a linear functional φ on V, i.e., $\varphi \in V^\#$. Each $\psi \in W^\#$ determines an element φ of $V^\#$ in this manner, so that if we write $\varphi = T^\# \psi$ we get a mapping $T^\#$ from $W^\#$ into $V^\#$. It is evident that $T^\#$ is linear. In other words, $T^\#$ is a linear operator from $W^\#$ into $V^\#$. Recapitulating, we have

$$(T^\# \psi)(f) = \psi(Tf) \text{ for all } f \in V \text{ and all } \psi \in W^\#.$$

Note also that $(\alpha T_1 + \beta T_2)^\# = \alpha T_1^\# + \beta T_2^\#$ for all T_1 and T_2 (from V into W) and all (real or complex) coefficients α, β. The operator $T^\#$ is called the *algebraic adjoint* of T.

Assume now that E and F are Riesz spaces with F Dedekind complete. As proved in Theorem 20.2, the space $\mathcal{L}_b(E, F)$ of all order bounded operators from E into F is a Dedekind complete Riesz space and hence every $T \in \mathcal{L}_b(E, F)$ can be written as $T = T^+ - T^-$ with T^+ and T^- positive. Given such an operator T, the restriction of $T^\#$ to $F^\sim$ will be denoted by $T^\sim$. The operator $T^\sim$ is called the *order adjoint* of T. We prove that $T^\sim$ maps $F^\sim$ into $E^\sim$. To this end, choose $0 \leq \psi \in F^\sim$ and let $\varphi = T^\sim \psi$ be its image. For any $f \in E$ we have

$$\varphi(f) = (T^\sim \psi)(f) = \psi(Tf) = \psi(T^+ f) - \psi(T^- f) = \varphi_1(f) - \varphi_2(f),$$

where φ_1 and φ_2 are evidently positive linear functionals on E. This shows that $\varphi = T^\sim \psi \in E^\sim$. It is evident that $T^\sim$ is positive if T is positive. Hence, if T is order bounded, i.e., $T \in \mathcal{L}_b(E, F)$, then $T^\sim$ is also order bounded, i.e., $T^\sim \in \mathcal{L}_b(F^\sim, E^\sim)$. It is even so that $T^\sim$ is order continuous. For the proof we may assume that T is positive (and so $T^\sim$ is positive). Let D be a downwards directed subset of $F^\sim$ satisfying $D \downarrow 0$. We have to show that $(T^\sim \psi : \psi \in D) \downarrow 0$ in $E^\sim$. For this purpose, choose an arbitrary $u \in E^+$ and observe that

$$\{(T^\sim \psi)(u) : \psi \in D\} = \{\psi(Tu) : \psi \in D\} \downarrow 0.$$

The following theorem holds, therefore.

Theorem 26.1. *If E and F are Riesz spaces with F Dedekind complete and $T : E \to F$ is order bounded, then the order adjoint $T^\sim : F^\sim \to E^\sim$ is order bounded and order continuous. If T is positive, then so is $T^\sim$. Furthermore, we have $\mid T^\sim \mid \leq \mid T \mid^\sim$.*

Proof. It remains to prove that $\mid T^\sim \mid \leq \mid T \mid^\sim$. It follows from $T \leq \mid T \mid$ and $-T \leq \mid T \mid$ that $T^\sim \leq \mid T \mid^\sim$ and $-T^\sim \leq \mid T \mid^\sim$. Hence

$$\mid T^\sim \mid = (T^\sim) \vee (-T^\sim) \leq \mid T \mid^\sim. \qquad \blacksquare$$

It is a remarkable result, due to J.Synnatzschke, that if $\psi \in F_n^\sim$ (i.e., ψ is an order continuous linear functional on F), then $\mid T^\sim \mid (\psi) = \mid T \mid^\sim (\psi)$. For the proof we need a lemma.

Lemma 26.2. *Let E and F be Riesz spaces with F Dedekind complete and let $T : E \to F$ be an order bounded operator. Then, for every $0 \leq u \in E$ and every $0 \leq \psi \in F^\sim$, we have*

$$\psi(\mid Tu \mid) \leq \{\mid T^\sim \mid (\psi)\}(u).$$

Proof. In view of Theorem 20.8 we have

$$\psi(\mid Tu \mid) = \max\{\chi(Tu) : \mid \chi \mid \leq \psi\}.$$

Hence, there exists an element χ in $F^\sim$ satisfying $\mid \chi \mid \leq \psi$ and $\psi(\mid Tu \mid) = \chi(Tu)$. It follows that

$$\psi(\mid Tu \mid) = \chi(Tu) = (T^\sim \chi)(u) \leq \mid T^\sim \chi \mid (u) \leq$$

$$\{\mid T^\sim \mid (\mid \chi \mid)\}(u) \leq \{\mid T^\sim \mid (\psi)\}(u). \qquad \blacksquare$$

Theorem 26.3. *Let E and F be Riesz spaces with F Dedekind complete and let $T : E \to F$ be an order bounded operator. Then*

$$\mid T^\sim \mid (\psi) = \mid T \mid^\sim (\psi) \text{ for every } \psi \in F_n^\sim.$$

Proof. From $\mid T^{\sim} \mid \leq \mid T \mid^{\sim}$ it follows that $\mid T^{\sim} \mid (\psi) \leq \mid T \mid^{\sim}(\psi)$ for every $\psi \in F^{\sim}$. For the inequality in the converse direction, let $0 \leq \psi \in F_n^{\sim}$ and $0 \leq u \in E$ be given. Note first that (by Theorem 20.10)

$$\mid T \mid (u) = \sup(\Sigma_{j=1}^n \mid Tu_j \mid : \text{all } u_j \in E^+ \text{ and } \Sigma_1^n u_j = u).$$

Hence, using now that ψ is order continuous, we see that

$$\{\mid T \mid^{\sim}(\psi)\}(u) = \psi\{\mid T \mid (u)\} =$$

$$\psi\{\sup(\Sigma_{j=1}^n \mid Tu_j \mid : u_j \in E^+ \text{ and } \Sigma_1^n u_j = u)\} =$$

$$\sup(\Sigma_{j=1}^n \psi(\mid Tu_j \mid) : u_j \in E^+ \text{ and } \Sigma_1^n u_j = u).$$

Therefore, in view of the last lemma,

$$\{\mid T \mid^{\sim}(\psi)\}(u) \leq \sup[\Sigma_{j=1}^n \{\mid T^{\sim} \mid (\psi)\}(u_j) : u_j \in E^+ \text{ and } \Sigma_1^n u_j = u] =$$

$$= \{\mid T^{\sim} \mid (\psi)\}(u).$$

This holds for every $u \in E^+$, and so $\mid T \mid^{\sim}(\psi) \leq \mid T^{\sim} \mid (\psi)$. Combining this with $\mid T^{\sim} \mid (\psi) \leq \mid T \mid^{\sim}(\psi)$, we obtain the desired result. ∎

The order bounded operator T maps E into F and its order adjoint $T^{\sim}$ is order continuous and maps $F^{\sim}$ into $E^{\sim}$, i.e., $T \in \mathcal{L}_b(E, F)$ and $T^{\sim} \in \mathcal{L}_n(F^{\sim}, E^{\sim})$. We shall denote the restriction of $T^{\sim}$ to $F_n^{\sim}$ by T'. Hence, T' is an order continuous operator mapping $F_n^{\sim}$ into $E^{\sim}$, i.e., $T' \in \mathcal{L}_n(F_n^{\sim}, E^{\sim})$. What we have proved in the last theorem can be expressed briefly by saying that $\mid T' \mid = \mid T \mid'$. Observing now that the mapping Φ from $\mathcal{L}_b(E, F)$ into $\mathcal{L}_n(F_n^{\sim}, E^{\sim})$, defined by $\Phi(T) = T'$ for every $T \in \mathcal{L}_b(E, F)$, is a positive linear operator from a Riesz space into a Dedekind complete Riesz space, this result can be written as

$$\mid \Phi(T) \mid = \Phi(\mid T \mid),$$

holding for every $T \in \mathcal{L}_b(E, F)$. It is clear, therefore, that Φ is a Riesz homomorphism (see Theorem 19.2). We state this as a theorem.

Theorem 26.4. *(J.Synnatzschke, 1972). Let E and F be Riesz spaces with F Dedekind complete and, for any order bounded operator T from E into F, let T' be the restriction of the order adjoint $T^{\sim}$ to $F_n^{\sim}$. Then the mapping transforming T into T' is a Riesz homomorphism, i.e.,*

$$(T_1 \vee T_2)' = T_1' \vee T_2' \text{ and } (T_1 \wedge T_2)' = T_1' \wedge T_2'$$

for all T_1 and T_2 in $\mathcal{L}_b(E, F)$. In particular $\mid T \mid' = \mid T' \mid$ for all $T \in \mathcal{L}_b(E, F)$.

We turn to the normed case for some moments. First, let V and W be normed spaces (not necessarily Riesz spaces) and let $T : V \to W$ be a norm bounded operator. The Banach dual space W^* of W is a linear subspace of the algebraic dual $W^{\#}$. The algebraic adjoint $T^{\#}$ of T maps $W^{\#}$ into $V^{\#}$;

the restriction of $T^{\#}$ to W^{*} is now called the *Banach adjoint* of T and is denoted by T^{*}. Hence

$$(T^{*}\psi)(f) = \psi(Tf) \text{ for every } f \in V \text{ and every } \psi \in W^{*}.$$

Obviously, $T^{*}\psi$ is a linear functional on V, so $T^{*}\psi \in V^{\#}$. We prove that $T^{*}\psi \in V^{*}$. Furthermore, T^{*} is linear (since T^{*} is a restriction of the linear operator $T^{\#}$) and T^{*} is norm bounded with $\| T^{*} \| \leq \| T \|$. We recall the proof. From

$$| (T^{*}\psi)(f) | = | \psi(Tf) | \leq \| \psi \| \cdot \| Tf \| \leq \| \psi \| \cdot \| T \| \cdot \| f \|$$

for all $f \in V$ it follows that $T^{*}\psi$ is a bounded linear functional on V with norm at most equal to $\| \psi \| \cdot \| T \|$, i.e., $T^{*}\psi \in V^{*}$ and $\| T^{*}\psi \| \leq \| T \| \cdot \| \psi \|$. This holds for all $\psi \in W^{*}$, so T^{*} is norm bounded with norm $\| T^{*} \|$ at most equal to $\| T \|$. More can be proved (see section 37), but for our present purposes this is enough.

The following variant of Synnatzschke's theorem can be proved now for normed Riesz spaces. Note first that any Banach lattice having order continuous norm is Dedekind complete (see Theorem 17.8).

Theorem 26.5. *Let E and F be Banach lattices such that F has order continuous norm (so that, therefore, F is Dedekind complete) and let T be an order bounded operator from E into F. As before, let T' be the restriction of $T^{\sim}$ to $F_{n}^{\sim}$ and let T^{*} be the Banach adjoint of T. Then $T^{*} = T'$ and, for all order bounded T_{1} and T_{2} from E into F, we have*

$$(T_{1} \vee T_{2})^{*} = T_{1}^{*} \vee T_{2}^{*} \quad and \quad (T_{1} \wedge T_{2})^{*} = T_{1}^{*} \wedge T_{2}^{*}.$$

Proof. The operator T^{*} is the restriction of $T^{\#}$ to F^{*} and T' is the restriction of $T^{\sim}$ (and hence the restriction of $T^{\#}$ as well) to $F_{n}^{\sim}$. Note now that $F^{*} = F^{\sim} = F_{n}^{\sim}$ by Theorem 25.10 (because F is a Banach lattice having order continuous norm). Furthermore, Synnatzschke's theorem holds for all order bounded T_{1} and T_{2} from E into F. Observing now that $T_{1}' = T_{1}^{*}, T_{2}' = T_{2}^{*}$ and $(T_{1} \vee T_{2})' = (T_{1} \vee T_{2})^{*}$, the desired result follows.   ∎

For our final remarks in this section assume again that E and F are Riesz spaces with F Dedekind complete and $T : E \to F$ is an order bounded operator, i.e., $T \in \mathcal{L}_{b}(E, F)$. We have seen that the order adjoint $T^{\sim}$ of T satisfies $T^{\sim} \in \mathcal{L}_{b}(F^{\sim}, E^{\sim})$ and $T^{\sim}$ is order continuous, so $T^{\sim} \in \mathcal{L}_{n}(F^{\sim}, E^{\sim})$. In particular, the restriction T' of $T^{\sim}$ to $F_{n}^{\sim}$ satisfies $T' \in \mathcal{L}_{n}(F_{n}^{\sim}, E^{\sim})$. If T itself is order continuous or σ-order continuous, we can say more.

Theorem 26.6. *If under the above hypotheses T is order continuous, then $T^\sim$ is order continuous and maps $F_n^\sim$ into $E_n^\sim$. In other words, $T' \in \mathcal{L}_n(F_n^\sim, E_n^\sim)$. Similarly, if T is σ-order continuous, then $T^\sim$ is order continuous and maps $F_c^\sim$ into $E_c^\sim$.*

Proof. Let $T : E \to F$ be order continuous. To show that $T^\sim\psi \in E_n^\sim$ for every $\psi \in F_n^\sim$, we may assume that T is positive and we have to prove that $D \downarrow 0$ in E implies $\{(T^\sim\psi)(u) : u \in D\} \downarrow 0$ for $0 \le \psi \in F_n^\sim$. Since

$$\{(T^\sim\psi)(u) : u\epsilon D\} = \{\psi(Tu) : u \in D\} \downarrow 0$$

on account of both T and ψ being order continuous, the desired result follows. ∎

CHAPTER 14
Signed Measures and the Radon-Nikodym Theorem

27. The Space of Signed Measures

Recall that a non-empty collection Γ of subsets of the set X is called a ring of subsets of X if $A \in \Gamma, B \in \Gamma$ implies $A \cup B \in \Gamma$ and $A \backslash B \in \Gamma$ (see Example 3.4(i)). It follows then that finite unions and finite intersections of sets in Γ are sets in Γ. The ring Γ is called an algebra of subsets of X if X itself is a member of Γ. The algebra Γ is called a σ-algebra if any countable union of sets in Γ is again a set in Γ (see Example 3.4(ii)). The mapping ν from the algebra Γ into $\mathbb{R}$ is called a (real) finitely additive signed measure (or a charge) on Γ if $\nu(A_1 \cup A_2) = \nu(A_1) + \nu(A_2)$ holds for all disjoint A_1 and A_2 in Γ (see Example 4.3(7)). If Γ is a σ-algebra and $\nu(\cup_1^\infty A_n) = \Sigma_1^\infty \nu(A_n)$ holds for every disjoint sequence $(A_n : n = 1, 2, \ldots)$ in Γ, then ν is called a *σ-additive signed measure* on Γ. In this section we shall briefly say "signed measure" when a σ-additive signed measure is meant. If there is only finite additivity, this will be explicitly mentioned. .

Let now ν be such a signed measure on the σ-algebra Γ of subsets of X and assume first that $(A_n : n = 1, 2, \ldots)$ is a disjoint sequence of sets in Γ. Then $\Sigma \mid \nu(A_n) \mid$ converges. For the proof, let B be the union of the sets A_n for which $\nu(A_n) > 0$ and C the union of the remaining sets A_n. It follows easily that

$$\nu(B) - \nu(C) = \Sigma \mid \nu(A_n) \mid .$$

Next, assume that $(B_n : n = 1, 2, \ldots)$ is an increasing sequence of sets in Γ and $B = \cup_1^\infty B_n$. Then $\nu(B_n)$ converges to $\nu(B)$ as $n \to \infty$. This follows by observing that

$$\nu(B) = \nu\{B_1 \cup (B_2 \backslash B_1) \cup (B_3 \backslash B_2) \cup \ldots\} =$$

$$\nu(B_1) + \Sigma_2^\infty \nu(B_k \backslash B_{k-1}) = \lim_{n \to \infty} \{\nu(B_1) + \Sigma_2^n \nu(B_k \backslash B_{k-1})\} =$$

$$\lim_{n \to \infty} \nu\{B_1 \cup (B_2 \backslash B_1) \cup \ldots \cup (B_n \backslash B_{n-1})\} = \lim_{n \to \infty} \nu(B_n).$$

We shall say that the set $A \in \Gamma$ is *ν–positive* if $\nu(B) \geq 0$ for any $B \in \Gamma$ that is a subset of A, and $A \in \Gamma$ is *strongly ν–positive* if A is ν-positive and $\nu(A) > 0$. For sets A_1 and A_2 in Γ, it does not necessarily follow from $\nu(A_1) \geq 0$ and

$\nu(A_2) \geq 0$ that $\nu(A_1 \cup A_2) \geq 0$. However, it is evident that if A_1 and A_2 are ν-positive, then $A_1 \cup A_2$ is ν-positive and, if one at least of A_1 and A_2 is strongly ν-positive, then $A_1 \cup A_2$ is strongly ν-positive. Hence, by induction, $\cup_1^n A_k$ is strongly ν-positive if all $A_k(k = 1, \ldots, n)$ are ν-positive and one at least of these sets is strongly ν-positive. Assume now that $(A_n : n = 1, 2, \ldots)$ is a sequence of ν-positive sets and one at least of these sets, say for $n = n_0$, is strongly ν-positive. Then $B_n = \cup_{k=1}^n A_k$ is strongly ν-positive for $n \geq n_0$ and the union $A = \cup_1^\infty A_n$ satisfies $A = \cup_1^\infty B_n$ with $\nu(B_n) \uparrow \nu(A)$. Hence (since $\nu(B_n) > 0$ for $n \geq n_0$) we have $\nu(A) > 0$. It follows easily that A is strongly ν-positive. The definitions and results for ν-negative and strongly ν-negative sets are similar.

It will be proved now that any set $A \in \Gamma$ satisfying $\nu(A) < 0$ has a strongly ν-negative subset.

Lemma 27.1. *Let ν be a signed measure on the σ-algebra Γ and let $A \in \Gamma$ satisfy $\nu(A) < 0$. Then A has a strongly ν-negative subset.*

Proof. We define a sequence $(A_n : n = 0, 1, 2, \ldots)$ of subsets of A as follows. Let $A_0 = \emptyset$ and let

$$\alpha_1 = \sup(\nu(B) : B \in \Gamma, B \subseteq A).$$

Then $\alpha_1 \geq 0$ (since $\emptyset \subseteq A$). If $\alpha_1 < \infty$, then choose $A_1 \in \Gamma, A_1 \subseteq A$ such that $\nu(A_1) \geq \alpha_1/2$. If $\alpha_1 = \infty$, then choose $A_1 \in \Gamma, A_1 \subseteq A$ such that $\nu(A_1) \geq 1$. Next, let

$$\alpha_2 = \sup(\nu(B) : B \in \Gamma, B \subseteq A \backslash A_1)$$

and choose $A_2 \in \Gamma, A_2 \subseteq A \backslash A_1$ such that $\nu(A_2) \geq \min\{1, \alpha_2/2\}$. Generally, if $A_0, A_1, \ldots, A_n$ have been chosen, let

$$\alpha_{n+1} = \sup\{\nu(B) : B \in \Gamma, B \subseteq (A \backslash \cup_1^n A_k)\}$$

and choose $A_{n+1} \in \Gamma, A_{n+1} \subseteq A \backslash \cup_1^n A_k$ such that $\nu(A_{n+1}) \geq \min\{1, \alpha_{n+1}/2\}$. We thus obtain a disjoint sequence (A_n) of sets in Γ, each one a subset of A and such that $\nu(A_n) \geq 0$ for all n. Hence, writing $S = \cup_1^\infty A_n$, we have

$$0 \leq \Sigma_1^\infty \nu(A_n) = \nu(S) < \infty,$$

which shows that $\nu(A_n) \to 0$, and so $\alpha_n \to 0$.

We prove now that $A \backslash S$ is strongly ν-negative. Observe first that

$$\nu(A \backslash S) = \nu(A) - \nu(S) \leq \nu(A) < 0.$$

Furthermore, if $C \in \Gamma$ and C is a subset of $A \backslash S$, then C is a subset of all sets $A \backslash \cup_1^n A_k$, so $\nu(C) \leq \alpha_{n+1}$ for $n = 0, 1, 2, \ldots$, i.e., $\nu(C) \leq 0$. It follows that $A \backslash S$ is strongly ν-negative.   ∎

It is not difficult to derive from this lemma that the point set X is the disjoint union of a ν-positive set S_1 and a ν-negative set S_2. The decomposition $X = S_1 \cup S_2$ is then called a *Hahn decomposition of X with respect to ν.* We present the proof.

Theorem 27.2. *(Hahn decomposition). Let ν be a signed measure on the σ-algebra Γ of subsets of X. Then X is the disjoint union of a ν-positive set and a ν-negative set.*

Proof. If ν assumes only non-negative or only non-positive values, the result is trivial. Hence, we may assume without loss of generality that ν assumes both positive and negative values. Therefore, by the lemma above, X has at least one strongly ν-negative subset. Let

$$\beta = \inf(\nu(A) : A \in \Gamma, A \text{ strongly } \nu\text{-negative}),$$

and let $(A_n : n = 1, 2, \ldots)$ be a sequence of strongly ν-negative sets satisfying $\nu(A_n) \to \beta$. Then, if $B_n = \cup_1^n A_k$ for every n, the sequence $(B_n : n = 1, 2, \ldots)$ is an increasing sequence of strongly ν-negative sets, so (in view of our earlier observations) the set $S_2 = \cup A_n = \cup B_n$ is strongly ν-negative and $\nu(B_n)$ converges decreasingly to $\nu(S_2)$. Since also $\beta \leq \nu(B_n) \leq \nu(A_n)$ for every n (the first inequality because of the definition of β), we must have $\nu(B_n) \downarrow \beta$, so $\nu(S_2) = \beta$(and hence β is a finite number). Finally, we prove that $S_1 = X\backslash S_2$ is ν-positive, and $X = S_1 \cup S_2$ is then the desired decomposition. Assuming that S_1 is not ν-positive, there exists a subset C of S_1 such that $\nu(C) < 0$. Then C has a strongly ν-negative subset D, so $S_2 \cup D$ is then strongly ν-negative and

$$\nu(S_2 \cup D) = \nu(S_2) + \nu(D) < \nu(S_2) = \beta.$$

This contradicts the definition of β. Therefore, S_1 is ν-positive. ∎

The Hahn decomposition of the signed measure ν is unique except for ν-null sets, where the set $A \in \Gamma$ is defined to be a *ν-null set* if not only $\nu(A) = 0$ but also $\nu(B) = 0$ for any $B \in \Gamma, B \subseteq A$. To prove the uniqueness of the Hahn decomposition, let $X = S_1 \cup S_2 = \tilde{S_1} \cup \tilde{S_2}$ be Hahn decompositions cf X with S_1 and $\tilde{S_1}$ the ν-positive sets. The symmetric differences $(S_1 \backslash \tilde{S_1}) \cup (\tilde{S_1} \backslash S_1)$ and $(S_2 \backslash \tilde{S_2}) \cup (\tilde{S_2} \backslash S_2)$ are both equal to $(S_1 \cap \tilde{S_2}) \cup (\tilde{S_1} \cap S_2)$. Any set $A \in \Gamma$ contained in $S_1 \cap \tilde{S_2}$ satisfies $\nu(A) \geq 0$(since $A \subseteq S_1$) as well as $\nu(A) \leq 0$(since $A \subseteq \tilde{S_2}$), so $\nu(A) = 0$. Similarly if $A \subseteq \tilde{S_1} \cap S_2$.

Any σ-additive signed measure is of course also a finitely additive signed measure. Any σ-additive signed measure assuming only non-negative values on the σ-algebra Γ is an (ordinary) finite measure on Γ such as for example the Lebesgue measure in a bounded interval of $\mathbb{R}$ or, more generally, a more abstract finite measure as in Example 9.5. It is not difficult to see, once we have the Hahn decomposition available, that every σ-additive signed measure

ν on Γ can be written as $\nu = \mu_1 - \mu_2$, where μ_1 and μ_2 are non-negative finite measures on Γ. The proof is simple. If $X = S_1 \cup S_2$ is a Hahn decomposition of X with respect to ν, where S_1 is ν-positive and S_2 is ν-negative, we set

$$\mu_1(A) = \nu(A \cap S_1) \text{ and } \mu_2(A) = -\nu(A \cap S_2)$$

for every $A \in \Gamma$. It is evident that μ_1 and μ_2 are σ-additive and assume only non-negative values. Also

$$\nu(A) = \nu\{(A \cap S_1) \cup (A \cap S_2)\} =$$

$$\nu(A \cap S_1) + \nu(A \cap S_2) = \mu_1(A) - \mu_2(A)$$

for every $A \in \Gamma$. This decomposition $\nu = \mu_1 - \mu_2$ is called the *Jordan Decomposition* of ν.

We proceed with another simple result. From the formula

$$\nu(A) = \nu(A \cap S_1) + \nu(A \cap S_2),$$

where $\nu(A \cap S_1)$ assumes only non-negative values and vanishes for $A \subseteq S_2$, whereas $\nu(A \cap S_2)$ assumes only non-positive values and vanishes for $A \subseteq S_1$, it follows that $\nu(A)$ attains its maximum for $A = S_1$ and its minimum for $A = S_2$. Hence

$$\sup(|\,\nu(A)\,| : A \in \Gamma) = \max\{\nu(S_1), -\nu(S_2)\}.$$

The supremum on the left, for a finitely additive signed measure ν, was denoted by $\|\,\nu\,\|$ in Example 4.2(7), and any ν for which $\|\,\nu\,\|$ is finite was said to be bounded. We see now, therefore, that every σ-additive ν on Γ is bounded in this sense. As shown in Example 4.2(7), the set E_b of all bounded finitely additive signed measures on the σ-algebra Γ is a Riesz space with respect to the natural definitions of addition, multiplication by real numbers and order ($\nu_1 \leq \nu_2$ if $\nu_1(A) \leq \nu_2(A)$ for every $A \in \Gamma$). It is evident now that the set M_σ of all σ-additive signed measures on Γ is a linear subspace of E_b. Actually, M_σ is a Riesz subspace of E_b. To prove this, it is sufficient to show that for every $\nu \in M_\sigma$ the supremum $\nu \vee 0$ with respect to the mentioned ordering is also a member of M_σ. We assert that $\nu \vee 0$ (where 0 denotes the measure identically zero) is exactly the measure μ_1 in the Jordan decomposition $\nu = \mu_1 - \mu_2$. The measure μ_1 is evidently an upper bound of ν and 0, so it remains to prove that any upper bound $\nu^{\sim}$ of ν and 0 satisfies $\nu^{\sim} \geq \mu_1$. For any $A \subseteq S_1$ we have $\nu^{\sim}(A) \geq \nu(A) = \mu_1(A)$ and for any $A \subseteq S_2$ we have $\nu^{\sim}(A) \geq 0 = \mu_1(A)$. Hence, if $A \in \Gamma$ is arbitrary, then

$$\nu^{\sim}(A) = \nu^{\sim}(A \cap S_1) + \nu^{\sim}(A \cap S_2) \geq \mu_1(A \cap S_1) + \mu_1(A \cap S_2) = \mu_1(A).$$

This shows that $\mu_1 = \nu \vee 0$, and hence $\mu_2 = (-\nu) \vee 0 = -(\nu \wedge 0)$. From here on we shall write ν^+ and ν^- instead of μ_1 and μ_2, so the Jordan decomposition of ν can be written as $\nu = \nu^+ - \nu^-$.

The space M_σ is not only a Riesz subspace of E_b, but it is an ideal in E_b. This follows by observing that if $0 \leq \mu_1 \leq \mu_2$ with $\mu_1 \in E_b$ and $\mu_2 \in M_\sigma$, then $\mu_1 \in M_\sigma$. We prove that M_σ is a band in E_b. For this purpose, assume that D is an upwards directed set in M_σ^+ such that $D \uparrow \mu_0$, where $\mu_0 \in E_b$. Furthermore, assume that $(A_n : n = 1, 2, \ldots)$ is an increasing sequence of sets in Γ such that $\cup_1^\infty A_n = A_0$ in Γ. To show that μ_0 is σ-additive it is sufficient to prove that $\mu_0(A_n) \uparrow \mu_0(A_0)$. Given $\epsilon > 0$, let $\mu_1 \in D$ be such that

$$0 \leq (\mu_0 - \mu_1)(A_0 - A_1) < \epsilon.$$

Then $0 \leq (\mu_0 - \mu_1)(A_0 - A_n) < \epsilon$ holds for $n = 1, 2, \ldots$, and also $\mu_1(A_0 - A_n) \to 0$ as $n \to \infty$. It follows that $0 \leq \mu_0(A_0 - A_n) < 2\epsilon$ for n sufficiently large. This is the desired result. Hence, M_σ is a band in E_b. Recall now that the Riesz space E_b is Dedekind complete. The easy proof was given in Example 12.5(iv). Therefore, E_b has the projection property, and so $E_b = M_\sigma \oplus (M_\sigma)^d$. The members of $(M_\sigma)^d$ are sometimes called *purely finitely additive signed measures*. Combining the results obtained so far, the following theorem has been proved.

Theorem 27.3. *The space M_σ of all σ-additive signed measures on the σ-algebra Γ of subsets of X is a band in the Riesz space E_b of all bounded finitely additive signed measures on Γ. Every bounded finitely additive measure ν can be uniquely written, therefore, as $\nu = \nu_c + \nu_f$, where ν_c is σ-additive and ν_f is purely finitely additive. Furthermore, given the σ-additive ν, there exists a decomposition $X = S_1 \cup S_2 (S_1$ and S_2 disjoint) such that $\nu^-(A) = 0$ for every $A \subseteq S_1$ and $\nu^+(A) = 0$ for every $A \subseteq S_2$. This is called the Hahn decomposition of X with respect to ν and the decomposition $\nu = \nu^+ - \nu^-$ of the σ-additive ν is called the Jordan decomposition of ν.*

The space M_σ of σ-additive measures has some additional properties. To describe one of these we first recall a definition. If μ and ν are members of M_σ, then ν is said to be μ-*absolutely continuous* if any $A \in \Gamma$ satisfying $| \mu | (A) = 0$ also satisfies $| \nu | (A) = 0$.

Theorem 27.4. *(i) If μ and ν are members of M_σ, then $\mu \perp \nu$ (i.e., $| \mu | \wedge | \nu | = 0$) if and only if there exist disjoint sets A and B in Γ such that $A \cup B = X$ and $| \nu | (A) = | \mu | (B) = 0$.*

(ii) If μ and ν are members of M_σ, then ν is an element of the band generated by μ in M_σ if and only if ν is μ-absolutely continuous.

Proof. (i) Let $\mu \perp \nu$. Then $\lambda = | \mu | - | \nu | \in M_\sigma$ is the difference of two disjoint positive elements, so the decomposition $\lambda = | \mu | - | \nu |$ is minimal in the sense of Theorem 5.6, which implies that $\lambda^+ = | \mu |$ and $\lambda^- = | \nu |$. Then, in view of the Hahn decomposition theorem for λ, there exist disjoint sets A and B in Γ such that $A \cup B = X$ and $| \nu | (A) = | \mu | (B) = 0$.

Conversely, if there exist disjoint sets A and B in Γ such that $A \cup B = X$ and $\mid \nu \mid (A) = \mid \mu \mid (B) = 0$, then $\tau = \mid \mu \mid \wedge \mid \nu \mid$ satisfies

$$\tau(C) = \tau(C \cap B) + \tau(C \cap A) \leq \mid \mu \mid (B) + \mid \nu \mid (A) = 0$$

for every $C \in \Gamma$, so $\tau = 0$, i.e., $\mu \perp \nu$.

(ii) It is easy to see that if ν is an element in the band generated by μ, then $\mid \mu \mid (A) = 0$ implies $\mid \nu \mid (A) = 0$. Conversely, assume that $\mid \mu \mid (A) = 0$ implies $\mid \nu \mid (A) = 0$ and let $\mid \nu \mid = \nu_1 + \nu_2$ with ν_1 in the band B_μ generated by μ and $\nu_2 \in (B_\mu)^d$. Since $\mid \mu \mid \perp \nu_2$, there exist disjoint sets A and B in Γ such that $A \cup B = X$ and $\nu_2(A) = \mid \mu \mid (B) = 0$. Then $\mid \nu \mid (B) = 0$ by hypothesis, so $\nu_2(B) = 0$ on account of $0 \leq \nu_2 \leq \mid \nu \mid$. It follows that

$$\nu_2(X) = \nu_2(A \cup B) = \nu_2(A) + \nu_2(B) = 0,$$

so ν_2 is the zero measure. This implies that $\mid \nu \mid = \nu_1 + \nu_2 = \nu_1 \in B_\mu$.   ∎

Exercise 27.5. In Example 4.2(7) it was shown that if for any $\nu \in E_b$ the number $\| \nu \|$ is defined by

$$\| \nu \| = \sup(\mid \nu(A) \mid : A \in \Gamma),$$

then $\| \cdot \|$ is a norm in E_b. It is obvious that this is not a Riesz norm, since in general ν and $\mid \nu \mid$ do not have the same norm. If we modify the definition of $\| \nu \|$ to

$$\| \nu \| = \sup(\mid \nu \mid (A) : A \in \Gamma) \text{ for } \nu \in E_b,$$

then $\| \nu \|$ is a Riesz norm in E_b. It is evident that $\| \nu \| = \mid \nu \mid (X)$. Show that if E_b is equipped with this norm, then E_b has the Riesz-Fischer property, so E_b is a Banach lattice. Therefore, every band in E_b is a Banach lattice.

28. The Radon-Nikodym Theorem

Definitions and notations are the same as in the preceding section. Hence, as before, M_σ is the Riesz space of all σ-additive signed measures on the σ-algebra Γ of subsets of X and, for any given $\mu \in M_\sigma$, the band generated by μ in M_σ is denoted by B_μ. Assume now that $\mu \in M_\sigma, \mu \geq 0$, is fixed. There is a relation between B_μ and the space $L_1 = L_1(X, \Gamma, \mu)$ of all μ-summable (also called μ-integrable) real functions on X. It is evident that if $f \in L_1$ is given and, for any $A \in \Gamma$, the number $\nu(A)$ is defined by $\nu(A) = \int_A f d\mu$, then ν is a σ-additive signed measure on Γ. Also, if $f \geq 0$ in L_1, then ν assumes only non- negative values and $\mu(A) = 0$ implies $\nu(A) = 0$, and so ν is a member of the band B_μ. If $f \in L_1$ is arbitrary, then $f = f^+ - f^-$ with $f^+ = f \vee 0$ and $f^- = (-f) \vee 0$, which implies that

$$\nu(A) = \int_A f d\mu = \int_A f^+ d\mu - \int_A f^- d\mu = \nu_1(A) - \nu_2(A),$$

where ν_1 and ν_2 are non-negative members of B_μ, so ν itself is a member of B_μ as well. Furthermore, it is not difficult to see that $\nu_1 \wedge \nu_2 = 0$ holds, so $\nu = \nu_1 - \nu_2$ is a minimal decomposition of ν as a difference of positive elements (in the sense of Theorem 5.6), i.e., $\nu_1 = \nu^+$ and $\nu_2 = \nu^-$. Therefore, if f corresponds to ν, then f^+ and f^- correspond to ν^+ and ν^- respectively, and so $|f|$ corresponds to $|\nu|$.

We ask now whether, conversely, there corresponds to every $\nu \in B_\mu$ a function $f \in L_1(X, \Gamma, \mu)$. This is indeed the case, as shown by the following theorem, known as the *Radon-Nikodym theorem*. The theorem is due to J. Radon (1913) for the case that μ is Lebesgue measure and to O. Nikodym (1930) for the more general case.

Theorem 28.1. *(Radon-Nikodym theorem). If ν is a σ-additive measure on Γ which is μ-absolutely continuous (i.e., $\nu \in B_\mu$), then there exists a function $f \in L_1(X, \Gamma, \mu)$ such that $\nu(A) = \int_A f d\mu$ holds for every $A \in \Gamma$.*

Proof. Without loss of generality we may assume that μ is not the zero measure and $\nu \geq 0$. Let G be the set of all μ-summable (real) functions on X such that $\int_A f d\mu \leq \nu(A)$ holds for all $A \in \Gamma$ and let

$$\alpha = \sup\left(\int_X f d\mu : f \in G\right),$$

so $\alpha \leq \nu(X)$. Any $f \in G$ satisfying $\int_X f d\mu = \alpha$ will be called extremal in G. The set G is not empty since the zero function is in G. We prove that if $f_1, \ldots, f_n$ are in G, then $f_1 \vee \cdots \vee f_n$ is in G. It is sufficient to prove this for $n = 2$, so let $f = f_1 \vee f_2$ with f_1 and f_2 in G. Then the sets

$$A_1 = \{x \in X : f_1(x) \geq f_2(x)\} \text{ and } A_2 = X \backslash A_1$$

are members of Γ, so if $A \in \Gamma$ is arbitrary, we have

$$\int_A f d\mu = \int_{A \cap A_1} f_1 d\mu + \int_{A \cap A_2} f_2 d\mu \leq \nu(A \cap A_1) + \nu(A \cap A_2) = \nu(A).$$

It follows that $f \in G$. Furthermore, if $(f_n : n = 1, 2, \ldots)$ is a pointwise increasing sequence of functions in G converging to the function f, then

$$\int_A f d\mu = \lim_{n \to \infty} \int_A f_n d\mu \leq \nu(A)$$

for every $A \in \Gamma$, so $f \in G$.

We prove that there exists an extremal function in G. For the proof, let $(f_n : n = 1, 2, \ldots)$ be a sequence of functions in G such that $\int_X f_n d\mu$ converges to α. It may be assumed that the sequence is increasing pointwise (if necessary replace f_n by $f_1 \vee \cdots \vee f_n$ for every n), so $\int_X f_n d\mu \uparrow \alpha$. It follows that the function $f = \lim_{n \to \infty} f_n$ satisfies $f \in G$ and $\int_X f d\mu = \alpha$, so f is extremal in G.

In the next step it will be shown that the extremal function f satisfies $\int_A f d\mu = \nu(A)$ for every $A \in \Gamma$. In other words, the extremal function f satisfies the conditions as stated in the theorem. For the proof, we set

$$\tau(A) = \nu(A) - \int_A f d\mu \text{ for every } A \in \Gamma.$$

Evidently, τ is a non-negative σ-additive measure on Γ, satisfying $0 \leq \tau \leq \nu$. What we have to prove now is that τ is the zero measure. Assume that τ is not the zero measure. Then $\tau(X) > 0$, so the quotient $\tau(X)/\mu(X)$ is positive, which implies that $\tau(X) > \beta\mu(X)$ for some number $\beta > 0$. Let the signed measure τ_0 be defined by $\tau_0 = \tau - \beta\mu$. Note that $\tau_0(X) > 0$. Now, let $X = X_1 \cup X_2$ be a Hahn decomposition of X with respect to τ_0, where X_1 is τ_0-positive and X_2 is τ_0-negative. Then $\tau_0(X_1) > 0$, because $\tau_0(X_1) = 0$ would imply

$$\tau_0(X) = \tau_0(X_1) + \tau_0(X_2) = \tau_0(X_2) \leq 0,$$

which contradicts $\tau_0(X) > 0$. It follows that $\mu(X_1) > 0$, because $\mu(X_1) = 0$ would imply $\nu(X_1) = 0$(since ν is μ-absolutely continuous), so $\tau(X_1) = \nu(X_1) - \int_{X_1} f d\mu = 0$, and then $\tau_0(X_1) = \tau(X_1) - \beta\mu(X_1) = 0$, which contradicts $\tau_0(X_1) > 0$. Now let us consider the function $f + \beta\chi_{X_1}$, where χ_{X_1} is the characteristic function of X_1. For any $A \in \Gamma$ we have

$$\int_A (f + \beta\chi_{X_1})d\mu = \int_{A \cap X_1} (f + \beta)d\mu + \int_{A \cap X_2} f d\mu =$$

$$\nu(A \cap X_1) - \tau(A \cap X_1) + \beta\mu(A \cap X_1) + \int_{A \cap X_2} f d\mu =$$

$$\nu(A \cap X_1) - \tau_0(A \cap X_1) + \int_{A \cap X_2} f d\mu.$$

Observe now that $\tau_0(A \cap X_1) \geq 0$ since X_1 is τ_0-positive and observe also that $\int_{A \cap X_2} f d\mu \leq \nu(A \cap X_2)$. Hence

$$\int_A (f + \beta\chi_{X_1})d\mu \leq \nu(A \cap X_1) + \nu(A \cap X_2) = \nu(A).$$

Since this holds for every $A \in \Gamma$, it follows that $f + \beta\chi_{X_1} \in G$. But

$$\int_X (f + \beta\chi_{X_1})d\mu = \alpha + \beta\mu(X_1) > \alpha,$$

which contradicts the definition of α. It follows that τ is the zero measure. For a remark about uniqueness of f we refer to Exercise 28.4 at the end of the section.   ∎

A second (and more direct) proof of the Radon-Nikodym theorem, based on Freudenthal's spectral theorem, will be presented in the final paragraph of section 34.

Once more, let M_σ be the Riesz space of all σ-additive signed measures on the σ-algebra Γ of subsets of X, and for any $\mu \in M_\sigma$, let B_μ be the band generated by μ in M_σ. For $\mu \geq 0$ fixed in M_σ, the band B_μ is itself a Riesz space, Dedekind complete and equipped with the Riesz norm $\| \nu \| = | \nu | (X)$ for $\nu \in B_\mu$. As indicated in Exercise 27.5, B_μ is a Banach lattice with respect to this norm. Furthermore, as observed already, B_μ is related to the space $L_1 = L_1(X, \Gamma, \mu)$ of all μ-summable (real) functions on X. If to any $f \in L_1$ we assign the signed measure $\nu \in B_\mu$ defined by $\nu(A) = \int_A f d\mu$ for every $A \in \Gamma$, then this mapping of L_1 into B_μ is linear and such that $| f |$ has $| \nu |$ as its image, so the mapping is a Riesz homomorphism. It follows from the Radon-Nikodym theorem that the mapping is not only into B_μ but it is onto B_μ, i.e., for every $\nu \in B_\mu$ there exists a (μ-almost unique) function $f \in L_1$ such that $\nu(A) = \int_A f d\mu$ holds for every $A \in \Gamma$. Hence, B_μ and $L_1 = L_1(X, \Gamma, \mu)$ are Riesz isomorphic. These spaces are even norm isomorphic. If f and ν correspond, i.e., if $\nu(A) = \int_A f d\mu$ for every $A \in \Gamma$, then

$$\| \nu \| = | \nu | (X) = \int_X | f | \, d\mu = \| f \| .$$

The so obtained result is summed up in the following theorem.

Theorem 28.2. *Let M_σ be the Banach lattice of all σ-additive signed measures on the σ-algebra Γ of subsets of X and, for $0 \leq \mu \in M_\sigma$ fixed, let B_μ be the band generated by μ in M_σ. Then B_μ is Riesz isomorphic to the Banach lattice $L_1(X, \Gamma, \mu)$ of all (real) μ-summable functions on X. The isomorphism is defined by assigning to any $f \in L_1$ the element $\nu \in B_\mu$ such that $\nu(A) = \int_A f d\mu$ holds for every $A \in \Gamma$. If f and ν correspond, then their norms are equal as well, i.e., $\| f \| = \int_X | f | \, d\mu$ equals $\| \nu \| = | \nu | (X)$.*

The formula $\nu(A) = \int_A f d\mu$ in the Radon-Nikodym theorem may be written as $\int_X g d\nu = \int_X g f d\mu$ for the case that g is the characteristic function of some set $A \in \Gamma$. The thus rewritten formula is valid for every ν-summable g, as shown in the following theorem.

Theorem 28.3. *Once more, let $0 \leq \mu \in M_\sigma$ be fixed and let $0 \leq \nu \in B_\mu$. In view of the Radon-Nikodym theorem there exists a μ-summable $f \geq 0$ such that $\nu(A) = \int_A f d\mu$ for every $A \in \Gamma$. Then, for any ν-summable function g on X, we have*

$$\int_A g d\nu = \int_A g f d\mu \quad \text{for every} \ A \in \Gamma.$$

Proof. It is sufficient to present the proof for $A = X$. For reasons of simplicity we shall write the formula to be proved as $\int g d\nu = \int g f d\mu$. Assume first that $g = \chi_B$ for some $B \in \Gamma$. Then

$$\int g d\nu = \int \chi_B d\nu = \nu(B) = \int_B f d\mu = \int \chi_B f d\mu = \int g f d\mu.$$

It follows that the desired result holds if g is a step function, i.e., g is a finite linear combination of characteristic functions of sets in Γ. If g is non-negative and ν-summable, there exists a sequence $(s_n : n = 1, 2, \ldots)$ of step functions such that $0 \le s_n \uparrow g$ pointwise, so

$$\int g d\nu = \lim \int s_n d\nu = \lim \int s_n f d\mu = \int g f d\mu.$$

The result for g real and ν-summable follows trivially. ∎

Exercise 28.4. Show that the function f in the Radon-Nikodym theorem is μ-almost uniquely determined.

Hint: It has to be shown that if $0 \le g \in L_1$ and $\int_A g d\mu = 0$ for all $A \in \Gamma$, then $g = 0(\mu$-almost everywhere). If not, then $g(x) > 0$ for all x in some set A_0 of positive measure. Show that there exists a number $\epsilon > 0$ such that $g(x) \ge \epsilon$ for all x in some subset B of A_0 such that B has positive measure. Then $\epsilon \chi_B \le g$, so $0 < \epsilon \mu(B) \le \int_B g d\mu = 0$. Contradiction, so $g = 0$. In fact, we deal here with the result that if the L_1-norm of g is zero, then $g = 0$.

CHAPTER 15
Linear Functionals on Spaces of Measurable Functions

29. Linear Functionals on Spaces of Measurable Functions

As in the Examples 9.5 and 12.5(iii) we assume that μ is a σ-finite (non-negative and σ-additive) measure in the non-empty point set X and $L_0 = L_0(X, \mu)$ is the Dedekind complete Riesz space of all μ-measurable (real) functions on X. To be precise, μ is defined on a σ-algebra Γ the members of which are called the μ-measurable subsets of X and the real function f on X is called μ-measurable whenever the set $(x \in X : f(x) > \alpha)$ is μ-measurable for every real α. It follows then easily that the sets of points x where $f(x) \geq \alpha, f(x) < \alpha, f(x) \leq \alpha$ or $\beta \leq f(x) \leq \alpha$ (for α and β real) respectively are also μ-measurable. Recall that functions differing only on a set of measure zero are identified, so that the members of L_0 are in fact equivalence classes of measurable functions. Similarly for the members of Γ (as explained already in Example 3.4(ii)).

Let A be an ideal in L_0. In Example 9.5 we have introduced the notion of the carrier of A. This is a measurable subset $c(A)$ of X having the property that on its complement $n(A) = X \backslash c(A)$ every $f \in A$ vanishes (μ-almost everywhere), whereas $c(A)$ itself does not have any subset of positive measure on which all $f \in A$ vanish (μ-almost everywhere). It may be asked whether the characteristic function of $c(A)$ is a member of A. This is not always so, as the following example shows. Let μ be Lebesgue measure in $[0, 1]$ and let A be the ideal of all measurable functions f on $[0, 1]$ having the property that f vanishes on some interval $[0, \alpha_f]$, where $0 < \alpha_f \leq 1$ and α_f may vary with f. Then $c(A) = [0, 1]$ and the characteristic function of $[0, 1]$ is not a member of A. Generally, for Y a measurable subset of $c(A)$, we cannot be sure that the characteristic function χ_Y is a member of A. We shall prove now, however, that for any Y of this kind there exists a sequence $Y_n \uparrow Y$ such that $\chi_{Y_n} \in A$ for all $n = 1, 2, \ldots$.

Theorem 29.1. *If A is an ideal in $L_0 = L_0(X, \mu)$ and Y is a measurable subset of the carrier $c(A)$ of A, then there exists an increasing sequence $(Y_n : n = 1, 2, \ldots)$ of measurable sets of finite measure such that $Y_n \uparrow Y$ and $\chi_{Y_n} \in A$ for all n.*

Proof. We show first that if P is a measurable subset of $c(A)$ such that $\mu(P) > 0$, then P has a subset Q of positive measure such that $\chi_Q \in A$. To prove this, observe that not all $f \in A$ vanish (μ-almost everywhere) on P, so there exists a function $f \in A$ such that $f(x) \neq 0$ for all x in a subset P_0 of P with $\mu(P_0) > 0$. We may assume that $f(x) > 0$ for all $x \in P_0$ (if necessary, replace f by $|\,f\,|$). Now, let $\epsilon_n \downarrow 0$ (every ϵ_n a positive number) and $P_n = (x \in P_0 : f(x) \geq \epsilon_n)$. Then $P_n \uparrow P_0$, so $\mu(P_n) \uparrow \mu(P_0) > 0$, which implies that $\mu(P_n) > 0$ from some $n = n_0$ onwards. Set $Q = P_{n_0}$. It follows from $\epsilon_{n_0}\chi_Q \leq f \in A$ that $\epsilon_{n_0}\chi_Q \in A$, and therefore $\chi_Q \in A$.

For the proof itself, assume first that $\mu(Y)$ is finite. Now, let

$$\alpha = \sup(\mu(Z) : Z \subseteq Y, \chi_Z \in A).$$

There exists a sequence $(Z_n : n = 1, 2, \ldots)$ of subsets of Y such that $\chi_{Z_n} \in A$ and $\mu(Z_n) \to \alpha$, where it may be assumed that the sequence is increasing (if necessary, replace Z_n by $Z_1 \cup \ldots \cup Z_n$). Then $Z = \lim Z_n = \cup_1^\infty Z_n$ satisfies $\mu(Z) = \alpha$. We prove that Z is almost equal to Y. If not, the set $Y \backslash Z$ is a measurable subset of $c(A)$ having positive measure, so (in view of the remark at the beginning) $Y \backslash Z$ has a subset Q of positive measure β such that $\chi_Q \in A$. Then the characteristic function of $Q \cup Z_n$ is a member of A and

$$\mu(Q \cup Z_n) = \beta + \mu(Z_n) \uparrow \beta + \alpha \text{ as } n \to \infty,$$

so $\mu(Q \cup Z_n) > \alpha$ for n sufficiently large. This contradicts the definition of α. Hence, Z is almost equal to Y, and so $Z_n \uparrow Y$ with $\chi_{Z_n} \in A$ for all n. This may also be expressed by saying that for any $\epsilon > 0$ there exists a set $R \subseteq Y$ such that $\chi_R \in A$ and $\mu(Y \backslash R) < \epsilon$.

If $\mu(Y) = \infty$, we can write $Y = \cup_1^\infty \tilde{Y_n}$ with $\mu(\tilde{Y_n})$ finite for all n (since the measure μ is σ-finite), and we may assume that $\tilde{Y_n}$ is increasing as n increases, so $\tilde{Y_n} \uparrow Y$. As proved above, each $\tilde{Y_n}$ has a subset Y_n such that $\chi_{Y_n} \in A$ and $\mu(\tilde{Y_n} \backslash Y_n) < n^{-1}$. Once more, it may be assumed that Y_n increases with n, so $Y_\infty = \lim Y_n = \cup_1^\infty Y_n$ exists and we have $\tilde{Y_n} \backslash Y_\infty \subseteq \tilde{Y_n} \backslash Y_n$, which implies that $\mu(\tilde{Y_n} \backslash Y_\infty) < n^{-1}$ for $n = 1, 2, \ldots$. But

$$\tilde{Y_n} \backslash Y_\infty \uparrow Y \backslash Y_\infty, \text{ so } \mu(Y \backslash Y_\infty) = \lim\{\mu(\tilde{Y_n} \backslash Y_\infty)\} = 0.$$

This shows that $Y_n \uparrow Y$ (except for a set of measure zero). Therefore, the sequence $(Y_n : n = 1, 2, \ldots)$ has all the desired properties.   ∎

The theorem we have thus proved would not be of much interest were it not that by means of this theorem we can obtain precise information about the space $\tilde{A_n}$ of all order continuous linear functionals on a given ideal A in $L_0 = L_0(X, \mu)$. Recall that L_0 is super Dedekind complete (see the remark following after Theorem 17.6), so in particular L_0 is order separable. It follows that every ideal A in L_0 is order separable, that is to say, every downwards directed set D in A satisfying $D \downarrow 0$ contains a sequence $(f_n : n = 1, 2, \ldots)$ satisfying

$f_n \downarrow 0$. As an immediate consequence, we see that an order bounded linear functional on A is order continuous if and only if it is σ-order continuous.

Theorem 29.2. *If A is an ideal in $L_0 = L_0(X,\mu)$ and the order bounded linear functional φ on A is order continuous, i.e., if $\varphi \in A_n^{\sim}$, then there exists a (real) μ-measurable function t on X such that*

(i)   for every $f \in A$ the product tf is μ-summable over X,
(ii) $\varphi(f) = \int_X tfd\mu$ holds for every $f \in A$.

The function t thus corresponding to φ is uniquely determined μ-almost everywhere on the carrier $c(A)$ of A. If φ is positive, then $t(x) \geq 0$ for μ-almost every $x \in c(A)$.

Conversely, if t is a (real)μ-measurable function on X possessing the property that tf is μ-summable over X for every $f \in A$, then the linear functional φ on A, defined by

$$\varphi(f) = \int_X tfd\mu \text{ for every } f \in A, \tag{1}$$

is order bounded and order continuous. Trivially, φ is uniquely determined by t. If $t(x) \geq 0$ for μ-almost every $x \in c(A)$, then φ is positive.

Proof. Let us assume first that $\varphi \in A_n^{\sim}$. Then φ^+ and φ^- are positive members of $A_n^{\sim}$, so it is sufficient to show that each of φ^+ and φ^- has the desired properties. In other words, we may assume that $0 \leq \varphi \in A_n^{\sim}$ holds and we have to prove the existence of a μ-measurable function t, non-negative on $c(A)$ and such that (i) and (ii) hold. In view of the last theorem there exists a sequence $(Y_n : n = 1, 2, \ldots)$ of measurable subsets of $c(A)$, each Y_n of finite measure, such that $Y_n \uparrow c(A)$ and $\chi_{Y_n} \in A$ for $n = 1, 2, \ldots$. Therefore, if Z is a μ-measurable subset of some Y_n, we have $\chi_Z \in A$ and so $\varphi(\chi_Z)$ is well defined. We shall denote $\varphi(\chi_Z)$ briefly by $\nu(Z)$ and we first restrict ourselves to the subsets of one fixed Y_n, say for $n = n_0$. Given such a subset $Z \subseteq Y_{n_0}$ and given that $Z_k \uparrow Z$ as $k \to \infty$, we have $\varphi(\chi_{Z_k}) \uparrow \varphi(\chi_Z)$ since φ is positive and order continuous. In other words, we have $\nu(Z_k) \uparrow \nu(Z)$, which shows that ν is a finite, non-negative, σ-additive measure on the σ-algebra of μ-measurable subsets of Y_{n_0}. Furthermore, the measure ν is μ-absolutely continuous, i.e., $\mu(Z) = 0$ implies $\nu(Z) = 0$. Hence, in virtue of the Radon-Nikodym theorem, there exists a non-negative μ-measurable function t on Y_{n_0} such that $\nu(Z) = \int_Z td\mu$ holds for every μ-measurable subset Z of Y_{n_0}. The function t is μ-almost uniquely determined on Y_{n_0} (see Exercise 28.4). By varying n_0, the function t can now be extended μ-almost uniquely to $\cup_1^{\infty} Y_n = c(A)$ such that $\nu(Z) = \int_Z td\mu$ holds for every Z satisfying $Z \subseteq Y_n$ for some n. Setting $t(x) = 0$ for all $x \in X \backslash c(A)$, we obtain

$$\varphi(\chi_Z) = \nu(Z) = \int_X t\chi_Z d\mu$$

for every Z of this particular type. Hence, if s is now a step function $s = \Sigma_{k=1}^{p} a_k \chi_{Z_k}$ with all Z_k of this type, it is evident that $\varphi(s) = \int_X t s d\mu$. Finally, let $0 \le f \in A$ be given. It is well-known (and not difficult to prove) that there exists a sequence $(s_{\tilde{n}} : n = 1, 2, \ldots)$ of μ-measurable step functions such that $0 \le s_{\tilde{n}} \uparrow f$. Then, for each n, $s_n = s_{\tilde{n}} \chi_{Y_n}$ is a step function of the particular type we need and $0 \le s_n \uparrow f$ still holds. Hence $\varphi(s_n) = \int_X t s_n d\mu$ for all n. Furthermore $\varphi(s_n) \uparrow \varphi(f)$ since φ is positive and order continuous, and

$$\int_X t s_n d\mu \uparrow \int_X t f d\mu$$

by the well-known theorem on integration of monotone sequences. It follows, as desired, that $\varphi(f) = \int_X t f d\mu$.

Finally, for any arbitrary $f \in A$, we have

$$\varphi(f) = \varphi(f^+) - \varphi(f^-) = \int_X t f^+ d\mu - \int_X t f^- d\mu = \int_X t f d\mu.$$

Conversely, assume now that the real μ-measurable function t on X has the property that tf is μ-summable over X for every $f \in A$. Then $|\,tf\,|$ is μ-summable and so $t^+ f$ and $t^- f$ are μ-summable. It is evident that φ, as defined in (1), is a linear functional on A and

$$\varphi(f) = \int_X t f d\mu = \int_X t^+ f d\mu - \int_X t^- f d\mu = \varphi_1(f) - \varphi_2(f)$$

for every $f \in A$, so φ is the difference of the positive linear functionals φ_1 and φ_2. This shows already that φ is order bounded. Furthermore, $f_n \downarrow 0$ in A implies $\varphi_1(f_n) = \int_X t^+ f_n d\mu \downarrow 0$, so φ_1 is order continuous. Similarly, φ_2 is order continuous. It follows that φ is order continuous.   ∎

Once more, let A be an ideal in $L_0 = L_0(X, \mu)$ having carrier $c(A)$. Since the set $X \backslash c(A)$ is of no importance for what follows, we may just as well assume that $c(A) = X$. As already shown, there corresponds to any $\varphi \in A_{\tilde{n}}$ a function $t \in L_0$ such that $\varphi(f) = \int t f d\mu$ for every $f \in A$. Conversely, if the function $t \in L_0$ has the property that tf is μ-summable over X for every $f \in A$, then the linear functional φ on A, defined by $\varphi(f) = \int t f d\mu$, satisfies $\varphi \in A_{\tilde{n}}$. The functional $\varphi \in A_{\tilde{n}}$ and the corresponding $t \in L_0$ determine each other uniquely (μ-almost everywhere as far as it concerns t). The set of all $t \in L_0$ thus corresponding to elements $\varphi \in A_{\tilde{n}}$ is obviously a linear subspace of L_0 which we shall denote by K_A. We prove that K_A is an ideal in L_0. If $t \in K_A$, then tf is summable (over X) for every $f \in A$, i.e., $|\,tf\,|$ is summable for every $f \in A$, and so $|\,t\,|\,f$ is summable for every $f \in A$. This shows that $|\,t\,| \in K_A$. Furthermore, it is evident that if $0 \le t_2 \le t_1$ and $t_1 \in K_A$, then $t_2 \in K_A$. In particular, $|\,t\,| \in K_A$ implies that t^+ and t^- are members of K_A, and so $t = t^+ - t^- \in K_A$. To summarize, we have proved that $t \in K_A$ if and only if $|\,t\,| \in K_A$ and $0 \le t_2 \le t_1 \in K_A$ implies that $t_2 \in K_A$. Hence, K_A is an ideal in L_0. The ideal K_A is sometimes called the *Köthe dual space of A* (named after G. Köthe who was one of the first to introduce this notion, first for the case of sequence spaces, i.e., X is then the set of natural numbers with discrete measure, so that L_0 is now the space of all real sequences).

As observed, the functional $\varphi \in A_n^{\sim}$ and the corresponding function $t \in K_A$ determine each other uniquely. Hence, denoting by $\Phi : A_n^{\sim} \to K_A$ the mapping which transforms φ into the corresponding t (i.e., $\Phi(\varphi) = t$), the inverse mapping Φ^{-1} exists (i.e., $\Phi^{-1}(t) = \varphi$). Evidently, both Φ and Φ^{-1} are positive linear operators. It follows then from Theorem 19.3 that Φ is a Riesz isomorphism. Therefore, the spaces $A_n^{\sim}$ and K_A are Riesz isomorphic. We collect these results in the following theorem.

Theorem 29.3. *Let A be an ideal in $L_0(X, \mu)$ having carrier X. The set K_A of all $t \in L_0$ such that tf is μ-summable over X for every $f \in A$ is an ideal in L_0. To any $\varphi \in A_n^{\sim}$ corresponds a (μ-almost) unique $t \in K_A$ such that $\varphi(f) = \int tf d\mu$ holds for every $f \in A$. Conversely, to any $t \in K_A$ corresponds a unique $\varphi \in A_n^{\sim}$ given by the same formula. The mapping $\Phi : A_n^{\sim} \to K_A$, defined by $\Phi(\varphi) = t$ (where t is the function corresponding to φ as mentioned) is a Riesz isomorphism from $A_n^{\sim}$ onto K_A. In particular, therefore, if $\Phi(\varphi) = t$, then $\Phi(|\varphi|) = |t|$.*

According to our assumptions the ideal A in L_0 has carrier $c(A) = X$. The Köthe dual K_A is likewise an ideal in L_0, but K_A does not necessarily have X as carrier. If, for example, $A = L_0(\mathbb{R}, \mu)$ with μ Lebesgue measure, then $A^{\sim} = \{0\}$. It follows then that the Riesz isomorphic image K_A of $A_n^{\sim}$ satisfies $K_A = \{0\}$, and therefore $c(K_A) = \{0\}$. On the other hand, if for example $A = L_\infty(X, \mu)$, then the unit function e, identically one on X, is a member of A, so for any $t \in K_A$ the product $te = t$ is summable over X, i.e., $t \in L_1(X, \mu)$. Conversely, if $t \in L_1$, then tf is μ-summable over X for every $f \in A$, so $t \in K_A$. This shows that $K_A = L_1$, and so $c(K_A) = X$. For μ Lebesgue measure in $\mathbb{R}$, any μ-measurable subset Y of $\mathbb{R}$ is the carrier of the Köthe dual K_A for some ideal A in $L_0(\mathbb{R}, \mu)$. Indeed, let A be the space of all $f \in L_0$ having the property that f is (μ-almost) bounded on Y, i.e., $f\chi_Y \in L_\infty(\mathbb{R}, \mu)$. In the next example we shall discuss the case that $A = L_p = L_p(X, \mu)$ for some p satisfying $1 \le p \le \infty$. For q determined by $p^{-1} + q^{-1} = 1$ (with $q = 1$ or ∞ if $p = \infty$ or $p = 1$ respectively), every $t \in L_q$ has the property that tf is μ-summable over X for every $f \in L_p$ (by Hölder's inequality), which shows that L_q is a subset of the Köthe dual of L_p. Since the carrier of L_q is already the whole set X, it follows that the Köthe dual of L_p has its carrier equal to X. More precisely, we shall prove that the Köthe dual of L_p is exactly L_q.

Example 29.4. As before, we assume that μ is a σ-finite measure in the point set X. Let p be a number satisfying $1 \le p \le \infty$ and define the number

q by $p^{-1} + q^{-1} = 1$. The space $L_p = L_p(X,\mu)$ is an ideal in $L_0(X,\mu)$. If equipped with the usual norm $\| f \|_p = (\int | f |^p \, d\mu)^{1/p}$ for $1 \leq p < \infty$ and

$$\| f \|_\infty = \mathrm{esssup}(| f(x) |: x \in X) \text{ for } p = \infty,$$

the space L_p is a Banach lattice, and so (by Theorem 25.8) the order dual $L_p^\sim$ and the norm dual L_p^* coincide. Recall, furthermore, that if $\varphi \in L_p^\sim = L_p^*$, then φ and $| \varphi |$ have equal norms (by Lemma 25.7). Finally, for $1 \leq p < \infty$ (but not for $p = \infty$, unless in trivial cases) the norm in L_p is order continuous, so every $\varphi \in L_p^\sim$ is now order continuous, i.e., $L_p^\sim = (L_p)_n^\sim$. For convenience of notation we shall denote the Köthe dual of L_p by $K(L_p)$. It will be proved that $t \in K(L_p)$ if and only if $t \in L_q$ and, furthermore, that if φ corresponds to t under the Riesz isomorphism between $(L_p)_n^\sim$ and $K(L_p)$, then $\| \varphi \| = \| t \|_q$. Several cases have to be distinguished.

(i) The case that $L_p = L_\infty$. As shown already above, we have $K(L_\infty) = L_1$, and so $\varphi \in (L_\infty)_n^\sim = (L_\infty)_n^*$ if and only if there exists a function $t \in L_1$ such that $\varphi(f) = \int tf d\mu$ for every $f \in L_\infty$, in which case also $| \varphi | (f) = \int | t | f d\mu$ for every $f \in L_\infty$. It is evident that

$$\| \varphi \| = \|| \varphi ||| = \sup\left(\int | t | f d\mu : \| f \|_\infty \leq 1 \right) \leq \int | t | \, d\mu = \| t \|_1 .$$

In the converse direction we have

$$\int | t | f d\mu = | \varphi | (f) \leq \|| \varphi ||| \cdot \| f \|_\infty \leq \|| \varphi ||| = \| \varphi \|$$

for every f satisfying $\| f \|_\infty \leq 1$. In particular, choosing f to be identically one on X, we see that $\int | t | \, d\mu \leq \| \varphi \|$, i.e., $\| t \|_1 \leq \| \varphi \|$. It follows that $\| \varphi \| = \| t \|_1$, as desired.

(ii) The case that $1 < p < \infty$. Assume first that $\varphi \in (L_p)_n^\sim = L_p^\sim$. Then $\varphi(f) = \int tf d\mu$ for some $t \in K(L_p)$ and all $f \in L_p$. We may assume that φ is not the null functional and so t is not identically zero. It has to be proved that $t \in L_q$ and $\| t \|_q = \| \varphi \|$. Since $| \varphi | (f) = \int | t | f d\mu$ for all $f \in L_p$ and since φ and t have the same norm as $| \varphi |$ and $| t |$ respectively, we may immediately assume now that φ and t are positive. Now, let $(s_n : n = 1, 2, \ldots)$ be a sequence of step functions on X, each s_n not identically zero and vanishing outside a set of finite measure, such that $0 \leq s_n(x) \uparrow t^q(x)$ for all $x \in X$. For each n, let $g_n = s_n^{1/p}$. Then $g_n \in L_p$ and $\| g_n \|_p = \| s_n \|_1^{1/p}$, so

$$\int tg_n d\mu = \varphi(g_n) \leq \| \varphi \| \cdot \| g_n \|_p = \| \varphi \| \cdot \| s_n \|_1^{1/p} .$$

Combining this with the observation that $s_n = s_n^{1/p} \cdot s_n^{1/q} \leq s_n^{1/p} \cdot t = tg_n$, it follows that

$$\| s_n \|_1 \leq \int tg_n d\mu \leq \| \varphi \| \cdot \| s_n \|_1^{1/p},$$

so (dividing by $\| s_n \|_1^{1/p} > 0$) we see that $\| s_n \|_1^{1/q} \leq \| \varphi \|$. This implies that $\int s_n d\mu = \| s_n \|_1 \leq \| \varphi \|^q$. Since $0 \leq s_n \uparrow t^q$, it follows in view of the theorem on integration of monotone sequences that $\int s_n d\mu \uparrow \int t^q d\mu$, and so $\int t^q d\mu \leq \| \varphi \|^q$, i.e., $t \in L_q$ and $\| t \|_q \leq \| \varphi \|$.

In the converse direction we have

$$| \varphi(f) | = | \int t f d\mu | \leq \| f \|_p \cdot \| t \|_q \leq \| t \|_q$$

for every $f \in L_p$ satisfying $\| f \|_p \leq 1$. Taking the supremum on the left for $\| f \|_p \leq 1$, we obtain the inequality $\| \varphi \| \leq \| t \|_q$. Combining this with the earlier inequality, it follows that $\| \varphi \| = \| t \|_q$, as desired.

(iii) The case that $L_p = L_1$. Assume that $\varphi \in (L_1)_n^{\sim} = L_1^{\sim}$. Then $\varphi(f) = \int t f d\mu$ for some $t \in K(L_1)$ and all $f \in L_1$. As in case (ii) we may assume that φ is positive and t is non-negative and not identically zero. It follows that

$$(2) \qquad 0 \leq \int t f d\mu = \varphi(f) \leq \| \varphi \| \cdot \| f \|$$

for every f satisfying $0 \leq f \in L_1$. Now choose a number $\alpha > \| \varphi \|$ and assume that $t(x) \geq \alpha$ for all x in a set X_α of positive measure. Let Y be a subset of X_α such that $0 < \mu(Y) < \infty$. The characteristic function χ_Y of Y satisfies $\chi_Y \in L_1$, so

$$\int t \chi_Y d\mu = \int_Y t d\mu \geq \alpha \mu(Y) > \| \varphi \| \cdot \mu(Y) = \| \varphi \| \cdot \| \chi_Y \|_1,$$

which contradicts the inequality in (2). Hence $t(x) \geq \alpha > \| \varphi \|$ can hold only for the points x in a set of measure zero, i.e., $t(x) \leq \| \varphi \|$ holds μ-almost everywhere. In other words, $t \in L_\infty$ and $\| t \|_\infty \leq \| \varphi \|$.

In the converse direction we have

$$| \varphi(f) | = | \int t f d\mu | \leq \| f \|_1 \cdot \| t \|_\infty \leq \| t \|_\infty$$

for every $f \in L_1$ satisfying $\| f \|_1 \leq 1$. Taking the supremum on the left for $\| f \|_1 \leq 1$, we obtain $\| \varphi \| \leq \| t \|_\infty$. Combining this with the earlier inequality it follows that $\| \varphi \| = \| t \|_\infty$, as desired.

Exercise 29.5. (i) Let μ be a σ-finite measure in the (non-empty) point set X. The Riesz space of all (real) μ-measurable functions on X is denoted by $L_0 = L_0(X, \mu)$. We have seen that for $X = \mathbb{R}$ and μ Lebesgue measure the order dual $L_0^{\sim}$ satisfies $L_0^{\sim} = \{0\}$, so in particular $(L_0)_n^{\sim} = \{0\}$, which implies that the Köthe dual of L_0 consists of the zero function only. For $X = (1, 2, \ldots)$ and μ the discrete measure in X (i.e., each point has measure one) the situation is different. In this case $L_0 = (s)$, i.e., L_0 is the space of all (real) sequences $f = (f_1, f_2, \ldots)$. Show that the linear functional φ on $L_0 = (s)$ satisfies $\varphi \in L_0^{\sim}$ if and only if $\varphi(f) = \Sigma_1^\infty t_n f_n$, where $(t_n : n = 1, 2, \ldots)$

is a (real) sequence such that $t_n \neq 0$ for only finitely many n. Show that $L_0^{\tilde{}} = (L_0)_n^{\tilde{}}$ and the Köthe dual $K(L_0)$ consists of all sequences $t = (t_1, t_2, \ldots)$ with $t_n \neq 0$ for only finitely many n. Hence, $K(L_0)$ has carrier X.

(ii) Different ideals in L_0 may have the same Köthe dual. For an example choose the space (c_0) of all null sequences (i.e., $(f_1, f_2, \ldots) \in (c_0)$ whenever $f_n \to 0$ as $n \to \infty$). Show that (c_0) is a Banach lattice with respect to the uniform norm. Show that the norm is order continuous and the Köthe dual of (c_0) is the space ℓ_1. Hence, ℓ_∞ and (c_0) have the same Köthe dual.

Hint: For (i), denote by $e_n (n = 1, 2, \ldots)$ the sequence having one in the n-th place and zeros elsewhere. Given $\varphi \in L_0^{\tilde{}}$, let $\varphi(e_n) = t_n$. Assume for a moment that $t_n \neq 0$ for infinitely many n, say $t_n > 0$ for infinitely many n. Let now $f = (f_1, f_2, \ldots) \in L_0$ satisfy $f_n = t_n^{-1}$ for these n and $f_n = 0$ for all other n. It would follow that $\varphi(f) = \infty$. Hence, $t_n \neq 0$ holds for only finitely many n.

CHAPTER 16
Embedding into the Bidual

30. Annihilators and Inverse Annihilators

Let V be a vector space and W a (fixed) linear subspace of the algebraic dual $V^{\#}$ of V.

Definition 30.1. For any (non-empty) subset A of V the annihilator A° of A is defined by

$$A^{\circ} = (\varphi \in W : \varphi(f) = 0 \text{ for all } f \in A).$$

For any (non-empty) subset B of W the inverse annihilator ${}^{\circ}B$ of B is defined by

$$ {}^{\circ}B = (f \in V : \varphi(f) = 0 \text{ for all } \varphi \in B).$$

It is evident that A° and ${}^{\circ}B$ are linear subspaces of W and V respectively. Furthermore, if $A \subseteq V$ and A_1 is the linear subspace of V generated by A, then $A_1^{\circ} = A^{\circ}$, so we may as well assume immediately that A is a linear subspace of V (recall that A_1 consists of all finite linear combinations $\alpha_1 f_1 + \cdots + \alpha_n f_n$ of elements of A with real coefficients $\alpha_n, \ldots, \alpha_n$). Similarly, if $B \subseteq W$ and B_1 is the linear subspace generated by B, then ${}^{\circ}B_1 = {}^{\circ}B$. We mention several simple properties following from the definitions.

(i) $A_1 \subseteq A_2 \subseteq V$ implies that $A_1^{\circ} \supseteq A_2^{\circ} \supseteq V^{\circ} = \{0\}$.
(ii) $B_1 \subseteq B_2 \subseteq W$ implies that ${}^{\circ}B_1 \supseteq {}^{\circ}B_2 \supseteq {}^{\circ}W$.

The subset B of W satisfies ${}^{\circ}B = \{0\}$ if and only if it follows from $\varphi(f) = 0$ for all $\varphi \in B$ that $f = 0$. In other words, $f_1 \neq f_2$ in V implies that $\varphi(f_1) \neq \varphi(f_2)$ for some $\varphi \in B$. This is often expressed by saying that B *separates the points of* V. It is evident that if $B_1 \subseteq B_2$ in W and B_1 separates the points of V, then so does the larger space B_2.

From here on we shall assume that V is a Riesz space E and W is the order dual of E, so $V = E$ and $W = E^{\sim}$.

Theorem 30.2. *(i) If A is an ideal in E, then A° is a band in $E^\sim$.*

(ii) If B is an ideal in $E^\sim$, then $^\circ B$ is an ideal in E. If $[B]$ is the band generated by the ideal B in $E^\sim$, then $^\circ[B] = {}^\circ B$.

(iii) If B is an ideal contained in $E_n^\sim$, then $^\circ B$ is a band in E.

Proof. (i) Let A be an ideal in E. We prove first that $\varphi \in A^\circ$ implies that $|\varphi| \in A^\circ$. For this it is sufficient to show that $\varphi \in A^\circ$ implies $|\varphi|(u) = 0$ for $0 \le u \in A$. Let, therefore, $0 \le u \in A, \varphi \in A^\circ$ and $|f| \le u$. Then $f \in A$, so $\varphi(f) = 0$ by hypothesis. It follows that

$$|\varphi|(u) = \sup(|\varphi(f)| : |f| \le u) = 0.$$

For the proof that A° is an ideal in $E^\sim$, assume now that $\varphi \in A^\circ, \psi \in E^\sim$ and $|\psi| \le |\varphi|$. It has to be proved that $\psi \in A^\circ$. We have already proved that $|\varphi| \in A^\circ$, and so $|\psi| \in A^\circ$ follows trivially. Then, for any $u \in A^+$, we have $|\psi(u)| \le |\psi|(u) = 0$, so $\psi(u) = 0$, i.e., $\psi \in A^\circ$. Hence, A° is an ideal in $E^\sim$.

Let now D be an upwards directed set in $(E^\sim)^+$ such that $D \uparrow \varphi_0$ and $\varphi \in A^\circ$ for all $\varphi \in D$, i.e., $\varphi(u) = 0$ for every $\varphi \in D$ and every $u \in A^+$. Then

$$\varphi_0(u) = \sup(\varphi(u) : \varphi \in D) = 0$$

for every $u \in A^+$, so it follows that $\varphi_0 \in A^\circ$. This shows that A° is a band in $E^\sim$.

(ii) Let B be an ideal in $E^\sim$. We prove first that $f \in {}^\circ B$ implies that $|f| \in {}^\circ B$. For this it is sufficient to show that $f \in {}^\circ B$ implies that $\varphi(|f|) = 0$ for $0 \le \varphi \in B$. Let, therefore, $0 \le \varphi \in B, f \in {}^\circ B$ and $|\psi| \le \varphi$. Then $\psi \in B$, so $\psi(f) = 0$ by hypothesis. It follows (by Theorem 20.8) that

$$\varphi(|f|) = \sup(|\psi(f)| : |\psi| \le \varphi) = 0.$$

For the proof that $^\circ B$ is an ideal in E, assume now that $f \in {}^\circ B, g \in E$ and $|g| \le |f|$. It has to be proved that $g \in {}^\circ B$. We have already proved that $|f| \in {}^\circ B$, and so $|g| \in {}^\circ B$ follows trivially. Then, for any $\varphi \in B^+$, we have $|\varphi(g)| \le \varphi(|g|) = 0$, so $\varphi(g) = 0$, i.e., $g \in {}^\circ B$. Hence, $^\circ B$ is an ideal in E.

Let $[B]$ be the band generated by the ideal B in $E^\sim$. Since $B \subseteq [B]$ implies $^\circ B \supseteq {}^\circ[B]$, it remains to prove that $^\circ B \subseteq {}^\circ[B]$. For this purpose, let $0 \le u \in {}^\circ B$ be given, so $\varphi(u) = 0$ for all $\varphi \in B$. Furthermore, let $0 \le \varphi_0 \in [B]$. There exists an upwards directed set D in B^+ such that $D \uparrow \varphi_0$, and hence

$$\varphi_0(u) = \sup(\varphi(u) : \varphi \in D) = 0.$$

This shows that $u \in {}^\circ[B]$. Hence $^\circ B = {}^\circ[B]$.

(iii) Let B be an ideal contained in $E_n^\sim$. To show that the ideal $^\circ B$ is a band in E, assume that D is an upwards directed set of positive elements in

$^\circ B$ such that $D \uparrow u_0$ for some $u_0 \in E$. Since all φ in B are order continuous by hypothesis, we have

$$\varphi(u_0) = \sup(\varphi(u) : u \in D) = 0$$

for every positive $\varphi \in B$. This shows that $u_0 \in {}^\circ B$. Hence, $^\circ B$ is a band in E. $\blacksquare$

If it is given that $^\circ B = \{0\}$ for some subset B of $E^\sim$ or of $E_n^\sim$, then this has consequences for the structure of E or $E_n^\sim$.

Theorem 30.3. *(i) If there exists a subset B of $E^\sim$ such that $^\circ B = \{0\}$, then E is Archimedean.*

(ii) If B is an ideal contained in $E_n^\sim$ such that $^\circ B = \{0\}$, then B is order dense in $E_n^\sim$, i.e., $E_n^\sim$ is now the band generated by B in $E^\sim$.

Proof. (i) Let $^\circ B = \{0\}$ for some subset B of $E^\sim$. Then, as observed above, we have $^\circ(E^\sim) = \{0\}$. For the proof that E is Archimedean, assume that $0 \le nv \le u$ in E for $n = 1, 2, \ldots$. It has to be shown that $v = 0$. For any $\varphi \ge 0$ in $E^\sim$ we have $0 \le n\varphi(v) \le \varphi(u)$ for $n = 1, 2, \ldots$, so $\varphi(v) = 0$. It follows that $\varphi(v) = 0$ for every $\varphi \in E^\sim$, which implies that $v = 0$ (on account of $^\circ(E^\sim) = \{0\}$).

(ii) Let B be an ideal in $E_n^\sim$ satisfying $^\circ B = \{0\}$. Observe first that in view of the result in part (i) the space E is Archimedean. To show that B is order dense in $E_n^\sim$ it is sufficient to prove that the disjoint complement of B in the space $E_n^\sim$ contains only the zero functional (see Theorem 23.3(i)). Hence, let $0 \le \varphi \in E_n^\sim$ and $\varphi \perp \psi$ for all $\psi \in B$. We have to prove that $\varphi = 0$. Since E is Archimedean and φ, as well as every $\psi \in B$, is order continuous, Theorems 24.5 and 24.6 may be applied. It follows then from $\varphi \perp \psi$ for all $\psi \in B$ that the carrier C_φ of φ satisfies $C_\varphi \perp C_\psi$ for all $\psi \in B$, i.e., $C_\varphi \subseteq N_\psi$ for all $\psi \in B$. This implies that every positive u in C_φ satisfies $\psi(u) = 0$ for all $\psi \in B$, i.e., $u \in {}^\circ B$. Hence $u = 0$ on account of our hypothesis that $^\circ B = \{0\}$. It has thus been shown that $C_\varphi = \{0\}$. Finally, once more since E is Archimedean and φ is order continuous, the null ideal N_φ is the disjoint complement of C_φ (see the remark preceding Theorem 24.2), so $N_\varphi = (C_\varphi)^d = \{0\}^d = E$. It follows that $\varphi = 0$. It has been proved thus that the ideal B is order dense in $E_n^\sim$. Since $E_n^\sim$ is Archimedean, this is equivalent to the statement that the band generated in $E_n^\sim$ by B is equal to $E_n^\sim$ itself (see Theorem 23.4). $\blacksquare$

It is evident that if f is a positive element of the Riesz space E, then $\varphi(f) \ge 0$ holds for every positive $\varphi \in E^\sim$. The converse holds if $E^\sim$ separates the points of E, as proved in the following theorem.

Theorem 30.4. *The following conditions for the Riesz space E are equivalent.*

(i) $E^\sim$ separates the points of E, i.e., $^\circ(E^\sim) = \{0\}$.

(ii) If $f \in E$ has the property that $\varphi(f) \geq 0$ for every $\varphi \geq 0$ in $E^{\sim}$, then $f \geq 0$.

Proof. (i)$\Rightarrow$ (ii) Assume that (i) holds and let $\varphi(f) \geq 0$ for every $\varphi \geq 0$ in $E^{\sim}$. For any fixed $\varphi \geq 0$ in $E^{\sim}$ and any $g \in E$ we have

$$0 \leq \varphi(g^{+}) = \max(\psi(g) : 0 \leq \psi \leq \varphi)$$

by Theorem 20.8, so in particular, for $g = -f$,

$$0 \leq \varphi(f^{-}) = \varphi\{(-f)^{+}\} = \max(-\psi(f) : 0 \leq \psi \leq \varphi).$$

This shows that $\varphi(f^{-}) = -\psi(f)$ for some ψ satisfying $0 \leq \psi \leq \varphi$. Hence $0 \leq \varphi(f^{-}) = -\psi(f) \leq 0$ (the last inequality because $\psi(f) \geq 0$ by hypothesis), so $\varphi(f^{-}) = 0$. Then $\varphi(f^{-}) = 0$ for every $\varphi \in E^{\sim}$, so $f^{-} = 0$ in view of (i). It follows that $f = f^{+} - f^{-} = f^{+} \geq 0$.

(ii)$\Rightarrow$ (i) Let $\varphi(f) = 0$ for all $\varphi \in E^{\sim}$. Then, by (ii), $f \geq 0$ as well as $-f \geq 0$. Hence $f = 0$. ∎

31. Embedding into the Order Bidual

In the present section we shall assume that B is an ideal in the order dual $E^{\sim}$ of the Riesz space E. The cases that $B = E^{\sim}$ or $B = E_n^{\sim}$ will be of special importance. For any given $f \in E$ we define the real linear functional $f^{\star}$ on B by $f^{\star}(\varphi) = \varphi(f)$ for all $\varphi \in B$. We prove immediately that $f^{\star}$ is an order bounded linear functional on B, i.e., $f^{\star} \in B^{\sim}$. For the proof, observe that if the given $f \in E$ is positive, then $f^{\star}(\varphi) = \varphi(f) \geq 0$ for every $\varphi \geq 0$ in B, so $f^{\star}$ is now a positive linear functional on B. Hence, if $f \in E$ is arbitrary, then $f = f^{+} - f^{-}$, so $f^{\star} = (f^{+})^{\star} - (f^{-})^{\star}$, i.e., $f^{\star}$ is the difference of two positive linear functionals on B, so $f^{\star} \in B^{\sim}$. We can prove more, as the following lemma shows.

Lemma 31.1. *For all f and g in E we have*

$$(f \vee g)^{\star} = f^{\star} \vee g^{\star} \quad and \quad (f \wedge g)^{\star} = f^{\star} \wedge g^{\star},$$

so in particular $\mid f^{\star} \mid = \mid f^{\star} \mid$ for every $f \in E$.

Proof. It is sufficient to prove only the formula for $f \vee g$, since the result for $f \wedge g$ follows by observing that $(f \vee g) + (f \wedge g) = f + g$. Furthermore, on account of $(f - g)^{\star} = f^{\star} - g^{\star}$, it is sufficient to prove the formula only for $g = 0$, i.e., it is sufficient to prove that $(f^{+})^{\star} = (f^{\star})^{+}$ for every $f \in E$. To this end, let $0 \leq \varphi \in B$. Then

$$(f^{\star})^{+}(\varphi) = \sup(f^{\star}(\psi) : 0 \leq \psi \leq \varphi) = \sup(\psi(f) : 0 \leq \psi \leq \varphi).$$

The last supremum is equal to $\varphi(f^+)$ by Theorem 20.8. Hence

$$(f^\star)^+(\varphi) = \varphi(f^+) = (f^+)^\star(\varphi).$$

This holds for every $\varphi \geq 0$ in B, and so it holds for every $\varphi \in B$. It follows that $(f^+)^\star = (f^\star)^+$. $\qquad\blacksquare$

It has been proved thus that the mapping $\sigma : E \to B^\sim$, defined by $\sigma(f) = f^\star$ for all $f \in E$, is a Riesz homomorphism. Actually, σ maps E into the space $B_n^\sim$ of all order continuous linear functionals on B. For the proof, let $f \in E$ be given and let D be a downwards directed set in B satisfying $D \downarrow 0$. It is sufficient to show that for any $f \in E$ we have

$$\{(f^\star)^+(\varphi) : \varphi \in D\} \downarrow 0.$$

This follows by observing that

$$(f^\star)^+(\varphi) = (f^+)^\star(\varphi) = \varphi(f^+) \text{ and } (\varphi(f^+) : \varphi \in D) \downarrow 0.$$

Some questions arise. As we have seen, the mapping $\sigma : E \to B_n^\sim$ is a Riesz homomorphism, i.e., σ preserves finite suprema and infima. Now, under what conditions does σ preserve arbitrary suprema and infima? And when is the Riesz homomorphism a Riesz isomorphism? Answers are contained in the following lemma.

Lemma 31.2. *(i) The Riesz homomorphism σ preserves arbitrary suprema and infima (i.e., σ is order continuous) if and only if the ideal B is contained in $E_n^\sim$ (i.e., if and only if the members of B are order continuous linear functionals).*

(ii) The Riesz homomorphism σ is a Riesz isomorphism if and only if B separates the points of E, i.e., if and only if ${}^\circ B = \{0\}$.

Proof. (i) Assume first that $B \subseteq E_n^\sim$. It is sufficient to show now that if D is an upwards directed set in E^+ such that $D \uparrow u_0$ holds in E, then $(\sigma(u) : u \in D) \uparrow \sigma(u_0)$ holds in $B_n^\sim$. In other words, using the same notations as before, we have to show that $(u^\star : u \in D) \uparrow u_0^\star$. For this purpose, let $0 \leq \varphi \in B$ be given. Then, since φ is order continuous, we have

$$(u^\star(\varphi) : u \in D) = (\varphi(u) : u \in D) \uparrow \varphi(u_0) = u_0^\star(\varphi).$$

This holds for every positive φ in B, so the desired result follows.

Conversely, if σ preserves arbitrary suprema and infima, then (in particular) $D \downarrow 0$ in E implies $(u^\star : u \in D) \downarrow 0$ in $B_n^\sim$, i.e.,

$$(u^\star(\varphi) : u \in D) \downarrow 0 \text{ for every } 0 \leq \varphi \in B.$$

Hence, for every $\varphi \geq 0$ in B, we have $(\varphi(u) : u \in D) \downarrow 0$, i.e., φ is order continuous. This shows that $B \subseteq E_n^{\sim}$.

(ii) The Riesz homomorphism is a Riesz isomorphism if and only if it follows from $f^* = 0$ that $f = 0$, i.e., if and only if it follows from $\varphi(f) = 0$ for all $\varphi \in B$ that $f = 0$, i.e., if and only if $^{\circ}B = \{0\}$.   ∎

The next problem is to find out how large the image $\sigma(E)$ of E is under the mapping $\sigma : E \to B_n^{\sim}$. To do this, observe first that $\sigma(E)$ is a Riesz subspace of $B_n^{\sim}$ and

$$^{\circ}\{\sigma(E)\} = \{\varphi \in B : f^*(\varphi) = 0 \text{ for all } f^* \in \sigma(E)\} =$$

$$\{\varphi \in B : \varphi(f) = 0 \text{ for all } f \in E\} = \{0\}.$$

Hence, the ideal J generated by $\sigma(E)$ in $B_n^{\sim}$ satisfies $^{\circ}J = \{0\}$, which shows (see Theorem 30.3(ii)) that J is order dense in $B_n^{\sim}$. In other words, the band generated by J in $B_n^{\sim}$ is $B_n^{\sim}$ itself. Equivalently, the band generated by J in $B^{\sim}$ is $B_n^{\sim}$. Since $\sigma(E)$ and J generate the same band, we may just as well say that the band generated by $\sigma(E)$ in $B^{\sim}$ is $B_n^{\sim}$.

If σ is a Riesz isomorphism, i.e., if $^{\circ}B = \{0\}$, we shall identify E and its image $\sigma(E)$. In other words, we shall regard E as a Riesz subspace of $B_n^{\sim}$ and we shall say that E is *embedded* in $B_n^{\sim}$ as a Riesz subspace. The case that $B = E_n^{\sim}$ and $^{\circ}(E_n^{\sim}) = \{0\}$ is of special interest. In this case, therefore, the following holds. Note that E is now Archimedean (see Theorem 30.3(i)).

Theorem 31.3. *If E is a Riesz space satisfying $^{\circ}(E_n^{\sim}) = \{0\}$, the mapping $\sigma : E \to (E_n^{\sim})_n^{\sim}$ is a Riesz isomorphism. The space E is thus embedded into $(E_n^{\sim})_n^{\sim}$ as a Riesz subspace having the property that the band generated by E in $(E_n^{\sim})_n^{\sim}$ is $(E_n^{\sim})_n^{\sim}$. Furthermore, the mapping preserves arbitrary suprema and infima.*

If E is a Riesz space satisfying $^{\circ}(E_n^{\sim}) = \{0\}$ as in the last theorem and if in this case the image $\sigma(E)$ is the whole space $(E_n^{\sim})_n^{\sim}$, we shall write $E = (E_n^{\sim})_n^{\sim}$, where the equality sign indicates that the homomorphism is an isomorphism. Any Riesz space E satisfying $E = (E_n^{\sim})_n^{\sim}$ is called a *perfect Riesz space*. The name was introduced originally for the case that E is a sequence space, i.e., E is an ideal in the sequence space (s) of all real sequences $f = (f_1, f_2, \ldots)$ with $f_1, f_2, \ldots$ real numbers. In this case $E_n^{\sim}$ is the Köthe dual of E, i.e., $E_n^{\sim}$ is the space of all sequences $t = (t_1, t_2, \ldots)$ such that $\Sigma f_n t_n$ converges for every $f \in E$ (see Theorem 29.3). It is evident that $^{\circ}(E_n^{\sim}) = \{0\}$ and, therefore, E is embedded in $(E_n^{\sim})_n^{\sim}$. If the Köthe dual of $E_n^{\sim}$ is exactly E again (and not larger), then E is perfect. In other words, if $\Sigma f_n t_n$ converges for all $t \in E_n^{\sim}$ only if $f \in E$, then E is perfect.

More generally, if μ is a σ-finite measure in the (non-empty) point set X and E is an ideal in the space $L_0 = L_0(X, \mu)$ of all μ-measurable (real) functions on X such that the carrier of E is X, then $E_n^\sim$ is the Köthe dual of E, i.e., $E_n^\sim$ is the ideal in L_0 consisting of all $t \in L_0$ such that ft is μ-summable over X for every $f \in E$ (see again Theorem 29.3). As we have already seen in several examples it is by no means always so that $^0(E_n^\sim) = \{0\}$. Precisely, $^0(E_n^\sim) = \{0\}$ holds if and only if the carrier of $E_n^\sim$ is X, as we shall prove now.

Let first $^0(E_n^\sim) = \{0\}$, i.e.,

$$\text{if } f \in E \text{ and } \int ft d\mu = 0 \text{ for all } t \in E_n^\sim, \text{ then } f = 0. \tag{1}$$

It has to be proved that the carrier of $E_n^\sim$ is X. Assume that this does not hold. Then there exists a subset Y of X of positive measure such that all $t \in E_n^\sim$ vanish on Y. Since the carrier of E is X, there exists a subset Z of Y of positive measure such that the characteristic function χ_Z of Z is a member of E, i.e., $\chi_Z \in E$ (see Theorem 29.1). Now, in view of $Z \subseteq Y$, all $t \in E_n^\sim$ vanish on Z, so $\int \chi_Z t d\mu = 0$ for all $t \in E_n^\sim$. This implies by (1) that $\chi_Z = 0$. Contradiction, since $\mu(Z) > 0$. Hence, $E_n^\sim$ has X as its carrier.

Assume, conversely, that $E_n^\sim$ has carrier X. We have to prove that (1) holds, so let $f \in E$ satisfy $\int ft d\mu = 0$ for each $t \in E_n^\sim$. It has to be shown that $f = 0$. If not, at least one of f^+ and f^-, for example f^+, is not the zero function. Then there exist a subset Y of X of positive measure and a number $\epsilon > 0$ such that $f^+(x) \geq \epsilon$ for all $x \in Y$. Observe that f^- vanishes on Y. Now, since $E_n^\sim$ has carrier X, there exists a subset Z of Y of positive measure such that $\chi_Z \in E_n^\sim$. Hence

$$0 = \int f\chi_Z d\mu = \int_Z f d\mu = \int_Z f^+ d\mu \geq \epsilon\mu(Z) > 0.$$

Contradiction. Hence, (1) holds.

Assume that $^0(E_n^\sim) = \{0\}$, so (by the last theorem)E is embedded in $(E_n^\sim)_n^\sim$. If $\int ft d\mu$ exists (as a finite number) for all $t \in E_n^\sim$ only if $f \in E$, then E is perfect. Example: All spaces $L_p(X, \mu), 1 \leq p \leq \infty$, are perfect. In particular, all sequence spaces $\ell_p(1 \leq p \leq \infty)$ are perfect. The sequence space (c_0) of all null sequences is not perfect because $(c_0)_n^\sim = \ell_1$ (see Exercise 29.5(ii)), so $\{(c_0)_n^\sim\}_n^\sim = \ell_\infty$. We proceed with an example in which E is a Riesz subspace of (s) but not an ideal in (s).

Example 31.4. Let E be the Riesz space (c) of all (real) converging number sequences. Evidently, $E = (c)$ is a Riesz subspace of (s) but not an ideal in (s). Observe that any $f \in E = (c)$ is of the form $f = \{\lim f(n)\} \cdot e + f^\sim$, where $e = (1, 1, 1, \ldots)$ and $f^\sim \in (c_0)$. Let $\varphi \in E^\sim$. Writing $\varphi(e) = t^\sim(0)$, we have $\varphi(f) = t^\sim(0) \cdot \lim f(n) + \varphi(f^\sim)$. As we have seen already, there exists a sequence $t = (t(1), t(2), \ldots) \in \ell_1$ such that $\varphi(f^\sim) = \Sigma_1^\infty t(n) f^\sim(n)$ for every $f^\sim \in (c_0)$. Hence, for any $\varphi \in E^\sim$ there exists a sequence

$$\{t^\sim(0), t(1), t(2), \ldots\}, \text{ satisfying } \Sigma_{n=1}^\infty |t(n)| < \infty,$$

such that

$$\varphi(f) = t\tilde{}(0) \cdot \lim f(n) + \Sigma_1^\infty t(n) f\tilde{}(n) =$$

$$t\tilde{}(0) \cdot \lim f(n) + \Sigma_1^\infty t(n)\{f(n) - \lim f(n)\} =$$

$$\{t\tilde{}(0) - \Sigma_1^\infty t(n)\} \cdot \lim f(n) + \Sigma_1^\infty t(n)f(n).$$

Hence, writing $t(0) = t\tilde{}(0) - \Sigma_1^\infty t(n)$, we get

$$\varphi(f) = t(0) \cdot \lim f(n) + \Sigma_1^\infty t(n)f(n) \text{ with } \Sigma_1^\infty \mid t(n) \mid \text{ converging,}$$

holding for every $f \in E = (c)$. Any $\varphi \in E\tilde{}$ satisfying $t(0) = 0$ is obviously order continuous, i.e., $\varphi \in E_n\tilde{}$. Observe here that if $(f_k : k = 1, 2, \ldots)$ is a sequence in (c) satisfying $f_k \downarrow 0$, then $\lim_{n\to\infty} f_k(n) \downarrow 0$ as $k \to \infty$ does not necessarily hold (Example: f_k is the element in (c) having the first k coordinates zero and the remaining coordinates one). Conversely, if $\varphi \in E_n\tilde{}$, then $t(0) = 0$. To see this, let $(f_k : k = 1, 2, \ldots)$ be the sequence in the last example. Then $f_k \downarrow 0$, so $\varphi(f_k) \to 0$(as $k \to \infty$). But $\varphi(f_k) = t(0) + \Sigma_{k+1}^\infty t(n)$, so $t(0) = 0$. It has been proved thus that $E_n\tilde{}$ is (isomorphic to) ℓ_1, from which it follows that $(E_n\tilde{})_n\tilde{} = \ell_\infty$. This shows that $E = (c)$ is not perfect. The ideal generated by $E = (c)$ in $(E_n\tilde{})_n\tilde{} = \ell_\infty$ is ℓ_∞ itself, so this ideal is perfect.

CHAPTER 17
Freudenthal's Spectral Theorem

32. Projection Bands and Components

Let μ be a σ-finite measure in the (non-empty) point set X. As well-known, any finite linear combination of characteristic functions of measurable subsets of X is called a μ-*step function*. It is not difficult to see that for any $0 \le f \in L_\infty(X, \mu)$ there exists a sequence $(s_n : n = 1, 2, \ldots)$ of μ-step functions such that $0 \le s_n \uparrow f$ holds uniformly. A similar result holds in an Archimedean Riesz space possessing a strong unit. This result, due to H.Freudenthal (1936), is known as *Freudenthal's spectral theorem*. Some preliminary theorems about projection bands will be useful. Recall that the band B in the Archimedean Riesz space E is called a projection band if $E = B \oplus B^d$ (see section 11). In this case B^d is likewise a projection band (we have $B = B^{dd}$ because E is Archimedean, so $E = B^d \oplus B^{dd}$). As in Theorem 11.4, the band projection on the projection band B is sometimes denoted by P_B. It is obvious that $P_{B^d} = I - P_B$, where I is the identity operator in E (i.e., $If = f$ for each $f \in E$, and so $P_{B^d} f = f - P_B f$). The following holds now.

Theorem 32.1. *If B_1 and B_2 are projection bands in the Archimedean Riesz space E, then $B_3 = B_1 \cap B_2$ is also a projection band and the corresponding band projections satisfy*

$$P_1 P_2 = P_2 P_1 = P_3.$$

Furthermore, we have $P_3 u = (P_1 u) \wedge (P_2 u)$ for any $u \in E^+$.

Proof. For any $u \in E^+$ we have $P_2 u \in B_2$. Hence, it follows from $0 \le P_1 P_2 u \le P_2 u$ that $P_1 P_2 u \in B_2$. Since it is evident that $P_1 P_2 u \in B_1$, we see already that $P_1 P_2 u \in B_1 \cap B_2 = B_3$. To show now that $B_3 = B_1 \cap B_2$ is a projection band such that $P_3 = P_1 P_2$ holds, it is sufficient to prove that $(u - P_1 P_2 u) \in B_3^d$, because this will show then that any $u \in E^+$ is the sum of an element in B_3 and an element in B_3^d (so $E = B_3 \oplus B_3^d$). For the proof, note that

$$u - P_1 P_2 u = (u - P_2 u) + (P_2 u - P_1 P_2 u),$$

where $u - P_2 u$ is the band projection of u on B_2^d and $P_2 u - P_1 P_2 u$ is the band projection of $P_2 u$ on B_1^d. Hence

$$u - P_2u \in B_2^d \subseteq (B_1 \cap B_2)^d \text{ and } P_2u - P_1P_2u \in B_1^d \subseteq (B_1 \cap B_2)^d.$$

It follows by addition that $u - P_1P_2u \in (B_1 \cap B_2)^d = B_3^d$.

Finally, let again $u \in E^+$ be given and denote $(P_1u) \wedge (P_2u)$ by w_0. On account of $P_3u \leq P_1u$ and $P_3u \leq P_2u$ we have $P_3u \leq w_0$. On the other hand, since $w_0 \in B_1 \cap B_2 = B_3$ and $0 \leq w_0 \leq u$, and since

$$P_3u = \sup(w : 0 \leq w \leq u, w \in B_3),$$

we have $w_0 \leq P_3u$. Hence $P_3u = w_0 = (P_1u) \wedge (P_2u)$.   ∎

Theorem 32.2. *If B_1 and B_2 are projection bands in the Archimedean Riesz space E with corresponding band projections P_1 and P_2, then the following conditions are equivalent.*

(i) $B_1 \subseteq B_2$.
(ii) $P_1P_2 = P_2P_1 = P_1$.
(iii) $P_1 \leq P_2$.

Proof. (i)$\rightarrow$ (ii). If $B_1 \subseteq B_2$, then $B_3 = B_1 \cap B_2 = B_1$, so

$$P_1P_2 = P_2P_1 = P_3 = P_1.$$

(ii) $\rightarrow$ (iii). For each $u \in E^+$ we have $P_1u = P_1P_2u \leq P_2u$.

(iii) $\rightarrow$ (i). For $0 \leq u \in B_1$ we have $0 \leq u = P_1u \leq P_2u \in B_2$, and so $B_1 \subseteq B_2$.   ∎

Theorem 32.3. *If B_1 and B_2 are projection bands in the Archimedean Riesz space E, then $B_3 = B_1 + B_2$ is also a projection band and the corresponding band projections satisfy*

$$P_3 = P_1 + P_2 - P_1P_2.$$

Furthermore, we have $P_3u = (P_1u) \vee (P_2u)$ for any $u \in E^+$.

Proof. The band $B_1^d \cap B_2^d$ is a projection band with band projection

$$(I - P_1)(I - P_2) = I - P_1 - P_2 + P_1P_2.$$

Therefore, $B_0 = (B_1^d \cap B_2^d)^d$ is a projection band with band projection $P_0 = P_1 + P_2 - P_1P_2$. It is evident that B_0 contains $B_1^{dd} = B_1$ as well as $B_2^{dd} = B_2$, so $B_0 \supseteq (B_1 + B_2)$. From

$$P_0u = P_1u + P_2u - P_1P_2u,$$

holding for every $u \in E^+$, it follows that $P_0u \in (B_1 + B_2)$, and so $B_0 \subseteq (B_1 + B_2)$. Therefore, $B_1 + B_2 = B_0$. It has been proved thus that $B_3 = B_1 + B_2$ is a projection band with band projection $P_3 = P_1 + P_2 - P_1P_2$. Furthermore,

$$P_3u = P_1u + P_2u - (P_1u) \wedge (P_2u) = (P_1u) \vee (P_2u)$$

for each $u \in E^+$.   ∎

Corollary 32.4. *Let E be an Archimedean Riesz space.*

(i) If B_1 and B_2 are projection bands in E with corresponding band projections P_1 and P_2 and if $B_2 \subseteq B_1$, then $B_1 \cap B_2^d$ is a projection band with band projection $P_1 - P_2$.

(ii) If the projection bands B_1, B_2, B_3, B_4 in E with corresponding band projections P_1, P_2, P_3, P_4 satisfy $B_4 \subseteq B_3 \subseteq B_2 \subseteq B_1$, then

$$(P_1 - P_2)u \perp (P_3 - P_4)u \text{ for every } u \in E^+.$$

Proof. (i) Note first that $P_1 P_2 = P_2$ in view of $B_2 \subseteq B_1$. The band projections on B_1 and B_2^d are P_1 and $I - P_2$ respectively, so $B_1 \cap B_2^d$ is a projection band with band projection

$$P_1(I - P_2) = P_1 - P_1 P_2 = P_1 - P_2.$$

(ii) We have $0 \le (P_1 - P_2)u \in B_2^d$ and $0 \le (P_3 - P_4)u \in B_3$. By hypothesis B_3 is a subset of B_2, so B_3 and B_2^d are disjoint sets. It follows that the elements $(P_1 - P_2)u$ and $(P_3 - P_4)u$ are disjoint. ∎

As in section 7, the principal ideal generated by an element f in the Riesz space E will be denoted by A_f. Recall that

$$A_f = (g \in E : \mid g \mid \le \mid \alpha f \mid \text{ for some } \alpha \in \mathbb{R}).$$

In Exercise 7.7 it was asked to prove that for any u and v in E^+ we have

$$A_{u \wedge v} = A_u \cap A_v \text{ and } A_{u \vee v} = A_{u+v} = A_u + A_v.$$

We indicate how to prove that $A_u \cap A_v$ is a subset of $A_{u \wedge v}$. If $0 \le w \in A_u \cap A_v$, then $w \le \alpha u$ and $w \le \beta v$ for appropriate non-negative α and β. Hence $w \le \gamma u$ and $w \le \gamma v$ for $\gamma = \alpha + \beta$, so $w \le \gamma(u \wedge v)$. For the proof that A_{u+v} is a subset of $A_u + A_v$, let $0 \le w \in A_{u+v}$, so $0 \le w \le \alpha(u+v)$ for some $0 \le \alpha \in \mathbb{R}$. Since $\alpha(u + v) \in A_u + A_v$ and $A_u + A_v$ is an ideal (see Theorem 7.6), it follows that $w \in A_u + A_v$.

The band generated by $f \in E$ will be denoted by B_f. If u and v are positive elements in E such that $u \in B_v$, then

$$u = \sup\{u \wedge (nv) : n = 1, 2, \ldots\}$$

by Corollary 11.6. This shows that for any $0 \le u \in B_v$ there exists a sequence $(s_n : n = 1, 2, \ldots)$ in the ideal A_v such that $0 \le s_n \uparrow u$. It follows now easily that $B_{u \wedge v} = B_u \cap B_v$ for arbitrary u and v in E^+. We briefly indicate how to prove that $B_u \cap B_v$ is a subset of $B_{u \wedge v}$. If $0 \le w \in B_u \cap B_v$, there exist sequences $0 \le s_n \uparrow w$ and $0 \le t_n \uparrow w$ such that $s_n \in A_u$ and $t_n \in A_v$ for all n. Then $0 \le (s_n \wedge t_n) \uparrow w$ with $(s_n \wedge t_n) \in A_u \cap A_v = A_{u \wedge v}$ for all n. It follows that $w \in B_{u \wedge v}$.

From $(u+v)/2 \leq u \vee v \leq u+v$ it follows that $B_{u\vee v} = B_{u+v}$, but it is not always so that $B_{u+v} = B_u + B_v$. This is shown by the example in section 7 which precedes Exercise 7.7 where, for appropriate u and v, the sum $B_u + B_v$ is a proper subset of B_{u+v}, simply already because $B_u + B_v$ fails to be a band. If, however, the space E is Archimedean and B_u, B_v are projection bands, then B_{u+v} is a projection band and $B_{u+v} = B_u + B_v$. To see this, observe first that $B_u + B_v$ is now a projection band (by Theorem 32.3 above). It is sufficient, therefore, to prove that B_{u+v} is a subset of $B_u + B_v$. Any $0 \leq s \in A_{u+v} = A_u + A_v$ is obviously a member of the band $B_u + B_v$. Hence, for any $0 \leq w \in B_{u+v}$ there exists a sequence $0 \leq s_n \uparrow w$ with $s_n \in (B_u + B_v)$ for all n. It follows that $w \in B_u + B_v$.

For convenience we shall say that $f \in E$ is a *projection element* whenever the set B_f is a projection band. Combining the last results with those earlier in this section, we can state the following theorem.

Theorem 32.5. *If u and v are positive projection elements in the Archimedean Riesz space E, then $u \wedge v, u \vee v$ and $u + v$ are projection elements and the corresponding projection bands satisfy*

$$B_{u\wedge v} = B_u \cap B_v \quad \text{and} \quad B_{u\vee v} = B_{u+v} = B_u + B_v.$$

Furthermore, the band projections satisfy

$$P_{u\wedge v} = P_u P_v = P_v P_u \quad \text{and} \quad P_{u\vee v} = P_{u+v} = P_u + P_v - P_u P_v.$$

Finally, u and v are disjoint if and only if $P_u P_v = P_v P_u = 0$ or, equivalently, $P_{u\vee v} = P_{u+v} = P_u + P_v$.

Exercise 32.6. Let E be a Riesz space having the principal projection property (so E is Archimedean; see Theorem 11.9). Let $0 \leq v \in B_u$ and write $w = P_v u$. Show that $B_w = B_v$.

Hint: Note first that if p and q are positive elements in E, then $p \perp q$ if and only if $B_p \perp B_q$. It follows from $w = P_v u \in B_v$ that $B_w \subseteq B_v$. We have $u = P_v u + z = w + z$ with $w \in B_v$ and $z \perp B_v$, so $v \perp B_z$. Show that it follows from $u = w + z$ and $w \perp z$ that $B_u = B_w \oplus B_z$. In view of $0 \leq v \in B_u$ we have $v = v_1 + v_2$ with $v_1 \in B_w$ and $v_2 \in B_z$. On the other hand $0 \leq v_2 \leq v$ and $v \perp B_z$, so $v_2 \perp B_z$. Therefore $v_2 = 0$, and so $v = v_1 \in B_w$. It follows that $B_v \subseteq B_w$.

If B is a projection band in the Riesz space E (i.e., $E = B \oplus B^d$) and $f \in E$, then the projection $P_B f$ is called the component of f in B (as defined in section 11). Note that if $f \geq 0$, then

$$(P_B f) \wedge (f - P_B f) = 0$$

because $f - P_B f$ is the component of f in the disjoint complement B^d of B. The notion of a component makes sense also if there is no projection band around. Let $e \in E^+$. Any element $p \in E^+$ satisfying $p \wedge (e - p) = 0$ is called a *component of* e. It is obvious that p is a component of e if and only if $e - p$ is so. Furthermore, the null element of E and e itself are components of e. We shall prove first that the set C_e of all components of e (with the ordering inherited from E) is a lattice. To this end, let p and q be components of e. Then

$$0 \leq e - (p \vee q) = e + \{(-p) \wedge (-q)\} = (e - p) \wedge (e - q),$$

so (by one of the distributive laws)

$$(p \vee q) \wedge \{e - (p \vee q)\} = (p \vee q) \wedge \{(e - p) \wedge (e - q)\} =$$

$$\{p \wedge (e - p) \wedge (e - q)\} \vee \{q \wedge (e - p) \wedge (e - q)\} = 0 \vee 0 = 0.$$

This shows that $p \vee q$ is a member of C_e. For the proof that $p \wedge q$ is likewise a member it is sufficient to show that $e - (p \wedge q)$ is a component of e. Now,

$$e - (p \wedge q) = (e - p) \vee (e - q),$$

and this is a component in virtue of what was proved a moment ago. It is obvious that the set C_e has the null element as smallest element and e as its largest element. Furthermore, since $p \vee (e - p) = e$ for any $p \in C_e$, the element $e - p$ is the complement of p in the sense as defined in section 2. All this leads up to the following theorem.

Theorem 32.7. *For $e \in E^+$, the set C_e of all components of e is a Boolean algebra (with respect to the ordering inherited from E). If E has the principal projection property, then $p \in C_e$ if and only if $p = P_f e$ for some principal projection P_f.*

Proof. Only the last statement needs proof, so assume that E has the principal projection property. For any $f \in E$ the element $P_f e$ satisfies $(P_f e) \wedge (e - P_f e) = 0$, so $P_f e$ is a component of e. Conversely, if p is a component of e, then $p \wedge (e - p) = 0$, so p is a member of the principal band B_p and $(e - p) \perp B_p$. This shows that $p = P_p e$.   ∎

Once more, let E be a Riesz space and assume now that $0 < e \in E$ (the case that $e = 0$ is excluded since this would not lead to anything of interest). Any $s \in E$ for which there exist pairwise disjoint components $p_1, \ldots, p_n$ of e and real numbers $\alpha_1, \ldots, \alpha_n$ such that $s = \Sigma_1^n \alpha_k p_k$ is called an *e-step function*. It is permitted here that one or several of the p_k or the coefficients α_k satisfies $p_k = 0$ or $\alpha_k = 0$. This way of writing s (with disjoint $p_1, \ldots, p_n$) is called a *standard representation* of s. Such a standard representation is not unique. Evidently, every e-step function is a member of the principal ideal A_e. Without loss of generality we may, if convenient, assume that $p_1 + \cdots + p_n = e$ in the standard representation $s = \Sigma_1^n \alpha_k p_k$ because, if not, then $p_1 + \cdots + p_n = \sup(p_1, \ldots, p_n)$ is a component of e, so $p_{n+1} = e - \sup(p_1, \ldots, p_n)$

is a component. Hence, we may add a term $\alpha_{n+1}p_{n+1}$ with $\alpha_{n+1} = 0$. In this case we shall say, for convenience again, that $s = \Sigma_1^{n+1}\alpha_k p_k$ is a full representation of s as an e-step function. If s and t are e-step functions with full representations

$$s = \Sigma_{k=1}^n \alpha_k p_k \text{ and } t = \Sigma_{j=1}^m \beta_j q_j,$$

then

$$s = \Sigma_{k=1}^n \Sigma_{j=1}^m \alpha_k (p_k \wedge q_j) \text{ and } t = \Sigma_{k=1}^n \Sigma_{j=1}^m \beta_j (p_k \wedge q_j)$$

are likewise full standard representations of s and t as e-step functions. Note here that $\Sigma_k \Sigma_j (p_k \wedge q_j) = e$ in view of

$$\Sigma_j (p_k \wedge q_j) = \sup_j (p_k \wedge q_j) = p_k \wedge (\sup q_j) = p_k \wedge e = p_k$$

for every k. Hence

$$s + t = \Sigma_k \Sigma_j (\alpha_k + \beta_j)(p_k \wedge q_j).$$

This shows that $s + t$ is an e-step function. The set of all e-step functions is, therefore, a linear subspace of E. We can say more. If s and t are e-step functions, the proof presented above shows that they can be written as linear combinations of the same disjoint components, i.e., if these components are $p_1, \ldots, p_n$, then $s = \Sigma_1^n \alpha_k p_k$ and $t = \Sigma_1^n \beta_k p_k$. If $\alpha_k \leq \beta_k$ for $k = 1, \ldots, n$, then $s \leq t$. To see that the converse holds, it is sufficient to prove that if $v = \Sigma_1^n \gamma_k p_k \geq 0$ with $p_k \neq 0$ for all k and all p_k mutually disjoint, then $\gamma_k \geq 0$ for $k = 1, \ldots, n$. For this purpose, write $v = v_1 + v_2$ with $v_1 = \Sigma' \gamma_k p_k$ containing all terms with $\gamma_k > 0$ (or $v_1 = 0$ if $\gamma_k \leq 0$ for all k) and $v_2 = \Sigma'' \gamma_k p_k$ containing all terms with $\gamma_k \leq 0$. Then $v_1 \geq 0, -v_2 \geq 0$ and $v_1 \wedge (-v_2) = 0$. It follows that $v_1 = v^+$ and $-v_2 = v^-$ (see Theorem 5.6). Since it is given here that $v \geq 0$, we have $v^- = 0$, i.e., all $\gamma_k \leq 0$ must be zero. In other words, $\gamma_k \geq 0$ holds for all k. Thus it has been proved that if $s = \Sigma_1^n \alpha_k p_k$ and $t = \Sigma_1^n \beta_k p_k$ (all $p_k \neq 0$ and mutually disjoint), then $s \leq t$ if and only if $\alpha_k \leq \beta_k$ for all k. This immediately implies that if again $s = \Sigma_1^n \alpha_k p_k$ and $t = \Sigma_1^n \beta_k p_k$, then $s \vee t$ and $s \wedge t$ are e-step functions with representations

$$s \vee t = \Sigma_1^n \{\max(\alpha_k, \beta_k)\} p_k \text{ and } s \wedge t = \Sigma_1^n \{\min(\alpha_k, \beta_k)\} p_k.$$

In particular $|s| = \Sigma_1^n |\alpha_k| p_k$. Therefore, the set of all e-step functions is a Riesz subspace of E.

33. Freudenthal's Spectral Theorem

Definitions and notations are as in the preceding section. Hence, E is a Riesz space and e is an element of E satisfying $e > 0$. We assume furthermore that E has the principal projection property, i.e., every principal band is a projection band. It will be proved that every f in the principal ideal A_e can be approximated e-uniformly by e-step functions, from below as well as from above. More precisely, there exist sequences $(s_n : n = 1, 2, \ldots)$ and $(t_n : n = 1, 2, \ldots)$ of e-step functions such that $s_n \uparrow f$ and $t_n \downarrow f$ hold e-uniformly. Since $f \in A_e$, there exist real numbers a and b such that $ae \leq f \leq (b-1)e$. For each real α we shall denote by B_α the band generated by $(\alpha e - f)^+$ and by P_α the band projection on B_α. For $\alpha \geq b$ we have $\alpha e - f \geq be - f \geq e$, so $(\alpha e - f)^+ = \alpha e - f \geq e$. It follows that B_α is equal to the band B_e generated by e, so $P_\alpha = P_e$. In particular $P_\alpha e = e$ and $P_\alpha f = f$ for $\alpha \geq b$. For $\alpha \leq a$ we have $\alpha e \leq f$, so $\alpha e - f \leq 0$, i.e., $(\alpha e - f)^+ = 0$. It follows that $B_\alpha = \{0\}$, so $P_\alpha = 0$ for $\alpha \leq a$. The following simple lemma is of importance for the approximation proof.

Lemma 33.1. *(i) For α real and $ae \leq f \leq (b-1)e$ we have*

$$P_\alpha f \leq \alpha P_\alpha e \quad and \quad \alpha(e - P_\alpha e) \leq f - P_\alpha f.$$

(ii) For $\alpha \leq \beta$ (with α and β real) and $ae \leq f \leq (b-1)e$ we have

$$\alpha(P_\beta - P_\alpha)e \leq (P_\beta - P_\alpha)f \leq \beta(P_\beta - P_\alpha)e.$$

Proof. (i) From $P_{g^+}(g) = g^+ \geq 0$, holding for all $g \in E$, it follows by choosing $g = (\alpha e - f)$ that $P_\alpha(\alpha e - f) \geq 0$, and so $P_\alpha f \leq \alpha P_\alpha e$.

Similarly, since $g - P_{g^+}(g) = g - g^+ = -g^- \leq 0$ for every $g \in E$, the choice $g = (\alpha e - f)$ shows that $\alpha e - f - P_\alpha(\alpha e - f) \leq 0$, and so $\alpha(e - P_\alpha e) \leq f - P_\alpha f$.

(ii) Observe first that the band generated by $(\alpha e - f)^+$ is included in the band generated by $(\beta e - f)^+$, and so $P_\alpha P_\beta = P_\beta P_\alpha = P_\alpha$ (see Theorem 32.2). Applying P_β to the inequality $\alpha(e - P_\alpha e) \leq f - P_\alpha f$, we obtain

$$\alpha(P_\beta e - P_\alpha e) \leq (P_\beta - P_\alpha)f.$$

Similarly, it follows from $P_\beta f \leq \beta P_\beta e$ that

$$(P_\beta - P_\alpha)f = (P_\beta - P_\alpha)P_\beta f \leq (P_\beta - P_\alpha)(\beta P_\beta e) = \beta(P_\beta - P_\alpha)e. \quad \blacksquare$$

We shall prove now Freudenthal's spectral theorem.

Theorem 33.2. *(Freudenthal's spectral theorem). Let E be a Riesz space having the principal projection property and let $0 < e \in E$. Then, for any f in the principal ideal A_e, there exist sequences $(s_n : n = 1, 2, \ldots)$ and $(t_n : n = 1, 2, \ldots)$ of e-step functions such that $s_n \uparrow f$ and $t_n \downarrow f$ hold e-uniformly.*

Proof. As observed already there exist real numbers a, b such that $ae \leq f \leq (b-1)e$. Since the e-step functions form a Riesz subspace of E, it is sufficient to prove that for any number $\epsilon > 0$ there exist e-step functions s and t such that $s \leq f \leq t$ and $t - s \leq \epsilon e$. For this purpose, let $a = \alpha_0 < \alpha_1 < \cdots < \alpha_n = b$ be a partition of $[a, b]$ with mesh less than ϵ, i.e., $0 < \alpha_k - \alpha_{k-1} < \epsilon$ for $k = 1, \ldots, n$. As in the lemma above, let $P_k (k = 0, \ldots, n)$ be the band projection onto the band generated by $(\alpha_k e - f)^+$ and let $v_k = (P_{\alpha_k} - P_{\alpha_{k-1}})e$ for $k = 1, \ldots n$. The elements v_k are mutually disjoint components of e (see Corollary 32.4) such that

$$v_1 + \cdots + v_n = (P_{\alpha_n} - P_{\alpha_0})e = (P_b - P_a)e = P_b e = e.$$

From the lemma established above it follows that

$$\Sigma_1^n \alpha_{k-1} v_k \leq \Sigma_1^n (P_{\alpha_k} - P_{\alpha_{k-1}})f \leq \Sigma_1^n \alpha_k v_k.$$

Since the term in the middle is equal to $(P_b - P_a)f = P_b f = f$, it follows that the e- step functions $s = \Sigma_1^n \alpha_{k-1} v_k$ and $t = \Sigma_1^n \alpha_k v_k$ satisfy $s \leq f \leq t$. Note finally that

$$t - s = \Sigma_1^n (\alpha_k - \alpha_{k-1})v_k \leq \epsilon \Sigma_1^n v_k = \epsilon e. \qquad \blacksquare$$

Writing $p_\alpha = P_\alpha e$ for abbreviation, the approximating sums s and t in the spectral theorem can be written as

$$s = \Sigma_1^n \alpha_{k-1}(p_{\alpha_k} - p_{\alpha_{k-1}}) \text{ and } t = \Sigma_1^n \alpha_k (p_{\alpha_k} - p_{\alpha_{k-1}})$$

respectively. It is evident, therefore, that if we now choose the numbers $\tilde{\alpha_k}$ ($k = 1, \ldots, n$) such that $\alpha_{k-1} \leq \tilde{\alpha_k} \leq \alpha_k$ for $k = 1, \ldots, n$, then the sum $\tilde{s} = \Sigma_1^n \tilde{\alpha_k}(p_{\alpha_k} - p_{\alpha_{k-1}})$ satisfies $s \leq \tilde{s} \leq t$, and so it follows from $s \leq f \leq t$ and $t - s \leq \epsilon e$ that $|f - \tilde{s}| \leq \epsilon e$, i.e.,

$$|f - \Sigma_1^n \tilde{\alpha_k}(p_{\alpha_k} - p_{\alpha_{k-1}})| \leq \epsilon e.$$

A suggestive notation is to write now

$$f = \int_a^b \alpha \, dp_\alpha.$$

This is justified by the analogy with the procedure of H.Lebesgue for defining the integral of a bounded realvalued measurable function. Note that if E is the Riesz space of all real functions on a point set X and e is the unit function on X (i.e., $e(x) = 1$ for all $x \in X$), then the spectral theorem states that every bounded real function can be approximated uniformly (in the familiar sense) by step functions, from below as well as from above.

We extend the spectral theorem to elements in the band B_e generated by e. The e-uniform convergence will be replaced by order convergence. Precisely, the following theorem holds.

Theorem 33.3. *As before, let E be a Riesz space having the principal projection property and let $0 \le e \in E$. Then, for any $u \ge 0$ in the band B_e generated by e there exists a sequence $(s_n : n = 1, 2, \ldots)$ of e-step functions such that $0 \le s_n \uparrow u$.*

Proof. Let $0 \le u \in B_e$ be given. As proved in Corollary 11.6, we have $u = \sup_n(u \wedge ne)$. Hence, writing $u_n = u \wedge ne$ for $n = 1, 2, \ldots$, every u_n is a member of the ideal A_e generated by e and $0 \le u_n \uparrow u$. In view of the last theorem there exists for each n a positive e-step function s_n such that $0 \le u_n - s_n \le n^{-1}e$. Without loss of generality we may assume that $s_n \uparrow$ (if necessary, replace s_n by $s_1 \vee \ldots \vee s_n$). Observe now that u_n converges in order to u and $u_n - s_n$ converges in order to 0. It follows that $s_n = u_n - (u_n - s_n)$ converges in order to u. Since s_n is increasing this shows that $s_n \uparrow u$.   ∎

As a special example we present a second proof of the Radon-Nikodym theorem in which use is made of Freudenthal's spectral theorem. The proof is direct, not by contradiction as the one in section 28. We recall some notations and results from section 27. Let Γ be a σ-algebra of subsets of the (non-empty) point set X and let M_σ be the Riesz space of all σ-additive signed measures on Γ.

(i) It was proved in Theorem 27.4(i) that if μ and ν are members of M_σ, then $\mu \perp \nu$ (i.e., $\mid \mu \mid \wedge \mid \nu \mid = 0$) if and only if there exist disjoint sets A and B in Γ such that $A \cup B = X$ and $\mid \nu \mid (A) = \mid \mu \mid (B) = 0$.

(ii) In Theorem 27.4(ii) it was shown that ν is an element of the band generated by μ if and only if ν is μ-absolutely continuous (i.e., any $A \in \Gamma$ satisfying $\mid \mu \mid (A) = 0$ satisfies $\mid \nu \mid (A) = 0$).

Let us assume now that $0 \le \mu \in M_\sigma$ is fixed. We determine the components of μ.

Lemma 33.4. *The measure π is a component of μ (i.e., $0 \le \pi \le \mu$ and $\pi \perp (\mu - \pi)$) if and only if there exists a set $A \in \Gamma$ such that*

$$\pi(Y) = \int_Y \chi_A d\mu$$

holds for every $Y \in \Gamma$ (where, as usual, the characteristic function of A is denoted by χ_A).

Proof. Assume first that π is a component of μ. In view of (i) above there exist disjoint sets A and B in Γ such that $A \cup B = X$ and $(\mu - \pi)(A) = \pi(B) = 0$. It follows that $\pi(Y) = \mu(Y)$ for every $Y \in \Gamma$ such that Y is a subset of A. Hence, if $Y \in \Gamma$ is now arbitrary, we have

$$\pi(Y) = \pi(Y \cap A) = \mu(Y \cap A) = \int_Y \chi_A d\mu.$$

Conversely, if A and B are disjoint sets in Γ such that $A \cup B = X$ and we define π by $\pi(Y) = \int_Y \chi_A d\mu$ for all $Y \in \Gamma$, then π is a σ-additive measure on Γ such that $0 \leq \pi \leq \mu$ and such that

$$(\mu - \pi)(Y) = \int_Y (1 - \chi_A)d\mu = \int_Y \chi_B d\mu$$

holds for all $Y \in \Gamma$. It is evident, therefore, that $\pi(B) = (\mu - \pi)(A) = 0$. Hence, in view of (i) above, we have $\pi \perp (\mu - \pi)$.

As a final remark we note that the set A in the present lemma is not uniquely determined. If $Z \in \Gamma$ satisfies $\mu(Z) = 0$, then Z or any subset of Z may be joined to or deleted from A without influencing the result. ∎

It follows from the lemma that for any μ-step function $\sigma = \Sigma_1^n \alpha_k \pi_k$ (where $\pi_1, \ldots, \pi_n$ are components of μ and $\alpha_1, \ldots, \alpha_n$ are real numbers) there exist sets $A_1, \ldots, A_n$ in Γ such that

$$\sigma(Y) = \int_Y \Sigma_1^n \alpha_k \chi_{A_k} d\mu$$

holds for every $Y \in \Gamma$. Hence, denoting the μ-measurable real step function $\Sigma_1^n \alpha_k \chi_{A_k}$ by s, we have $\sigma(Y) = \int_Y s d\mu$ for every $Y \in \Gamma$. It follows that $\sigma \geq 0$ holds in M_σ if and only if $s(x) \geq 0$ holds for μ-almost every $x \in X$ (because $s(x) < 0$ on a set $B \in \Gamma$ of positive measure would imply that $\sigma(B) < 0$). Note that $s(x) \geq 0$ almost everywhere is not the same as asserting that $\alpha_k \geq 0$ holds for all coefficients in s; this holds if and only if s is written with non-zero disjoint terms. We may conclude from these remarks, therefore, that if σ_1 and σ_2 are μ-step functions in M_σ, then $\sigma_1 \leq \sigma_2$ if and only if the corresponding real step functions s_1 and s_2 satisfy $s_1(x) \leq s_2(x)$ for μ-almost every $x \in X$.

The proof of the Radon-Nikodym theorem is now easy. Let μ be the same measure as before, so $0 \leq \mu \in M_\sigma$, and let ν be a σ-additive signed measure which is μ-absolutely continuous, so ν is a member of the band B_μ generated by μ in M_σ. We have to prove the existence of a (real) μ-measurable function f such that $\nu(Y) = \int_Y f d\mu$ holds for every $Y \in \Gamma$. Without loss of generality we may assume that ν is positive, i.e., $\nu \in B_\mu^+$. In view of the spectral theorem there exists a sequence $(\sigma_n : n = 1, 2, \ldots)$ of μ-step functions in B_μ such that $0 \leq \sigma_n \uparrow \nu$. For each σ_n there exists a non-negative μ-measurable step function s_n on X such that

$$\sigma_n(Y) = \int_Y s_n(x)d\mu \text{ for every } Y \in \Gamma.$$

From $0 \leq \sigma_n \uparrow \nu$ (in B_μ) it follows that $0 \leq \sigma_n(Y) \uparrow \nu(Y)$, so

$$\int_Y s_n(x)d\mu \uparrow \nu(Y) \text{ for every } Y \in \Gamma.$$

Since $0 \leq s_n(x) \uparrow$ holds for μ-almost every $x \in X$, the integral on the left converges to $\int_Y f(x)d\mu$, where f is the pointwise limit (μ-almost everywhere) of the increasing sequence $(s_n : n = 1, 2, \ldots)$. Hence

$$\nu(Y) = \int_Y f d\mu \text{ for every } Y \in \Gamma,$$

which is the desired result. This method of proving the Radon-Nikodym theorem is related to the method in a paper by K.Yosida (1940).

CHAPTER 18
Functional Calculas and Multiplication

34. Functional Calculus

As in the preceding section we assume that E is a Riesz space having the principal projection property. Let $0 < e \in E$. Since we shall restrict ourselves to properties holding in the principal ideal A_e generated by e we may also assume that $A_e = E$, i.e., e is a strong unit in E. Given the element $f \in E = A_e$, there exist real numbers $a, b (a < b)$ and a number $\delta > 0$ such that $ae \le f \le (b - \delta)e$. The interval $[a, b]$ is then sometimes called a *spectral interval* of f. Let $[a, b]$ be such a spectral interval of f and let

$$\mathcal{P} : a = \alpha_0 < \alpha_1 < \cdots < \alpha_m = b$$

be a partition of $[a, b]$. For any $\alpha \in [a, b]$ the band projection onto the band generated by $(\alpha e - f)^+$ will be denoted by P_α. Note that $P_{\alpha_0} = 0$ and $P_{\alpha_m} = I$ (the identity operator). The equality $P_{\alpha_m} = P_b = I$ holds because $be - f \ge \delta e$, so the band generated by $(be - f)^+ = be - f$ is the band generated by e, i.e., it is E. Writing

$$v_k = (P_{\alpha_k} - P_{\alpha_{k-1}})e \text{ for } k = 1, \ldots, m,$$

the elements v_k are pairwise disjoint components of e such that $\Sigma_1^m v_k = e$ and the e-step functions $s = \Sigma_1^m \alpha_{k-1} v_k$ and $S = \Sigma_1^m \alpha_k v_k$ satisfy $s \le f \le S$ (see the proof of Freudenthal's spectral theorem in the preceding section). The elements s and S are called the *lower sum* and *upper sum* belonging to f and the partition $\mathcal{P}$. If the partition points are sufficiently near to each other, then both s and S are near to f. Precisely stated, if $\alpha_k - \alpha_{k-1} \le \epsilon$ for $k = 1, \ldots, m$, then $0 \le S - s \le \epsilon e$, so

$$0 \le f - s \le \epsilon e \text{ and } 0 \le S - f \le \epsilon e.$$

In this case, observe also that if the numbers $\alpha_{\tilde{k}}$ are chosen such that $\alpha_{k-1} \le \alpha_{\tilde{k}} \le \alpha_k$ for $k = 1, \ldots, m$, then the e-step function $s^{\tilde{}} = \Sigma_1^m \alpha_{\tilde{k}} v_k$ satisfies $s \le s^{\tilde{}} \le S$, and so $| f - s^{\tilde{}} | \le \epsilon e$.

We shall consider now a somewhat more general situation. Let F be a real continuous function on the spectral interval $[a.b]$ and let, for $k = 1, \ldots, m$, the numbers m_k and M_k be the minimum and maximum respectively of F in the interval $[\alpha_{k-1}, \alpha_k]$ of the partition $\mathcal{P}$ of $[a, b]$. Similarly as before for the case that $F(\alpha) = \alpha$ for all $\alpha \in [a, b]$, but with a slightly different notation, let $s = \Sigma_1^m m_k v_k$ and $S = \Sigma_1^m M_k v_k$ be the corresponding lower and upper sums. For any $\epsilon > 0$ there exists a number $\alpha_0 > 0$ such that

$$| F(\alpha') - F(\alpha'') | \leq \epsilon \text{ if } | \alpha' - \alpha'' | \leq \alpha_0.$$

This holds because a continuous function on a closed interval is uniformly continuous. Hence, if the partition points of $\mathcal{P}$ are sufficiently near to each other, i.e., if $\alpha_k - \alpha_{k-1} \leq \alpha_0$ for $k = 1, \ldots, m$, then $0 \leq M_k - m_k \leq \epsilon$ for all k, and so

$$S - s = \Sigma_1^m (M_k - m_k)v_k \leq \epsilon \Sigma_1^m v_k = \epsilon e.$$

If the numbers $m_{\tilde{k}}(k = 1, \ldots, m)$ are chosen such that $m_{\tilde{k}} = F(\alpha_{\tilde{k}})$ for some $\alpha_{\tilde{k}}$ satisfying $\alpha_{k-1} \leq \alpha_{\tilde{k}} \leq \alpha_k$, then $m_k \leq m_{\tilde{k}} \leq M_k$ for all k and $s^{\tilde{}} = \Sigma_1^m m_{\tilde{k}} v_k$ is now called an *intermediate sum* (belonging to f and $\mathcal{P}$). It is evident that $s \leq s^{\tilde{}} \leq S$, so if the partition $\mathcal{P}$ is such that we have $S - s \leq \epsilon e$ for some $\epsilon > 0$ and $s_{\tilde{1}}, s_{\tilde{2}}$ are intermediate sums belonging to $\mathcal{P}$, then $| s_{\tilde{1}} - s_{\tilde{2}} | \leq \epsilon e$.

In the case that $F(\alpha) = \alpha$ for all α in the spectral interval $[a, b]$ of f, the upper and lower sums tend to a common limit if the maximal length of the partition intervals tends to zero, namely to the limit f (Freudenthal's spectral theorem). It may be asked, therefore, if also in the present more general case S and s tend to a common limit if the partition intervals tend to zero in length. We shall prove that this is indeed so if, besides having the principal projection property, E is also uniformly complete. In our case, since e is a strong unit in E, the space E is uniformly complete if and only if E is e-uniformly complete (see section 10). Furthermore, as proved in Theorem 12.8, E is uniformly complete if E is Dedekind σ-complete. Therefore, if E is Dedekind σ-complete, then E has the principal projection property and besides that E is uniformly complete. This is what we shall use now to derive some results related to the spectral theorem. It can be proved, but the proof is somewhat laborious, that Dedekind σ-completeness of E is not only sufficient but also necessary for E to have the principal projection property and to be uniformly complete. Before presenting now our first theorem we make some further remarks about partitions of $[a, b]$ and the corresponding upper and lower sums. As well-known, the partition $\mathcal{P}_2$ is said to be a *refinement* of the partition $\mathcal{P}_1$ if every partition point of $\mathcal{P}_1$ is also a partition point of $\mathcal{P}_2$. The corresponding lower sums s_1, s_2 and upper sums S_1, S_2 satisfy $s_1 \leq s_2 \leq S_2 \leq S_1$. If $\mathcal{P}_1$ and $\mathcal{P}_2$ are arbitrary, their common refinement is the partition $\mathcal{P}_3$ having as partition points the union of the partition points of $\mathcal{P}_1$ and $\mathcal{P}_2$. It follows that $s_1 \leq s_3 \leq S_3 \leq S_2$. Hence, since $\mathcal{P}_1$ and $\mathcal{P}_2$ are arbitrary, the inequality $s_1 \leq S_2$ shows that any lower sum is less than or equal to any upper sum, no matter to which partitions these sums belong. A further (notational) remark is that we shall sometimes write $s(\mathcal{P})$ and $S(\mathcal{P})$ to denote that s and S depend on $\mathcal{P}$.

Theorem 34.1. *Let e be a strong unit in the Dedekind σ-complete space E, let $f \in E$ be fixed and let $[a, b]$ be a spectral interval for f. Furthermore, let F be a real continuous function on $[a, b]$. For any partition $\mathcal{P}$ of $[a, b]$, let $s = s(\mathcal{P})$ and $S = S(\mathcal{P})$ be the corresponding lower and upper sums (for f and $\mathcal{P}$), as explained above. Then the set of all $s(\mathcal{P})$, for all possible partitions $\mathcal{P}$ of $[a, b]$, has a supremum in E which is at the same time the infimum of all possible $S(\mathcal{P})$. We shall denote this element by $F(f)$. The reasons for this notation will become clear later. For any number $\epsilon > 0$ there exists a number $\alpha_0 > 0$ such that if the maximal length of the partition intervals of $\mathcal{P}$ is less than α_0 and $s^\sim(\mathcal{P})$ is any intermediate sum for $\mathcal{P}$, then*

$$| F(f) - s^\sim(\mathcal{P}) | \le \epsilon e.$$

Proof. Choose a sequence of numbers $(\epsilon_n : n = 1, 2, \ldots)$ such that $\epsilon_n \downarrow 0$. For each ϵ_n there exists a partition $\mathcal{P}_n$ of $[a, b]$ such that $S_n - s_n \le \epsilon_n e$. It may be assumed here that each $\mathcal{P}_{n+1}$ is a refinement of $\mathcal{P}_n$ (if necessary, replace $\mathcal{P}_2$ by the common refinement of $\mathcal{P}_1$ and $\mathcal{P}_2$, and so on). Then $(s_n : n = 1, 2, \ldots)$ is an increasing sequence in E majorized by S_1 (on account of $s_n \le S_1$ for every n). Hence, since E is Dedekind σ-complete, the supremum of this sequence exists. Denoting this supremum by $F(f)$, ee have $s_n \uparrow F(f)$. Since $S_n - s_n \le \epsilon_n e \downarrow 0$, it follows that $S_n - s_n \to 0$, so $S_n = (S_n - s_n) + s_n \to F(f)$. Precisely, $S_n \downarrow F(f)$.

We prove that $F(f) = \sup(s(\mathcal{P}): \text{all } \mathcal{P})$. Since we know already that $F(f) = \sup_n s(\mathcal{P}_n)$, it is sufficient to prove that $F(f) \ge s(\mathcal{P})$ for any arbitrary $\mathcal{P}$. Let $(\mathcal{P}_n : n = 1, 2, \ldots)$ be the same sequence of partitions as above. Since $s(\mathcal{P}) \le S(\mathcal{P}_n)$ for all n and $S(\mathcal{P}_n) \downarrow F(f)$, it follows that $s(\mathcal{P}) \le F(f)$. Similarly, $F(f) \le S(\mathcal{P})$ for every $\mathcal{P}$. Hence

$$F(f) = \sup(s(\mathcal{P}): \text{all } \mathcal{P}) = \inf(S(\mathcal{P}): \text{all } \mathcal{P}).$$

Finally, given $\epsilon > 0$, there exists a number $\alpha_0 > 0$ such that if the maximal length of the partition intervals of $\mathcal{P}$ is less than α_0, then $0 \le S(\mathcal{P}) - s(\mathcal{P}) \le \epsilon e$. Any intermediate sum $s^\sim(\mathcal{P})$ lies between $s(\mathcal{P})$ and $S(\mathcal{P})$ and the same holds for $F(f)$. Hence

$$| F(f) - s^\sim(\mathcal{P}) | \le \epsilon e. \qquad \blacksquare$$

As observed in section 33, Freudenthal's spectral theorem is sometimes expressed by writing

$$f = \int_a^b \alpha \, dp_\alpha,$$

where $[a, b]$ is a spectral interval of f and where $p_\alpha = P_\alpha e$ with P_α the band projection onto the band generated by $(\alpha e - f)^+$. Similarly, we can write now

$$f^n = \int_a^b \alpha^n dp_\alpha \text{ for } n = 1, 2, \ldots$$

and, more generally,

$$F(f) = \int_a^b F(\alpha) dp_\alpha$$

for any continuous real function F on $[a, b]$.

This method of defining $F(f)$ for any $f \in E$ and any (real) continuous F on $[a, b]$ is an example of what is known as a *functional calculus*.

We mention some examples. With the same notations as before, if $F(\alpha) = 1$ for all $\alpha \in [a, b]$, then all lower sums, upper sums and intermediate sums are equal to $\Sigma_1^m v_k = e$, so $F(f) = e$ in this case. If $F(\alpha) = \alpha$ for all $\alpha \in [a, b]$, then $F(f) = f$ according to Freudenthal's theorem. If n is a natural number and $F(\alpha) = \alpha^n$ for all $\alpha \in [a, b]$, then the intermediate sums are of the form $\Sigma_1^m (\alpha_{\tilde{k}})^n v_k$, where $\alpha_{\tilde{k}}$ is some value in the interval $[\alpha_{k-1}, \alpha_k]$. These intermediate sums, therefore, converge now e-uniformly to what, for $n = 2, 3, \ldots$, we have called f^n by definition. The notation suggests that it should perhaps be possible to define a multiplication in E. In particular, since for f and g in E the elements f^2, g^2 and $(f + g)^2$ are now defined, one might try to define the product fg by

$$fg = \frac{1}{2}\{(f + g)^2 - f^2 - g^2\}.$$

It seems difficult, however, to prove that with this definition the distributive and associate laws hold. One of the main difficulties is caused by the fact that in the above formula for an intermediate sum belonging to $\mathcal{P}$ and f the elements v_k depend not only on $\mathcal{P}$ but also on f. To show that it is possible to define a multiplication in E with the usual properties we shall proceed differently. First some remarks dealing with the intermediate sums in the case that $f = e$. Any interval $[a, b]$ with 1 as an interior point is now a spectral interval for $f = e$ and the band generated by $(\alpha e - e)^+$ is E for $\alpha > 1$ and $\{0\}$ for $\alpha \leq 1$. Therefore, if $\mathcal{P}$ is a partition of $[a, b]$ with 1 as an interior point of $[\alpha_{k-1}, \alpha_k]$ for $k = k_0$, then the elements v_k in the corresponding lower, upper and intermediate sums satisfy $v_k = e$ for $k = k_0$ and $v_k = 0$ for $k \neq k_0$. Choosing now $\alpha_{\tilde{k}} = 1$ for $k = k_0$ and $\alpha_{\tilde{k}}$ arbitrary in $[\alpha_{k-1}, \alpha_k]$ for $k \neq k_0$, the corresponding intermediate sum $s^{\tilde{}}$ satisfies

$$s^{\tilde{}} = \Sigma \alpha_{\tilde{k}} v_k = 1 \cdot v_{k_0} = e.$$

Note that also

$$\Sigma(\alpha\tilde{k})^n v_k = 1 \cdot v_{k_0} = e \text{ for } n = 2, 3, \ldots.$$

Hence, since the sum on the left converges to e^n according to our notations, we have $e^n = e$ for $n = 2, 3, \ldots$. This suggests that the multiplication we are looking for should have the order unit e as its multiplicative unit. Then $fe = f \geq 0$ for f positive, so it is reasonable to require that the multiplication is *positive*, i.e., if f and g are positive, then fg is positive.

Assume that there exists such an associative and distributive positive multiplication in E having e as multiplicative unit. Fix $f_0 \in E$ and define the mapping $\pi : E \to E$ by $\pi g = f_0 g$ for all $g \in E$. It is obvious that π is linear, so π is an operator in E according to our conventions. Furthermore, since there exist real numbers α, β such that $\alpha e \leq f_0 \leq \beta e$, it follows for any $u \geq 0$ in E that

$$\alpha u = \alpha e u \leq f_0 u \leq \beta e u = \beta u,$$

i.e., $\alpha u \leq \pi u \leq \beta u$. This shows that $\alpha I \leq \pi \leq \beta I$, where I is the identity operator in E. Furthermore $\pi e = f_0 e = f_0$. Now, let us forget that π is derived from a multiplication and prove that if π is any operator in E such that $\pi e = f_0$ and $\alpha I \leq \pi \leq \beta I$, where f_0, α and β are given, then π is uniquely determined.

Lemma 34.2. *If $f_0 \in E$ is given and π is an operator in E such that $\pi e = f_0$ and there exist real numbers α, β such that $\alpha I \leq \pi \leq \beta I$ (where I is the identity operator in E), then π is uniquely determined.*

Proof. Note first that it follows from $\alpha u \leq \pi u \leq \beta u$ for any $u \geq 0$ that $| \pi u | \leq \gamma u$, where $\gamma = \max(\beta, -\alpha)$. We prove now that $g \perp h$ in E implies that $\pi g \perp h$. To this end, observe that it follows from $g^+ \perp h$ and $| \pi g^+ | \leq \gamma g^+$ that $\pi g^+ \perp h$. Similarly $\pi g^- \perp h$. Hence $\pi g \perp h$. Assume now that B is a band in E and $g \in B$. Then it follows from $| \pi g^+ | \leq \gamma g^+$ and $| \pi g^- | \leq \gamma g^-$ that πg^+ and πg^- are members of B, so $\pi g \in B$. We have proved thus that $g \in B$ implies $\pi g \in B$. This can be expressed by saying that π is *band preserving*. In the case that B is a projection band with corresponding band projection P and $g \in E$ is arbitrary, we have $g = g_1 + g_2$ with $g_1 = Pg \in B$ and $g_2 \perp B$, so

$$\pi g = \pi g_1 + \pi g_2 \text{ with } \pi g_1 \in B \text{ and } \pi g_2 \perp B.$$

It follows that $\pi Pg = \pi g_1$ as well as

$$P\pi g = P\pi g_1 + P\pi g_2 = P\pi g_1 = \pi g_1.$$

Hence $\pi Pg = P\pi g$ for every $g \in E$, i.e., $\pi P = P\pi$. This shows that π commutes with any arbitrary band projection.

Now, let p be a component of e and denote by P the band projection onto the principal band generated by p, so $p = Pe$. Then

$$\pi p = \pi P e = P \pi e = P f_0.$$

Thus, πp is uniquely determined by the given conditions. If $s = \Sigma_1^m \alpha_k p_k$ is an e- step function, where $p_1, \ldots, p_m$ are components of e with corresponding band projections $P_1, \ldots, P_m$, then $\pi s = \Sigma_1^m \alpha_k P_k f_0$. For any $g \in E$ there exists a sequence $(s_n : n = 1, 2, \ldots)$ of e-step functions such that $s_n \uparrow g$ holds e-uniformly. Hence

$$| \pi g - \pi s_n | = | \pi(g - s_n) | \leq \gamma(g - s_n) \downarrow 0,$$

so $\pi s_n \to \pi g$ holds e-uniformly. This shows, therefore, that πg is uniquely determined for any $g \in E$ by the given conditions.   ∎

The next result to prove is that there actually exists an operator π in E satisfying the conditions in the last lemma. This operator will then be denoted by π_{f_0} to indicate that it depends on the given f_0 for which $\pi e = f_0$. Writing simply f and πf instead of f_0 and π_{f_0}, the desired multiplication will be defined for any pair f, g in E by $fg = \pi_f g$.

Theorem 34.3. *Let $f \in E$ be given. Therefore, there exist real numbers α, β such that $\alpha e \leq f \leq \beta e$ holds. Then there exists a uniquely defined operator π in E such that $\pi e = f$ and $\alpha I \leq \pi \leq \beta I$, where I is the identity operator in E. If f is positive, we may choose $\alpha = 0$ and so π is then positive.*

Proof. We prove the existence of an operator π satisfying the given conditions by defining a mapping $\pi : E \to E$ step by step and showing that π is linear and satisfies $\alpha I \leq \pi \leq \beta I$. First we define πp for any component p of e by $\pi p = Pf$, where P is the band projection onto the band generated by p. In particular, $\pi e = f$. It is important for what follows to note that π is additive on the set of components of e because if $p = p_1 + p_2$, then p_1 and p_2 are necessarily disjoint, so the corresponding band projections satisfy $P = P_1 + P_2$, which implies that $\pi p = \pi p_1 + \pi p_2$. Now, let $s = \Sigma_1^m \alpha_k p_k$ be an arbitrary e-step function, written with disjoint terms and such that $\Sigma_1^m p_k = e$ (some of the α_k may now be zero). We define πs by $\pi s = \Sigma_1^m \alpha_k \pi p_k$. The mapping π is now uniquely defined on the set of e-step functions; the uniqueness is due to the additivity of π on the set of components of e. For the linearity of π it is sufficient to show that $\pi(s_1 + s_2) = \pi s_1 + \pi s_2$. This follows by writing s_1 and s_2 as linear combinations of the same disjoint components (see the discussion on e-step functions in section 32). It follows from the linearity that $s = \Sigma \alpha_k p_k$ implies $\pi s = \Sigma \alpha_k \pi p_k$, even in the case that the p_k are not disjoint. Let now $s = \Sigma \alpha_k p_k \geq 0$ with disjoint components and terms with $p_k = 0$ omitted. Then, as observed already earlier (near the end of section 32), all α_k are non-negative. Hence, on account of $f \leq \beta e$, we have

$$\pi s = \Sigma \alpha_k P_k f \leq \Sigma \alpha_k \beta P_k e = \beta \Sigma \alpha_k p_k = \beta s,$$

so $\pi \leq \beta I$. Similarly $\alpha I \leq \pi$. Observe furthermore that it follows from $\alpha s \leq \pi s \leq \beta s$ for $s \geq 0$ that $\mid \pi s \mid \leq \gamma s$ for $\gamma = \max(\beta, -\alpha)$. Hence, for s not necessarily positive, we have

$$\mid \pi s \mid \leq \mid \pi s^+ \mid + \mid \pi s^- \mid \leq \gamma s^+ + \gamma s^- = \gamma \mid s \mid .$$

Finally, let $g \in E$ be given. There exists a sequence $(s_n : n = 1, 2, \ldots)$ of e-step functions converging e-uniformly to g. The sequence is then an e-uniform Cauchy sequence, so it follows from

$$\mid \pi(s_m - s_n) \mid \leq \gamma \mid s_m - s_n \mid$$

that $(\pi s_n : n = 1, 2, \ldots)$ is an e-uniform Cauchy sequence. Since E is uniformly complete, the sequence $(\pi s_n : n = 1, \ldots)$ has an e-uniform limit. Call this limit πg and observe that πg depends on g but not on the approximating sequence. For $g \geq 0$ we may assume that $s_n \geq 0$ for all n, so $\beta s_n - \pi s_n \geq 0$ for all n. Since $\beta s_n - \pi s_n$ converges e-uniformly (and therefore also in order) to $\beta g - \pi g$, it follows that $\beta g - \pi g \geq 0$. Hence $\alpha g \leq \pi g \leq \beta g$ for every $g \geq 0$. The operator π has, therefore, all the desired properties. The preceding lemma (Lemma 34.2) shows that there exists only one operator having all these properties. $\blacksquare$

Remark 34.4. Some additional remarks follow. From the inequality $\mid \pi s \mid \leq \gamma \mid s \mid$, holding for any e-step function s, it follows immediately by e-uniform convergence that $\mid \pi g \mid \leq \gamma \mid g \mid$ holds for any $g \in E$. Hence, if $g_n \to g$ holds e-uniformly, then

$$\mid \pi g - \pi g_n \mid = \mid \pi(g - g_n) \mid \leq \gamma \mid g - g_n \mid \to 0$$

holds e-uniformly,i.e., πg_n converges e-uniformly to πg.

For the next observations we shall denote the operator π satisfying $\pi e = f$ by $\pi[f]$ to emphasize that π depends on f. It is obvious that for arbitrary f and g in E and α, β real numbers we have

$$\pi[\alpha f + \beta g] = \alpha \pi[f] + \beta \pi[g].$$

Let f_n converge e-uniformly to f in E. For any $\epsilon > 0$ there exists, therefore, a natural number n_0 such that $-\epsilon e \leq f - f_n \leq \epsilon e$ for all $n \geq n_0$. Now fix $h \in E$ (so $\mid h \mid \leq \alpha e$ for some real $\alpha > 0$). Then

$$\mid \pi[f]h - \pi[f_n]h \mid = \mid \pi[f - f_n]h \mid \leq \epsilon \alpha e \text{ for } n \geq n_0.$$

This shows that for $h \in E$ fixed $\pi[f_n]h$ converges e-uniformly to $\pi[f]h$.

Let us consider some simple examples.

(i) The case that $f = e$, so $\pi[e]e = e$. Recall that for any f the operator $\pi[f]$ commutes with any band projection. Let p be a component of e with corresponding band projection P, so $p = Pe$. Then

$$\pi[e]p = \pi[e]Pe = P\pi[e]e = Pe = p.$$

It follows that $\pi[e]s = s$ for any e-step function s and so, by e-uniform convergence, $\pi[e]g = g$ for any $g \in E$. Hence $\pi[e] = I$, the identity operator.

(ii) The case that $f = p_0$, where p_0 is a component of e with corresponding band projection P_0. Hence $\pi[p_0]e = p_0$. Let p be an arbitrary component of e with corresponding band projection P. Then

$$\pi[p_0]p = \pi[p_0]Pe = P\pi[p_0]e = Pp_0 = PP_0e = P_0Pe = P_0p,$$

where it has been used that band projections commute (see Theorem 32.1). It follows that $\pi[p_0]s = P_0s$ for any e-step function s and so, by e-uniform convergence, $\pi[p_0]g = P_0g$ for any $g \in E$. Hence $\pi[p_0] = P_0$.

It is important to note here that, similarly, $\pi[p]p_0 = P_0Pe = PP_0e$, so

$$\pi[p_0]p = \pi[p]p_0$$

for any pair p_0 and p of components of e.

35. Multiplication

Definitions and notations are as in the preceding section. After all the preliminary results obtained there we now introduce a multiplication in E by defining for any pair f, g in E the product fg by $fg = \pi[f]g$. The thus defined product has the usual properties of a product as we shall prove here.

(i) The distributive laws follow immediately from

$$f(\alpha g + \beta h) = \pi[f](\alpha g + \beta h) = \alpha\pi[f]g + \beta\pi[f]h = \alpha fg + \beta fh,$$

$$(\alpha f + \beta g)h = \pi[\alpha f + \beta g]h = (\alpha\pi[f] + \beta\pi[g])h =$$

$$\alpha\pi[f]h + \beta\pi[g]h = \alpha fh + \beta gh,$$

holding for all real numbers α, β.

(ii) The order unit e in E is a multiplicative unit element because $ef = \pi[e]f = If = f$ as well as $fe = \pi[f]e = f$ for any $f \in E$. It has been used here that $\pi[e] = I$, the identity operator, and $\pi[f]e = f$ by the definition of $\pi[f]$.

(iii) The multiplication is positive, i.e., if f and g are positive, then fg is positive. To see this, recall that in Theorem 34.3 it was shown that the operator $\pi[f]$ is positive for positive f. Hence, $fg = \pi[f]g$ is positive for $f \geq 0$ and $g \geq 0$.

(iv) The multiplication is commutative, i.e., $fg = gf$ for all f,g. For the proof, observe that $\pi[p_0]p = \pi[p]p_0$ for any two components p_0, p of e, i.e.,

$p_0 p = p p_0$ (see Remark 34.4). Hence, by the distributive laws, $s_0 s = s s_0$ for any two e-step functions s, s_0. Given $g \in E$, there exists a sequence ($s_n : n = 1, 2, \dots$) of e-step functions converging e-uniformly to g. It follows then that $s_0 s_n = \pi[s_0] s_n$ converges e-uniformly to $s_0 g$ and $s_n s_0 = \pi[s_n] s_0$ converges e-uniformly to $g s_0$. Hence, since $s_0 s_n = s_n s_0$ for all n, we get $s_0 g = g s_0$. Similarly, it follows now from $s_n^{\sim} \to f$ (e-uniformly) and $s_n^{\sim} g = g s_n^{\sim}$ that $fg = gf$.

(v) The multiplication is associative, i.e., $(fg)h = f(gh)$ for all f, g, h in E. Note that

$$(fg)h = \pi[fg]h \text{ and } f(gh) = \pi[f](gh) = \pi[f]\pi[g]h.$$

We have to show, therefore, that the operators $\pi[fg]$ and $\pi[f]\pi[g]$ are the same. It is easy to see that we may restrict ourselves to the case that f and g are positive. Let $0 \le f \le \alpha e$ and $0 \le g \le \beta e$, so $0 \le \pi[f] \le \alpha I$ and $0 \le \pi[g] \le \beta I$ (as shown in Theorem 34.3). It follows that $\pi[f]\pi[g]$ is a positive operator in E satisfying $\pi[f]\pi[g] \le \alpha\beta I$. Furthermore

$$fg = \pi[f]g \le \alpha I g = \alpha g \le \alpha\beta e,$$

so $\pi[fg] \le \alpha\beta I$ (once more in view of Theorem 34.3). The operators $\pi[fg]$ and $\pi[f]\pi[g]$ are, therefore, bounded from below as well as from above by a multiple of the identity operator. According to Lemma 34.2 an operator of this kind (called π in Lemma 34.2) is completely determined by its image πe. In the case we consider here we have

$$\pi[fg]e = fg \text{ and } \pi[f]\pi[g]e = \pi[f]g = fg.$$

Hence, $\pi[fg]$ is the same operator as $\pi[f]\pi[g]$.

(vi) $\mid fg \mid \le \mid f \mid \cdot \mid g \mid$ for all f, g. This follows immediately by writing $f = f^+ - f^-$ and $g = g^+ - g^-$.

(vii) If $g \perp h$, then $fg \perp h$ for all $f \in E$. For the proof, assume first that f, g and h are positive. Let $0 \le f \le \alpha e$. Then $0 \le fg \le \alpha eg = \alpha g$. Hence, since $\alpha g \perp h$, it follows that $fg \perp h$. For f, g and h arbitrary, we derive from $g \perp h$ that $\mid g \mid \perp \mid h \mid$, so $\mid f \mid \cdot \mid g \mid \perp \mid h \mid$. Hence, since $\mid fg \mid \le \mid f \mid \cdot \mid g \mid$, we have $\mid fg \mid \perp \mid h \mid$, so $fg \perp h$. It follows easily that

$$\mid fg \mid = \mid f \mid \cdot \mid g \mid \text{ for all } f, g.$$

As an immediate corollary we obtain the result that $g \perp h$ implies $gh = 0$. Indeed, from $g \perp h$ it follows that $gh = hg \perp h$. Similarly, it follows now from $h \perp gh$ that $gh \perp gh$, and so $gh = 0$. In particular, if p_1 and p_2 are disjoint components of e, then $p_1 p_2 = 0$.

(viii) For any component p of e we have $p^2 = p$ (and so $p^n = p$ for $n = 2, 3, \dots$). For the proof, denote the band projection corresponding to p by

P and recall that $\pi[p] = P$ (see Remark 34.4 in the preceding section). Hence $p^2 = \pi[p]p = Pp = p$.

(ix) If $s = \Sigma \alpha_k p_k$ and $t = \Sigma \beta_k p_k$ are e-step functions with all components p_k pairwise disjoint, then $st = \Sigma \alpha_k \beta_k p_k$. In particular $s^2 = \Sigma \alpha_k^2 p_k$ and, generally, $s^n = \Sigma \alpha_k^n p_k$ for $n = 2, 3, \ldots$. This follows immediately by observing that $p_k^2 = p_k$ for all k and $p_k p_j = 0$ for $k \neq j$.

(x) We have $g \perp h$ if and only if $gh = 0$. It was shown already in (vii) that $g \perp h$ implies $gh = 0$. It remains to show that $gh = 0$ implies $g \perp h$. Assume first that g and h are positive and denote $g \wedge h$ by k. Then $k \geq 0$ and $0 \leq k^2 \leq gh = 0$, so $k^2 = 0$. If s is any e-step function satisfying $0 \leq s \leq k$, then $0 \leq s^2 \leq sk \leq k^2 = 0$, so $s^2 = 0$. Now, s can be written as $s = \Sigma \alpha_j p_j$ with all $p_j > 0$, pairwise disjoint and all $\alpha_j \geq 0$, so $\Sigma \alpha_j^2 p_j = s^2 = 0$. It follows that $\alpha_j = 0$ for all j, which implies that $s = 0$. This holds for any e-step function s satisfying $0 \leq s \leq k$, so $k = 0$, i.e., $g \wedge h = 0$. In other words, $g \perp h$.

If g and h are arbitrary, it follows from $gh = 0$ that $g^+ h = g^- h$. On the other hand we have $g^+ \perp g^-$, so $g^+ h \perp g^- h$. Hence $g^+ h = g^- h = 0$. But then $g^+ h^+ = g^+ h^-$ as well as $g^+ h^+ \perp g^+ h^-$, and so $g^+ h^+ = g^+ h^- = 0$. Since g^+, h^+ and h^- are positive, it follows now from the first part of the proof that $g^+ \perp h^+$ and $g^+ \perp h^-$. Hence $g^+ \perp (h^+ - h^-)$, i.e., $g^+ \perp h$. Similarly $g^- \perp h$, and so $g \perp h$.

(xi) If $f^2 = 0$, then $f = 0$. More generally, if $f^n = 0$ for some natural number n, then $f = 0$. To see this, observe that $f^2 = 0$ implies by virtue of (x) that $f \perp f$, so $f = 0$. Let $f^n = 0$ for some natural number $n > 2$. If n is of the form $n = 2^m$ for some positive integer $m > 1$, then

$$(f^{2^{m-1}})^2 = f^{2^m} = f^n = 0,$$

so $f^{2^{m-1}} = 0$. Continuing, we obtain $f = 0$. If n is not a power of 2, for example $n = 5$, so $f^5 = 0$, then $f^8 = f^5 \cdot f^3 = 0$, so $f = 0$. In algebra, elements $f \neq 0$ satisfying $f^n = 0$ for some natural number $n > 1$ are called *nilpotent* elements. In the present case, therefore, there are no nilpotent elements in the multiplication.

(xii) Finally, note that $f^+ f^- = f^- f^+ = 0$ for any $f \in E$, and hence

$$f^2 = (f^+ - f^-)^2 = (f^+)^2 + (f^-)^2 \geq 0.$$

According to Freudenthal's spectral theorem any $f \in E$ can be approximated e- uniformly by e-step functions of the form $\Sigma \alpha_k p_k$ with the components p_k mutually disjoint. It was proved then at the beginning of the preceding section that, for n a natural number greater than one, the e-step functions $\Sigma \alpha_k^n p_k$ also converge e-uniformly and we have denoted the limit element by f^n without giving a good reason at that moment in the proceedings. In the functional calculus notation we have written $f^n = \int_a^b \alpha^n dp_\alpha$. That this notation is meaningful and justified follows now from property (x) of the introduced multiplication, in which it is shown that for $s = \Sigma \alpha_k p_k$ we have $s^n = \Sigma \alpha_k^n p_k$. We leave it to the reader to prove that, generally, it follows from e-uniform convergence of f_n to f and g_n to g that $f_n g_n$ converges e-uniformly to fg.

There exist Dedekind σ-complete Riesz spaces E possessing a strong order unit e such that in E there is defined from the beginning on a "natural" positive multiplication having e as multiplicative unit. An example is the space ℓ_∞ or, more generally, the space $L_\infty(X, \mu)$ for μ a σ-finite measure in X. The "natural" multiplication is here the pointwise multiplication: $(fg)(x) = f(x)g(x)$ for $x \in X$. The multiplication is distributive and the function identically one is a strong order unit as well as a multiplicative unit. Is this the same multiplication as the one introduced in the present section? The answer is that the multiplications are indeed the same. For the proof we shall denote the initially given "natural" multiplication of f by g by $f * g$. We assume that this is a positive multiplication, having e as unit (so $f * e = e * f = f$ for every $f \in E$) and such that the distributive laws hold in the sense that

$$f * (\alpha g + \beta h) = \alpha(f * g) + \beta(f * h),$$

$$(\alpha f + \beta g) * h = \alpha(f * h) + \beta(g * h)$$

for all f, g, h in E and real α, β. Fix $f \in E$ and define the mapping $\tau[f] : E \to E$ by $\tau[f](g) = f * g$, holding for all $g \in E$. It is immediately evident that $\tau[f]$ is a linear mapping, i.e., $\tau[f]$ is an operator in E. Furthermore, there exist real numbers α, β such that $\alpha e \le f \le \beta e$, which implies that for any $g \ge 0$ we have

$$\tau[f](g) = f * g \ge (\alpha e) * g = \alpha g, \text{ so } \tau[f] \ge \alpha I;$$

$$\tau[f](g) = f * g \le (\beta e) * g = \beta e, \text{ so } \tau[f] \le \beta I,$$

where I is the identity operator in E. Hence $\alpha I \le \tau[f] \le \beta I$ and $\tau[f](e) = f$. Since $\tau[f]$ satisfies these two conditions, it follows that $\tau[f]$ is the same as the (uniquely defined) operator $\pi[f]$ in Lemma 34.2 and Theorem 34.3. Therefore, $\tau[f](g) = \pi[f](g)$ for all $g \in E$, i.e., $f * g = fg$.

Exercise 35.1. Let e be a strong order unit in the Dedekind σ-complete Riesz space E and let p be a (non-zero) component of e.

(i) Show that p is a strong order unit in the band B generated by p.

(ii) Assume that the multiplication in E with e as multiplicative unit, as defined above, has been introduced and denote the product of f and g by fg. Show that if one at least of f and g is a member of B, then $fg \in B$.

(iii) Assume that the multiplication in B with p as multiplicative unit has also been introduced and denote the corresponding product of f and g by $f * g$. Show that $f * g = fg$ for all f and g in B.

Hint: For (ii) assume that $f \in B$, so $f = Pf$ with P the band projection onto B. Then, for any $g \in E$, we have

$$P(fg) = P\pi[g](f) = \pi[g](Pf) = \pi[g](f) = fg,$$

so $fg \in B$. For (iii), note that for all $f \in B$ we have $f * p = p * f = f$ as well as

$$fp = pf = \pi[f](Pe) = P\pi[f](e) = Pf = f.$$

Exercise 35.2. Once more, assume that e is a strong unit in the Dedekind σ-complete Riesz space E and let the multiplication in E with e as multiplicative unit be introduced. As usual, if $f \in E$ and there exists an element $g \in E$ such that $fg = gf = e$, then g is called an inverse of f and g is then denoted by f^{-1}. It is easy to see that f^{-1}, if existing, is uniquely determined.

(i) Show that if $f \geq 0$ and f^{-1} exists, then $f^{-1} \geq 0$. Since e is a strong unit, we have $0 \leq f \leq \alpha e$ for some positive number α. Show now that $f^{-1} \geq \alpha^{-1}e$. It follows that if $f \geq 0$ and f^{-1} exists, then $f \geq \beta e$ for some positive number β.

(ii) To prove that, conversely, any $f \in E$ satisfying $f \geq \beta e$ for some $\beta > 0$ has an inverse, we may assume without loss of generality that $\beta = 1$, so $e \leq f \leq \alpha e$ for some $\alpha > 1$. For $n = 0, 1, 2, \ldots$, let

$$s_n = \Sigma_{k=0}^n (e - \alpha^{-1}f)^k.$$

Note that $0 \leq e - \alpha^{-1}f \leq (1 - \alpha^{-1})e$ and $0 < 1 - \alpha^{-1} < 1$, so

$$0 \leq s_n \leq \Sigma_{k=0}^n (1 - \alpha^{-1})^k e.$$

Show that $(s_n : n = 0, 1, 2, \ldots)$ is an e-uniform Cauchy sequence. Since E is uniformly complete, there exists an element $s \in E$ such that $s_n \uparrow s$ holds e-uniformly, i.e., $s = \Sigma_0^\infty (e - \alpha^{-1}f)^k$. Then

$$(e - \alpha^{-1}f)s = \Sigma_1^\infty (e - \alpha^{-1}f)^k = s - e,$$

which implies that $\alpha^{-1}fs = e$. This shows that $\alpha^{-1}s$ is the inverse of f.

Hint: For (i), let $f \geq 0$ have an inverse which, for convenience of notation, we denote by g, so $fg = gf = e$. Then $g^+f - g^-f = e$ with g^+f and g^-f positive and disjoint. It follows that $g^+f = e^+ = e$, so both g^+ and g are inverses of f, i.e., $g = g^+ \geq 0$.

CHAPTER 19
Complex Operators

36. Complex Operators

If T is a (linear) operator mapping the real vector space V into the real vector space W, then T has a unique extension as a (linear) operator from $V + iV$ into $W + iW$ by defining $T(f + ig) = Tf + iTg$ for all f and g in V. The set $\mathcal{L}(V, W)$ of all (linear) operators mapping V into W can be considered thus as a real-linear subspace of the vector space of all complex (linear) operators from $V + iV$ into $W + iW$. Recall here that a subset K of a complex vector space G is said to be a real-linear subspace of G whenever $\alpha f + \beta g \in K$ for all f and g in K and all real numbers α, β. Any $T \in \mathcal{L}(V, W)$ which is thus extended is called a *real (linear) operator* from $V + iV$ into $W + iW$. Any arbitrary operator $T \in \mathcal{L}(V + iV, W + iW)$ has a unique decomposition $T = T_1 + iT_2$ with T_1 and T_2 real. To see this, decompose Tf for any $f \in V$ into its real and imaginary parts: $Tf = T_1 f + iT_2 f$. Obviously, the thus defined T_1 and T_2 are (linear) operators from V into W which can be extended from $V + iV$ into $W + iW$. It follows then that $Th = T_1 h + iT_2 h$ for any $h = f + ig \in V + iV$, and so $T = T_1 + iT_2$ holds on $V + iV$ with T_1 and T_2 real. This shows that the space $\mathcal{L}(V + iV, W + iW)$ is the complexification of $\mathcal{L}(V, W)$.

Assume now that E and F are Riesz spaces such that E is Archimedean and uniformly complete and F is Dedekind complete. Hence, both E and F have complexifications admitting an absolute value (see Theorem 13.4). Furthermore, an operator $T : E \to F$ is order bounded if and only if it is regular (see Theorem 20.2). The operator $T = T_1 + iT_2$ (with T_1 and T_2 real) from $E + iE$ into $F + iF$ is said to be *order bounded* if T maps order bounded sets into order bounded sets; an order bounded set in $E + iE$ is a subset D of $E + iE$ for which there exists an appropriate $f_0 \in E^+$ such that $| f | \le f_0$ for all $f \in D$. Similarly for $F + iF$. This definition is in agreement with the one in the real case. It is evident that the operator $T = T_1 + iT_2$ from $E + iE$ into $F + iF$ is order bounded if and only if T_1 and T_2 are order bounded operators from E into F (i.e., if and only if T_1 and T_2 are regular). The space $\mathcal{L}_b(E + iE, F + iF)$ of all order bounded operators from $E + iE$ into $F + iF$ is therefore the complexification of $\mathcal{L}_b(E, F)$. The operator T is called *positive* if T is real and the restriction of T to E is positive; $T = T_1 + iT_2$ is said to be order continuous (or σ-order continuous) whenever T_1 and T_2 are so. For any order bounded operator $T = T_1 + iT_2$ the absolute value $\mid T \mid$ is, by definition, the operator

$$\mid T \mid = \sup(T_1 \cos \theta + T_2 \sin \theta : 0 \leq \theta \leq 2\pi).$$

The supremum exists since $\mathcal{L}_b(E, F)$ is Dedekind complete (see Theorem 20.2). In the case that T is a member of $\mathcal{L}_b(E, F)$ the absolute value $\mid T \mid = T^+ \vee T^-$ satisfies

$$\mid T \mid (u) = \sup(\mid Tf \mid : f \in E, \mid f \mid \leq u)$$

for any $u \in E^+$. The corresponding formula is valid for complex T but all existing proofs are rather complicated. For that reason we shall restrict ourselves to the case that E is Dedekind σ-complete because in this case there is a simple proof available in which use is made of a complex variant of Freudenthal's spectral theorem, as follows.

Theorem 36.1. *Let e be a strictly positive element in the Dedekind σ-complete Riesz space E, so $0 < e \in E$. Furthermore, let $h = f + ig \in E + iE$ satisfy $\mid h \mid \leq e$. Then, for any $\epsilon > 0$, there exist disjoint components $e_1, \ldots, e_n$ of e and complex numbers $\gamma_1, \ldots, \gamma_n$ such that $\mid \gamma_k \mid \leq 1$ for $k = 1, \ldots, n$ and*

$$\mid h - \Sigma_1^n \gamma_k e_k \mid \leq \epsilon e$$

as well as

$$0 \leq \mid h \mid - \Sigma_1^n \mid \gamma_k \mid e_k \leq \epsilon e.$$

Proof. Let $\epsilon > 0$ be given and let p and q be e-step functions such that

$$0 \leq p \leq f^+, 0 \leq q \leq f^- \text{ and } f^+ - p \leq \epsilon e/4, f^- - q \leq \epsilon e/4.$$

Note that $p \wedge q = 0$. Hence, defining the e-step function s_1 by $s_1 = p - q$, we have $\mid s_1 \mid = p + q$ and

$$\mid f - s_1 \mid \leq \epsilon e/2 \text{ and } 0 \leq \mid f \mid - \mid s_1 \mid \leq \epsilon e/2.$$

Similarly, there exists an e-step function s_2 such that

$$\mid g - s_2 \mid \leq \epsilon e/2 \text{ and } 0 \leq \mid g \mid - \mid s_2 \mid \leq \epsilon e/2.$$

It follows (in virtue of the properties proved in Theorem 13.6) that the complex e-step function $s = s_1 + is_2$ satisfies

$$\mid s \mid = \mid s_1 + is_2 \mid = \mid\mid s_1 \mid + i \mid s_2 \mid\mid \leq \mid\mid f \mid + i \mid g \mid\mid = \mid f + ig \mid = \mid h \mid .$$

Furthermore,

$$| h - s | \leq | f - s_1 | + | g - s_2 | \leq \epsilon e,$$

and so (triangle inequality)

$$0 \leq | h | - | s | = \big| | h | - | s | \big| \leq | h - s | \leq \epsilon e.$$

It remains to prove that s can be written as a finite sum $\Sigma \gamma_k e_k$ with disjoint components e_k of e and with $| \gamma_k | \leq 1$ for all k. To see this, observe that s_1 can be written as $s_1 = \Sigma \alpha_j v_j$ with disjoint components v_j such that $\Sigma v_j = e$. Similarly, $s_2 = \Sigma \beta_m w_m$ with disjoint components w_m such that $\Sigma w_m = e$. Writing $e_{jm} = v_j \wedge w_m$, the elements e_{jm} are mutually disjoint components of e such that $\Sigma_{jm} e_{jm} = e$. It follows that

$$s_1 = \Sigma_{jm} \alpha_j e_{jm} \text{ and } s_2 = \Sigma_{jm} \beta_m e_{jm}$$

Writing $\gamma_{jm} = \alpha_j + i \beta_m$, we see that

$$s = s_1 + i s_2 = \Sigma_{jm} \gamma_{jm} e_{jm}$$

with disjoint components e_{jm}, so (by Corollary 13.7)

$$| s | = \Sigma | \gamma_{jm} | e_{jm}.$$

Now denote by P_{jm} the band projection onto the band generated by e_{jm}. Then $P_{jm}(| s |) = | \gamma_{jm} | e_{jm}$, so

$$| \gamma_{jm} | e_{jm} = P_{jm}(| s |) \leq P_{jm}(| h |) \leq P_{jm} e = e_{jm}.$$

It follows that $| \gamma_{jm} | \leq 1$ for all e_{jm} that are not the zero element. Renumbering the terms $\gamma_{jm} e_{jm}$ for which e_{jm} is not the zero element, we obtain the desired result.

Note that in this proof the Dedekind σ-completeness of E implies that E is uniformly complete (which guarantees that E has a "correct" complexification) and that E has the principal projection property (which guarantees that Freudenthal's spectral theorem holds in E).   ∎

In the lemma which follows now we prove the inequality $| Th | \leq | T | (| h |)$ in some special cases. Dedekind completeness of E is not yet needed.

Lemma 36.2. *Let E and F be Riesz spaces such that E is Archimedean and uniformly complete and F is Dedekind complete. Then the following holds.*

(i) If $T \in \mathcal{L}_b(E, F)$ is real and $h \in E + iE$, then $| Th | \leq | T | (| h |)$.

(ii) If $T \in \mathcal{L}_b(E + iE, F + iF)$ and $f \in E$, then $| Tf | \leq | T | (| f |)$. More generally, if $s = \Sigma_1^n \lambda_k f_k$ with all λ_k complex numbers and $f_1, /ldots, f_n$ mutually disjoint elements in E, then $| Ts | \leq | T | (| s |)$.

Proof. (i) Let $h = f + ig$ with f and g in E. Since T is real, Th has the decomposition $Th = Tf + iTg$, so

$$| Th | = \sup(| Tf\cos\theta + Tg\sin\theta |: 0 \le \theta \le 2\pi) \le$$

$$\sup | T | (| f\cos\theta + g\sin\theta |: 0 \le \theta \le 2\pi) \le | T | (| h |).$$

(ii) Let $T = T_1 + iT_2$ with T_1 and T_2 real. Since f is real, Tf has the decomposition $Tf = T_1 f + iT_2 f$, so

$$| Tf | = \sup(| T_1 f\cos\theta + T_2 f\sin\theta |: 0 \le \theta \le 2\pi) =$$

$$\sup_{\theta} | (T_1\cos\theta + T_2\sin\theta)(f) | \le$$

$$\sup_{\theta}\{| T_1\cos\theta + T_2\sin\theta | (| f |)\} \le | T | (| f |).$$

Now, let $s = \Sigma_1^n \lambda_k f_k$ with complex coefficients λ_k and the elements f_k pairwise disjoint in E. Note that $| s | = \Sigma_1^n | \lambda_k | \cdot | f_k |$ (see Corollary 13.7). It follows that

$$| Ts | = | \Sigma\lambda_k Tf_k | \le \Sigma | \lambda_k | \cdot | Tf_k | \le$$

$$| T | (\Sigma | \lambda_k | \cdot | f_k |) = | T | (| s |). \qquad\blacksquare$$

As we shall soon see, it will be easy to derive from this lemma that $| Th | \le | T | (| h |)$ holds for any $h \in E + iE$, provided E is Dedekind σ-complete. For the proof that

$$| T | (u) = \sup(| Th |: h\epsilon E + iE, | h | \le u)$$

holds for $u \in E^+$, we still need an auxiliary result in the converse direction in which we do not assume that E is Dedekind σ-complete.

Lemma 36.3. *Let D be a non-empty set the elements of which we shall denote by α. For each $\alpha \in D$, let $T_\alpha \in \mathcal{L}_b(E, F)$. Assume furthermore that $T_0 = \sup(T_\alpha : \alpha \in D)$ exists in $\mathcal{L}_b(E, F)$. Then, for any $u \in E^+$, we have*

$$T_0 u = \sup(T_{\alpha_1} u_1 + \cdots + T_{\alpha_n} u_n),$$

where $(\alpha_1, \ldots, \alpha_n)$ runs through all non-empty finite subsets of D and, for each subset of this type, $(u_1, \ldots, u_n)$ runs through all n-tuples of positive elements of E satisfying $u_1 + \cdots + u_n = u$.

Proof. If D consists of two elements α_1 and α_2, so that $T_0 = T_{\alpha_1} \vee T_{\alpha_2}$, the formula to be proved is

$$T_0 u = \sup(T_{\alpha_1} v + T_{\alpha_2} w : v \geq 0, w \geq 0, v + w = u).$$

This was shown to hold in Theorem 20.4, formula (3). If D has only a finite number of elements, the result follows by induction. In the general case, observe that the set of all finite suprema

$$T\{\alpha_1, \ldots, \alpha_n\} = \sup(T_{\alpha_1}, \ldots, T_{\alpha_n})$$

is upwards directed and T_0 is the supremum of the set. Hence

$$T_0 u = \sup T\{\alpha_1, \ldots, \alpha_n\}(u) =$$

$$\sup_{\{\alpha_1,\ldots,\alpha_n\}} (T_{\alpha_1} u_1 + \cdots + T_{\alpha_n} u_n : u_1, \ldots, u_n \quad \text{in } E^+, \Sigma u_k = u). \quad \blacksquare$$

We come to the main theorem.

Theorem 36.4. *Let E and F be Riesz spaces such that E is Dedekind σ-com-plete and F is Dedekind complete. Furthermore, let $T = T_1 + iT_2$ be a member of $\mathcal{L}_b(E + iE, F + iF)$ and let $u \in E^+$. Then*

$$\mid T \mid (u) = \sup(\mid T(f + ig) \mid : \mid f + ig \mid \leq u).$$

In particular, therefore, $\mid Th \mid \leq \mid T \mid (\mid h \mid)$ holds for every $h \in E + iE$.

Proof. The proof consists of two parts. In the first part Dedekind σ-completeness of E is not yet needed. Writing $T(\theta) = T_1 \cos \theta + T_2 \sin \theta$, we have $\mid T \mid = \sup(T(\theta); 0 \leq \theta \leq 2\pi)$, so it follows from the last lemma that

$$\mid T \mid (u) = \sup \Sigma_j T(\theta_j) u_j, \tag{1}$$

where the supremum is taken over all finite systems (θ_j, u_j) such that all u_j are positive and $\Sigma u_j = u$. Choosing one of these systems, let us write

$$p = \Sigma_j (\cos \theta_j) u_j \text{ and } q = \Sigma_j (\sin \theta_j) u_j.$$

Then

$$\mid p - iq \mid = \mid \Sigma(e^{-i\theta_j}) u_j \mid \leq \Sigma u_j = u,$$

$$T(p - iq) = \Sigma(e^{-i\theta_j}) T u_j = \Sigma e^{-i\theta_j} (T_1 u_j + iT_2 u_j), \tag{2}$$

so the real part of $T(p - iq)$ satisfies

$$\mathcal{R}e\{T(p - iq)\} = \Sigma(T_1 u_j \cos \theta_j + T_2 u_j \sin \theta_j) = T_1 p + T_2 q.$$

Keeping in mind that (2) holds, it follows that

$$| T_1 p + T_2 q \,|=| \, \mathcal{R}e\{T(p - iq)\} \,|\leq$$

$$| \, T(p - iq) \,|\leq \sup(| \, T(f + ig) \,|:| \, f + ig \,|\leq u). \tag{3}$$

Observing now that

$$T_1 p + T_2 q = \Sigma\{(T_1 \cos \theta_j) u_j + (T_2 \sin \theta_j) u_j\} = \Sigma T(\theta_j) u_j,$$

we derive from (3) that

$$\Sigma_j T(\theta_j) u_j \leq \sup(| \, T(f + ig) \,|:| \, f + ig \,|\leq u).$$

This holds for all systems (θ_j, u_j), so it follows from (1) above that

$$| \, T \,| \, (u) \leq \sup(| \, T(f + ig) \,|:| \, f + ig \,|\leq u).$$

For the inequality in the converse direction it is sufficient to prove that
$| \, Th \,|\leq| \, T \,| \, (| \, h \,|)$ holds for every $h = f + ig \in E + iE$. This is the part
of the proof in which we shall appeal to the complex version of Freudenthal's
spectral theorem. We first prove a weak variant of the inequality, as follows.
If $h \in E + iE$, we have

$$| \, Th \,|=| \, T_1 h + i T_2 h \,|\leq| \, T_1 h \,| + | \, T_2 h \,|\leq (| \, T_1 \,| + | \, T_2 \,|) \, | \, h \,|$$

(in virtue of Lemma 36.2(1) above). This implies that $| \, Th \,|\leq 2 \, | \, T \,| \, (| \, h \,|)$.It
follows that if h_n converges u-uniformly to h in $E + iE$ for some $u \in E^+$ (all
h_n and h in $E + iE$), then Th_n converges v-uniformly to Th in $F + iF$ for
$v = 2 \, | \, T \,| \, (u)$. Then $| \, Th_n \,|$ converges v-uniformly to $| \, Th \,|$ in F, because

$$\big|| \, Th \,| - | \, Th_n \,|\big| \, \leq | \, Th - Th_n \,| \, .$$

Hence, if $| \, Th_n \,|\leq w$ for some $w \in F$ and all n, then $| \, Th \,|\leq w$.

Finally, for the last part of the proof, fix $h \in E + iE$ and fix $\epsilon > 0$. By the
spectral theorem (Theorem 36.1) there exists an $| \, h \,|$-step function $s = \Sigma \gamma_j e_j$
(with disjoint components e_j) such that $| \, \gamma_j \,|\leq 1$ for all j and

$$| \, h - \Sigma \gamma_j e_j \,|\leq \epsilon \, | \, h \,| \quad 0 \leq| \, h \,| - \Sigma \, | \, \gamma_j \,| \, e_j \, \leq \, \epsilon \, | \, h \,| \, .$$

For $\epsilon_n \downarrow 0$ there exists for each ϵ_n such a step function s_n. Then s_n converges
$| \, h \,|$-uniformly to h, so that by the remark above $| \, Ts_n \,|$ converges v-uniformly
to $| \, Th \,|$ for $v = 2 \, | \, T \,| \, (| \, h \,|)$. It will be sufficient, therefore, to show that
$| \, Ts_n \,|\leq| \, T \,| \, (| \, h \,|)$ for all n, since this will imply (once more by the remark
above) that $| \, Th \,|\leq| \, T \,| \, (| \, h \,|)$. For any $s_n = \Sigma \gamma_j e_j$ we have (similarly as in
Lemma 36.2)

$$| \, Ts_n \,| \, =| \, \Sigma \gamma_j Te_j \,|\leq \Sigma \, | \, \gamma_j \,| \cdot | \, Te_j \,|\leq$$

$$| \, T \,| \, (\Sigma \, | \, \gamma_j \,| \, e_j) \leq| \, T \,| \, (| \, h \,|).$$

This is the desired result.   ■

As observed already, the theorem can be proved also if we assume about E only that E is Archimedean and uniformly complete. The first proofs in this general case were given by W.J. de Schipper and H.H. Schaefer (1973-74), independently of each other. In these proofs use is made of a so-called *representation theorem*, stating in this particular case that if A_e is the principal ideal generated in E by some positive non-zero element e, then there exists a compact Hausdorff topological space X such that A_e is Riesz isomorphic to the space $C(X)$ of all real continuous functions on X (and such that under the isomorphism the element e corresponds to the function identically one on X). Not only is it so, however, that the proof of this representation theorem (and of similar other representation theorems) is far from simple, but it also seems inevitable that one needs in these proofs a certain axiom about partially ordered sets which is known as Zorn's lemma, and about which we shall have to say more in the next chapter. Returning to the inequality $|\,Th\,| \leq |\,T\,|\,(|\,h\,|)$ for $h \in E + iE$, there has been found not long ago a proof without Zorn's lemma (J.J. Grobler, 1988), but this proof as well is not simple.

Assume again that E is uniformly complete and F is Dedekind complete. As defined already, the operator $T = T_1 + iT_2$ from $E + iE$ into $F + iF$ (with T_1 and T_2 real) is said to be order continuous (or σ-order continuous) whenever T_1 and T_2 are so. Since $|\,T_1\,|$ and $|\,T_2\,|$ are majorized by $|\,T\,|$ and $|\,T\,|$ is majorized by $|\,T_1\,| + |\,T_2\,|$, it follows immediately that T is order continuous (or σ-order continuous) if and only if $|\,T\,|$ is so. The sets $\mathcal{L}_n = \mathcal{L}_n(E, F)$ and $\mathcal{L}_c = \mathcal{L}_c(E, F)$ of all order continuous or σ-order continuous operators in $\mathcal{L}_b(E, F)$ are projection bands (as shown in section 22). Hence, in view of the definition of a projection band in the complex case (section 13), the bands $\mathcal{L}_n + i\mathcal{L}_n$ and $\mathcal{L}_c + i\mathcal{L}_c$ are projection bands in

$$\mathcal{L}_b + i\mathcal{L}_b = \mathcal{L}_b(E + iE, F + iF).$$

Let T_1 and T_2 be members of $\mathcal{L}_b$. Then $T_1 = T_{1,n} + T_{1,t}$ with $T_{1,n} \in \mathcal{L}_n$ and $T_{1,t} \in \mathcal{L}_t$, where $\mathcal{L}_t$ is the disjoint complement of $\mathcal{L}_n$ in $\mathcal{L}_b$. Similarly, we have $T_2 = T_{2,n} + T_{2,t}$. It follows that $T = T_1 + iT_2$ can be written as

$$T = T_1 + iT_2 = (T_{1,n} + iT_{2,n}) + (T_{1.t} + iT_{2,t}) = T_n + T_t$$

with $T_n \in \mathcal{L}_n + i\mathcal{L}_n$ and $T_t \in \mathcal{L}_t + i\mathcal{L}_t$. Since both $T_{1,t}$ and $T_{2,t}$ are members of $\mathcal{L}_t$, each of these operators is disjoint to $T_{1,n}$ and to $T_{2,n}$, so

$$|\,T_{1,t}\,| + |\,T_{2,t}\,| \text{ is disjoint to } |\,T_{1,n}\,| + |\,T_{2,n}\,|\,.$$

This implies that $|\,T_t\,|$ is disjoint to $|\,T_n\,|$, so

$$|\,T\,| = |\,T_n + T_t\,| = |\,T_n\,| + |\,T_t\,|\,.$$

If F is the space $\mathbb{R}$ of real numbers, then $F + iF$ is the space $\mathbb{C}$ of complex numbers. The space $\mathcal{L}_b(E, F)$ is now the order dual $E^\sim$ of E. Any complex linear functional φ on $E + iE$ is of the form $\varphi = \varphi_1 + i\varphi_2$, where φ_1 and φ_2

are the extensions to $E + iE$ of real linear functionals on E. It is clear that φ is order bounded if and only if φ_1 and φ_2 are order bounded on E, i.e.,

$$(E + iE)^\sim = E^\sim + iE^\sim.$$

The sets of all order continuous or σ-order continuous linear functionals on $E + iE$ are the projection bands $E_n^\sim + iE_n^\sim$ or $E_c^\sim + iE_c^\sim$ respectively.

37. Synnatschke's Theorem

As in the preceding section, we assume that the Riesz space E is Archimedean and uniformly complete and the Riesz space F is Dedekind complete. We recall Synnatschke's theorem, as proved in section 26. For any $T \in \mathcal{L}_b(E, F)$, the adjoint operator $T^\sim : F^\sim \to E^\sim$ is defined by

$$(T^\sim \psi)(f) = \psi(Tf) \text{ for } f \in E, \psi \in F^\sim,$$

and the restriction of $T^\sim$ to $F_n^\sim$ is denoted by T'. For $T \geq 0$ we have $T^\sim \geq 0$, and so $T' \geq 0$. Synnatschke's theorem states that for any T_1 and T_2 in $\mathcal{L}_b(E, F)$ we have

$$(T_1 \vee T_2)' = T_1' \vee T_2' \text{ and } (T_1 \wedge T_2)' = T_1' \wedge T_2',$$

so in particular $|\, T\, |' = |\, T'\, |$ for any $T \in \mathcal{L}_b(E, F)$. We shall extend the theorem to the complex case. For this purpose we need a lemma which is somewhat similar to Lemma 36.3.

Lemma 37.1. *Let D be a non-empty set the elements of which we shall denote by α. For each $\alpha \in D$, let $0 \leq T_\alpha \in \mathcal{L}_b(E, F)$. Assume that the set of all T_α is upwards directed and $T_0 = \sup(T_\alpha : \alpha \in D)$ exists, so $0 \leq T_\alpha \uparrow T_0$. Then $0 \leq T_\alpha' \uparrow T_0'$.*

Proof. It follows immediately from $0 \leq T_\alpha \uparrow T_0$ that $0 \leq T_\alpha' \uparrow \leq T_0'$. The element $S = \sup T_\alpha'$ exists in $\mathcal{L}_n(F_n^\sim, E^\sim)$ on account of the Dedekind completeness of $\mathcal{L}_n(F_n^\sim, E^\sim)$. We shall prove that $(S\psi)(f) = \psi(T_0 f)$ holds for all $0 \leq f \in E, 0 \leq \psi \in F_n^\sim$. This will imply then that the same holds for all $f \in E$ and all $\psi \in F_n^\sim$, and so $(S\psi)(f) = (T_0'\psi)(f)$ for all f and ψ, i.e., $S = T_0'$.

Assume, therefore, that $0 \leq f \in E$ and $0 \leq \psi \in F_n'$. Since $T_\alpha f \uparrow T_0 f$ and ψ is order continuous, we have

$$\psi(T_\alpha f) \uparrow \psi(T_0 f). \tag{1}$$

On the other hand $T_\alpha' \psi \uparrow S\psi$, so

$$(T_\alpha' \psi)(f) \uparrow (S\psi)(f). \tag{2}$$

The left hand sides of (1) and (2) are equal for each α, so the right hand sides are equal as well, i.e., $(S\psi)(f) = \psi(T_0 f)$. This completes the proof. ∎

If T_1 and T_2 are members of $\mathcal{L}_b(E, F)$ with adjoints T_1' and T_2' and if $T = T_1 + iT_2$, we shall say that $T_1' + iT_2'$ is the adjoint of T and we shall write $T^\sim = T_1' + iT_2'$. This name and the notation are justified because $(T^\sim\psi)(f) = \psi(Tf)$ holds for all $f \in E + iE$ and $\psi \in F^\sim + iF^\sim$. To see this, observe that the equality holds for $f \in E$ and $\psi \in F^\sim$ and then extend the equality first to the case that $f \in E + iE$ and $\psi \in F^\sim$ and finally extend it to the general case. As in the real case, the restriction of $T^\sim$ to $F_n^\sim + iF_n^\sim$ is denoted by T'. Hence, $T' = T_1' + iT_2'$.

We extend Synnatschke's theorem to the complex case.

Theorem 37.2. *If* $T \in \mathcal{L}_b(E + iE, F + iF)$, *then* $\mid T \mid' = \mid T' \mid$.

Proof. Let $T = T_1 + iT_2$ with T_1 and T_2 real. For any θ such that $0 \leq \theta \leq 2\pi$, write

$$T(\theta) = T_1 \cos\theta + T_2 \sin\theta \text{ and } T'(\theta) = T_1' \cos\theta + T_2' \sin\theta.$$

Note that $[T(\theta)]' = T'(\theta)$. From Synnatschke's theorem for the real case it follows that

$$\mid T(\theta) \mid' = \mid T'(\theta) \mid .$$

Now, for any finite set $\{\theta_1, \ldots, \theta_n\}$, write

$$T\{\theta_1, \ldots, \theta_n\} = \mid T(\theta_1) \mid \vee \cdots \vee \mid T(\theta_n) \mid,$$

$$T'\{\theta_1, \ldots, \theta_n\} = \mid T'(\theta_1) \mid \vee \cdots \vee \mid T'(\theta_n) \mid .$$

Once more by Synnatschke's theorem for the real case we have

$$[T\{\theta_1, \ldots, \theta_n\}]' = \mid T(\theta_1) \mid' \vee \cdots \vee \mid T(\theta_n) \mid' =$$

$$\mid T'(\theta_1) \mid \vee \cdots \vee \mid T'(\theta_n) \mid = T'\{\theta_1, \ldots, \theta_n\}.$$

The set of all $T\{\theta_1, \ldots, \theta_n\}$ is upwards directed and has $\mid T \mid$ as its supremum, i.e.,

$$0 \leq T\{\theta_1, \ldots, \theta_n\} \uparrow \mid T \mid .$$

Similarly, we have

$$0 \leq [T\{\theta_1, \ldots, \theta_n\}]' = T'\{\theta_1, \ldots, \theta_n\} \uparrow \mid T' \mid .$$

It follows now from the last lemma that $\mid T \mid' = \mid T' \mid$. ∎

CHAPTER 20
Results with the Hahn-Banach Theorem

38. The Hahn-Banach Theorem in Normed Vector Spaces

This chapter is devoted to several results the proofs of which are based on a theorem known as the Hahn-Banach theorem (due to H. Hahn, 1927, and S. Banach, 1929). Most of these results, as well as the Hahn-Banach theorem itself, are extension theorems dealing with extension of a (linear) operator from a linear subspace to the entire space, thereby preserving certain properties of the operator (such as, for example, positivity or norm boundedness). To prove the Hahn-Banach theorem it is necessary to accept a certain axiom about partially ordered sets, called *Zorn's lemma*. We formulate the axiom. The definition of a chain and of a maximal element in a partially ordered set are given in section 1.

Zorn's lemma. *If X is a (non-empty) partially ordered set having the property that every chain in X has an upper bound, then X contains a maximal element.*

For the benefit of readers who perhaps wonder where such an axiom comes from we observe that it can be proved that Zorn's lemma is equivalent to another axiom, called the *axiom of choice*. This says that if we have a non-empty collection of non-empty sets, then we can form a new set by picking one element from each of the given sets. More formally stated, if $\mathcal{D}$ is the collection of the given sets and F is their union, so $F = \cup(D : D \in \mathcal{D})$, then there exists a function $f : \mathcal{D} \to F$ such that $f(D) \in D$ for each $D \in \mathcal{D}$. The function f is an example of what is called a *choice function*.

To formulate the Hahn-Banach extension theorem, let V be a (real) vector space and E a (real) Riesz space. We recall that the function $p : V \to E$ is said to be *sublinear* whenever

(i) $p(f + g) \leq p(f) + p(g)$ for all f and g in V,

(ii) $p(\alpha f) = \alpha p(f)$ for every $f \in V$ and every real number $\alpha \geq 0$.

For $E = \mathbb{R}$ we have then a *sublinear functional*.

Theorem 38.1. *(Hahn-Banach) Let V be a real vector space and E a real Dedekind complete Riesz space. Furthermore, let $p : V \to E$ be sublinear. If W is a linear subspace of V and $S : W \to E$ is a (linear) operator such that $Sf \le p(f)$ holds for all $f \in W$, then there exists a (linear) operator $T : V \to E$ such that*

(i) $Tf = Sf$ for all $f \in W$ (i.e., T is an extension of S),
(ii) $Tf \le p(f)$ for all $f \in V$.

Proof. If $W = V$, there is nothing to prove. We may assume, therefore, that W is a proper linear subspace of V. Choose an element $f \in V$ such that f is not in W and denote by Y the linear subspace generated by W and f, so

$$Y = (g + \alpha f : g \in W \text{ and } \alpha \in \mathbb{R}).$$

Assume now that T is a (linear) extension of S to Y (not yet necessarily to the whole of V) such that $Th \le p(h)$ for all $h = g + \alpha f \in Y$. Observing that

$$Th = Tg + \alpha Tf = Sg + \alpha Tf,$$

we see that $Th \le p(h)$ is equivalent to

$$Sg + \alpha Tf \le p(g + \alpha f), \tag{1}$$

holding for all $g \in W$ and all $\alpha \in \mathbb{R}$. To show that there exists such an extension T of S, it is necessary to prove that Tf can be chosen in E such that (1) is satisfied for all $g \in W$ and $\alpha \in \mathbb{R}$. In other words, we have to prove that there exists an element $x \in E$ such that x satisfies

$$Sg + \alpha x \le p(g + \alpha f) \text{ for all } g \in W \text{ and all } \alpha \in \mathbb{R}. \tag{2}$$

For $\alpha > 0, (2)$ is equivalent to

$$Sg + x \le p(g + f)$$

for all $g \in W$. To see this, divide (2) by α and denote g/α by g again. For $\alpha < 0$, (2) is equivalent to

$$Sk - x \le p(k - f)$$

for all $k \in W$. To see this, divide (2) now by $-\alpha$ and denote $-g/\alpha$ by k. Both of these inequalities will be satisfied if x is chosen such that

$$Sk - p(k - f) \le x \le p(g + f) - Sg \tag{3}$$

holds for all k and g in W. To see that there exists an element $x \in E$ satisfying (3), note that for k and g in W we have

$$Sk + Sg = S(k + g) \le p(k + g) =$$
$$p\{(k - f) + (g + f)\} \le p(k - f) + p(g + f),$$

which implies that

$$Sk - p(k - f) \leq p(g + f) - Sg$$

holds for all k and g in W. Since E is Dedekind complete, the supremum s, for $k \in W$, on the left and the infimum t, for $g \in W$, on the right exist and satisfy $s \leq t$. Any $x \in E$ satisfying $s \leq x \leq t$ now satisfies (3), and therefore satisfies (2). It has been proved thus that S can be linearly extended to the subspace Y such that the extension is still majorized by p.

It is possible to repeat this extension procedure, thereby extending the given operator step by step. This, however, does not lead to an extension defined on the whole space V (unless in some special cases). We shall proceed differently, and this is where Zorn's lemma will be used. By (Z, T_Z) we shall denote a linear subspace Z of V such that $W \subseteq Z \subseteq V$ and a linear operator $T_Z : Z \to E$ such that $T_Z f = S f$ for all $f \in W$ and $T_Z f \leq p(f)$ for all $f \in Z$. Let $\mathcal{D}$ be the set of all (Z, T_Z). Obviously, $\mathcal{D}$ is not empty. The set $\mathcal{D}$ is partially ordered by defining that $(Z_1, T_{Z_1}) \leq (Z_2, T_{Z_2})$ whenever $Z_1 \subseteq Z_2$ and $T_{Z_1} = T_{Z_2}$ on Z_1. It is easy to see that if $\mathcal{C}$ is a chain in $\mathcal{D}$, then $\mathcal{C}$ has a least upper bound (Z_0, T_{Z_0}); the space Z_0 is simply the union of all Z in the chain. Hence, in view of Zorn's lemma, $\mathcal{D}$ has a maximal element (Z_m, T_{Z_m}). If Z_m would be a proper subspace of V, the operator T_{Z_m} could be further extended by means of the procedure described in the beginning of the present proof, so that then (Z_m, T_{Z_m}) would not be maximal. Hence $Z_m = V$ and T_{Z_m} satisfies all required conditions. The Hahn-Banach theorem has thus been proved.   ∎

In the original version of the theorem the Riesz space E was the space $\mathbb{R}$. This is an important case which deserves a special formulation.

Corollary 38.2. *Let p be a sublinear (real) functional on the real vector space V and let ψ be a linear functional on the linear subspace W such that $\psi(f) \leq p(f)$ for all $f \in W$. Then there exists a linear functional φ on V such that $\varphi(f) = \psi(f)$ for all $f \in W$ and $\varphi(f) \leq p(f)$ for all $f \in V$.*

If V is a normed vector space, then the norm $p(f) = \parallel f \parallel$ is an example of a sublinear functional p on V, having the additional property that $p(-f) = p(f)$ holds for all $f \in V$. Hence, if $p(f) = \alpha \parallel f \parallel$ for some (strictly) positive constant α, then p is sublinear and also $p(-f) = p(f)$ for all $f \in V$. Observe now that if ψ is a linear functional on the linear subspace W such that $\psi(f) \leq p(f) = \alpha \parallel f \parallel$ for all $f \in W$, then

$$-\psi(f) = \psi(-f) \leq p(-f) = p(f) = \alpha \parallel f \parallel$$

holds as well, so $\mid \psi(f) \mid \leq \alpha \parallel f \parallel$ for all $f \in W$. In other words, ψ is a norm bounded linear functional on W. Conversely, if $\mid \psi(f) \mid \leq \alpha \parallel f \parallel$ for all $f \in W$, then of course $\psi(f) \leq \alpha \parallel f \parallel$ for all $f \in W$. The following extension theorem follows therefore immediately.

Theorem 38.3. *If ψ is a norm bounded linear functional on the linear subspace W of the normed vector space V, then there exists a norm bounded linear functional φ on V such that $\varphi(f) = \psi(f)$ for all $f \in W$ and $\parallel \varphi \parallel = \parallel \psi \parallel$.*

Proof. Choose $\alpha = \parallel \psi \parallel$ in the remark above, so $p(f) = \parallel \psi \parallel \cdot \parallel f \parallel$ for all $f \in V$. Then the linear extension φ of ψ which exists by the above corollary has the property that $\mid \varphi(f) \mid \leq \parallel \psi \parallel \cdot \parallel f \parallel$ for all $f \in V$, which shows that $\parallel \varphi \parallel \leq \parallel \psi \parallel$. On the other hand we have $\parallel \varphi \parallel \geq \parallel \psi \parallel$ since φ is an extension of ψ. Hence $\parallel \varphi \parallel = \parallel \psi \parallel$. $\blacksquare$

The next theorem follows now easily.

Theorem 38.4. *For any given element $f_0 \neq 0$ of the normed vector space V there exists a norm bounded linear functional φ on V such that $\parallel \varphi \parallel = 1$ and $\varphi(f_0) = \parallel f_0 \parallel$.*

Proof. Denote the linear subspace of all (real) multiples of f_0 by W and define the linear functional ψ on W by $\psi(\alpha f_0) = \alpha \parallel f_0 \parallel$ for every $\alpha \in \mathbb{R}$. Then $\parallel \psi \parallel = 1$ and $\psi(f_0) = \parallel f_0 \parallel$. In view of the preceding theorem ψ has a linear extension φ to the whole of V such that $\parallel \varphi \parallel = \parallel \psi \parallel$. It follows that $\parallel \varphi \parallel = 1$ and $\varphi(f_0) = \parallel f_0 \parallel$. $\blacksquare$

We have actually proved here that if V is a normed vector space containing non-zero elements, then the Banach dual V^* (i.e., the space of all norm bounded linear functionals on V) contains non-zero elements as well. More precisely, it follows from the last theorem that V^* separates the points of V, i.e., if $f \in V$ and $\varphi(f) = 0$ for all $\varphi \in V^*$, then $f = 0$. The Banach dual V^* is introduced in section 18. For the notion of separation, see the beginning of section 30.

If V is a normed vector space and $\varphi \in V^*$, i.e., φ is a norm bounded linear functional on V, then (by definition) the norm of φ is given by

$$\parallel \varphi \parallel = \sup(\mid \varphi(f) \mid : f \in V, \parallel f \parallel \leq 1).$$

There is a dual formula for the norm of an element in V, as follows.

Corollary 38.5. *If V is a normed vector space and $f \in V$, then*

$$\parallel f \parallel = \sup(\mid \varphi(f) \mid : \varphi \in V^*, \parallel \varphi \parallel \leq 1).$$

Proof. For every $\varphi \in V^*$ satisfying $\parallel \varphi \parallel \leq 1$ we have

$$\mid \varphi(f) \mid \leq \parallel \varphi \parallel \cdot \parallel f \parallel \leq \parallel f \parallel .$$

Hence, $\sup \mid \varphi(f) \mid$ for $\parallel \varphi \parallel \leq 1$ is at most equal to $\parallel f \parallel$. In view of the last theorem there exists at least one $\varphi \in V^*$ such that $\parallel \varphi \parallel = 1$ and $\varphi(f) = \parallel f \parallel$. Hence, $\sup \mid \varphi(f) \mid$ for $\parallel \varphi \parallel \leq 1$ is equal to $\parallel f \parallel$. $\blacksquare$

In section 31 we have discussed the embedding of a Riesz space E into its order bidual $(E^\sim)^\sim$ or, more generally, into $B^\sim$ where B is an ideal in $E^\sim$. If B separates the points of E the embedding is an isomorphism, i.e., the embedding is then one-one (see Lemma 31.2). Something similar can be done in a normed vector space V. For any given $f \in V$ we define the linear functional f^* on V^* by $f^*(\varphi) = \varphi(f)$ for all $\varphi \in V^*$. Since

$$\mid f^*(\varphi) \mid = \mid \varphi(f) \mid \leq \parallel f \parallel \cdot \parallel \varphi \parallel$$

for every φ, the linear functional f^* is norm bounded and $\parallel f^* \parallel \leq \parallel f \parallel$. Hence $f^* \in (V^*)^*$. Briefly written, $f^* \in V^{**}$. It is evident from the definition that $(\alpha f + \beta g)^* = \alpha f^* + \beta g^*$ for real α and β, so the mapping from V into V^{**} which maps $f \in V$ into $f^* \in V^{**}$ is linear. The mapping is also norm preserving as follows from the last corollary by observing that

$$\parallel f^* \parallel = \sup(\mid f^*(\varphi) \mid : \varphi \in V^*, \parallel \varphi \parallel \leq 1) =$$

$$\sup(\mid \varphi(f) \mid : \varphi \in V^*, \parallel \varphi \parallel \leq 1) = \parallel f \parallel .$$

This shows in particular that $f \neq 0$ implies $f^* \neq 0$. Identifying now f^* and f for all $f \in V$, the space V is thus embedded into V^{**} as a linear subspace. In general V is a proper linear subspace of V^{**}. Example: $V = (c_0), V^* = \ell_1, V^{**} = \ell_\infty$. If $V = L_p = L_p(X, \mu)$ for some p satisfying $1 < p < \infty$, then $V^* = L_q$ for $p^{-1} + q^{-1} = 1$, so $V^{**} = L_p = V$. If $V^{**} = V$, the space is said to be *reflexive*.

If V and W are normed vector spaces and $T : V \to W$ is a norm bounded operator, then the Banach adjoint operator $T^* : W^* \to V^*$, as defined in section 26, satisfies

$$(T^*\psi)(f) = \psi(Tf) \text{ for all } f \in V, \psi \in W^*.$$

It was proved in section 26 that $\parallel T^* \parallel \leq \parallel T \parallel$. We complete this result by proving now that $\parallel T^* \parallel = \parallel T \parallel$. It is sufficient to show that $\parallel T \parallel \leq \parallel T^* \parallel$. We may assume that $\parallel T \parallel > 0$ and we recall that

$$\parallel T \parallel = \sup(\parallel Tf \parallel : f \in V, \parallel f \parallel = 1).$$

Hence, for $0 < \epsilon < \parallel T \parallel$, there exists an element $f \in V$ such that $\parallel f \parallel = 1$ and $\parallel Tf \parallel \geq \parallel T \parallel - \epsilon$. Now, in view of Theorem 38.4 above, there exists an element $\psi \in W^*$ such that $\parallel \psi \parallel = 1$ and $\psi(Tf) = \parallel Tf \parallel$. It follows that

$$0 < \parallel T \parallel - \epsilon \leq \parallel Tf \parallel = \psi(Tf) = (T^*\psi)(f) \leq \parallel T^*\psi \parallel \cdot \parallel f \parallel =$$

$$\parallel T^*\psi \parallel \leq \parallel T^* \parallel \cdot \parallel \psi \parallel = \parallel T^* \parallel .$$

The inequality $0 < \parallel T \parallel - \epsilon \leq \parallel T^* \parallel$ holds for every ϵ satisfying $0 < \epsilon < \parallel T \parallel$, so we see that $\parallel T \parallel \leq \parallel T^* \parallel$. This is the desired result. $\blacksquare$

39. The Hahn-Banach Theorem in Normed Riesz Spaces

In the present section some results in Riesz spaces will be obtained which depend on the Hahn-Banach theorem. First a definition. The sublinear mapping p from the Riesz space E into the Riesz space F is said to be *absolute* if $p(f) = p(|\,f\,|)$ for all $f \in E$ and p is said to be *monotone* if it follows from $0 \leq f \leq g$ in E that $p(f) \leq p(g)$. From the sublinearity of p it is evident that $p(0) = 0$(simply choose $\alpha = 0$ in $p(\alpha f) = \alpha p(f)$, holding for $\alpha \geq 0$). Hence, if p is absolute, then $p(f) = p(|\,f\,|) = p(-f)$, and so

$$0 = p(0) = p\{f + (-f)\} \leq p(f) + p(-f) = 2p(f)$$

for every $f \in E$. This shows that $p(f) \geq 0$ for every $f \in E$. As a special case we mention the case that $F = \mathbb{R}$ and E is a normed Riesz space, so the norm in E is now sublinear, absolute and monotone.

We present the first extension theorem in Riesz spaces.

Theorem 39.1. *Let E and F be Riesz spaces with F Dedekind complete and let $p : E \to F$ be sublinear, absolute and monotone. Furthermore, let G be a Riesz subspace of E and let $S : G \to F$ be a positive (linear) operator such that $Sf \leq p(f)$ holds for all $f \in G$. Then there exists a positive operator $T : E \to F$ such that $Tf = Sf$ for all $f \in G$ and $|\,Tf\,| \leq T(|\,f\,|) \leq p(f)$ for all $f \in E$.*

Proof. Define $p_1 : E \to F$ by $p_1(f) = p(f^+)$ for all $f \in E$. Then $p_1(\alpha f) = \alpha p_1(f)$ for $0 \leq \alpha \in \mathbb{R}$ and

$$p_1(f + g) = p\{(f + g)^+\} \leq p(f^+ + g^+) \leq$$

$$p(f^+) + p(g^+) = p_1(f) + p_1(g)$$

for all f and g in E. This shows that p_1 is sublinear. Furthermore, since S is positive, we have

$$Sf \leq S(f^+) \leq p(f^+) = p_1(f)$$

for all $f \in G$. Hence, in virtue of the Hahn-Banach theorem, there exists an operator $T : E \to F$ which extends S and is such that $Tf \leq p_1(f)$ holds for all $f \in E$. It follows that for $f \in E^+$ we have

$$T(-f) \leq p_1(-f) = p\{(-f)^+\} = p(0) = 0,$$

so $Tf \geq 0$. This shows that T is a positive operator. Finally, using now that T is positive, we derive for f arbitrary in E that

$$|\,Tf\,| \leq T(|\,f\,|) \leq p_1(|\,f\,|) = p(|\,f\,|) = p(f). \qquad \blacksquare$$

For the case that $F = \mathbb{R}$ and E is a normed Riesz space, the following extension theorem is an immediate corollary.

Theorem 39.2. *If ψ is a positive linear functional on the Riesz subspace G of the normed Riesz space E such that $\psi(f) \leq \parallel f \parallel$ holds on G, then there exists a positive linear functional φ on E such that $\varphi(f) = \psi(f)$ for all $f \in G$ and $\mid \varphi(f) \mid \leq \varphi(\mid f \mid) \leq \parallel f \parallel$ for all $f \in E$. It follows that every norm bounded positive linear functional ψ on G can be extended to a norm bounded positive linear functional φ on E such that $\parallel \varphi \parallel = \parallel \psi \parallel$.*

Note that the last theorem is similar to Theorem 38.3, where we dealt with arbitrary norm bounded linear functionals on a normed space. The theorem which follows now is similar to Theorem 38.4.

Theorem 39.3. *For any $u > 0$ in the normed Riesz space E there exists a positive linear functional φ on E such that $\parallel \varphi \parallel = 1$ and $\varphi(u) = \parallel u \parallel$.*

Proof. Define the positive linear functional ψ on the Riesz subspace of all real multiples αu of u by $\psi(\alpha u) = \alpha \parallel u \parallel$. Then $\parallel \psi \parallel = 1$ and $\psi(u) = \parallel u \parallel$. In view of the preceding theorem ψ has a positive extension φ to E such that $\parallel \varphi \parallel = \parallel \psi \parallel = 1$. $\qquad\blacksquare$

If E is a normed Riesz space with σ-order continuous norm and φ is a norm bounded linear functional on E (i.e., $\varphi \in E^*$), then $\varphi(u_n) \to 0$ for any sequence $u_n \downarrow 0$ in E. This is evident if φ is positive since then $\varphi(u_n) \leq \parallel \varphi \parallel \cdot \parallel u_n \parallel \downarrow 0$. For an arbitrary $\varphi \in E^*$ it follows from $E^* \subseteq E^\sim$ (see Theorem 25.8) that $\varphi = \varphi^+ - \varphi^-$ with φ^+ and φ^- positive, so

$$\varphi(u_n) = \varphi^+(u_n) - \varphi^-(u_n) \to 0.$$

Similarly, if E has order continuous norm and D is a downwards directed set in E satisfying $D \downarrow 0$, then $\varphi(D)$ is downwards directed with $\varphi(D) \downarrow 0$ for any positive $\varphi \in E^*$. We shall prove now that the converse holds as well. It will be convenient to formulate first a simple lemma about extension of a certain positive linear functional on the space (c) of convergent number sequences. We recall that (c) is the Riesz subspace of ℓ_∞ consisting of all sequences $f = (f_1, f_2, \ldots)$ in ℓ_∞ for which $\lim f_n$ exists as $n \to \infty$.

Lemma 39.4. *The positive linear functional ψ of norm one on (c), defined by $\psi(f) = \lim f_n$ for any $f = (f_1, f_2, \ldots) \in (c)$, can be extended to a positive linear functional Ψ on ℓ_∞ such that $\parallel \Psi \parallel = 1$.*

Proof. Follows from Theorem 39.2 above. $\qquad\blacksquare$

Theorem 39.5. *The following conditions in the normed Riesz space E are equivalent.*

(i) For every sequence $u_n \downarrow 0$ in E we have $\varphi(u_n) \downarrow 0$ for every positive $\varphi \in E^*$.

(ii) For every sequence $u_n \downarrow 0$ in E we have $\| u_n \| \downarrow 0$ (i.e., E has σ-order continuous norm).

Proof. It is evident that (ii) implies (i). For the proof in the converse direction, assume that we have a sequence $0 \leq u_n \downarrow 0$ in E and $\varphi(u_n) \downarrow 0$ for every positive $\varphi \in E^*$, but $\| u_n \| \geq \alpha > 0$ for some α and all n. In view of the last theorem proved above (Theorem 39.3) there exists for each n a positive linear functional φ_n on E such that $\| \varphi_n \| = 1$ and $\varphi_n(u_n) = \| u_n \|$. For any arbitrary $f \in E$ we have $| \varphi_n(f) | \leq \| f \|$, so the number sequence $(\varphi_1(f), \varphi_2(f), \ldots)$ is a member of ℓ_∞. Let Ψ be the positive linear functional on ℓ_∞ introduced in the last lemma and define $\varphi_0(f)$ for each $f \in E$ by

$$\varphi_0(f) = \Psi\{\varphi_1(f), \varphi_2(f), \ldots\}.$$

Then φ_0 is a positive linear functional on E and

$$| \varphi_0(f) | \leq \Psi\{| \varphi_1(f) |, | \varphi_2(f) |, \ldots\} \leq \Psi\{\| f \|, \| f \|, \ldots\} = \| f \|$$

for each $f \in E$, where it is used now that $\Psi(t) = \lim t_n$ for any converging number sequence $t = (t_1, t_2, \ldots)$. It follows that $\varphi_0 \in E^*$ and $\| \varphi_0 \| \leq 1$. Now fix a natural number k. We shall make an estimate for $\varphi_0(u_k)$. Observing that $\varphi_0(u_n) \leq \varphi_0(u_k)$ for $n \geq k$ since $u_n \downarrow$ and φ_0 is positive, we see that

$$\varphi_0(u_k) = \Psi\{\varphi_1(u_k), \varphi_2(u_k), \ldots\} \geq$$

$$\Psi\{\varphi_1(u_k), \ldots, \varphi_k(u_k), \varphi_{k+1}(u_{k+1}), \ldots\} =$$

$$\Psi\{\varphi_1(u_k), \ldots, \| u_k \|, \| u_{k+1} \|, \| u_{k+2} \|, \ldots\} \geq$$

$$\Psi\{\varphi_1(u_k), \ldots, \alpha, \alpha, \alpha, \ldots\} = \alpha.$$

It has been shown thus that $\varphi_0(u_k) \geq \alpha > 0$ for all $k = 1, 2, \ldots$, which contradicts the hypothesis that $\varphi_0(u_k) \downarrow 0$ as $k \to \infty$. It follows, therefore, that $\| u_n \| \downarrow 0$. This proof, more elementary than earlier proofs, is due to E. de Jonge and A.C.M. van Rooij (1977). ∎

For the case of order continuity it is convenient to make some preliminary remarks. If X is an arbitrary (non-empty) set, the normed Riesz space of all (real) bounded functions f on X, equipped with the uniform norm, i.e.,

$$\| f \| = \sup(| f(x) | : x \in X),$$

is sometimes denoted by $\ell_\infty(X)$. In general there does not exist a Riesz subspace similar to (c) in the case that X is countable. If, however, there exists a lattice such that X is a directed subset of the lattice, the situation becomes more like in the countable case. Assume, for example, that X is downwards directed. We shall say now that the element $t \in \ell_\infty(X)$ converges to the number α whenever for each $\epsilon > 0$ there exists a point $x_\epsilon \in X$ such that $|\alpha - t(x)| \le \epsilon$ holds for all $x \le x_\epsilon$. Notation: $\alpha = \lim(t(x) : x \in X)$. It is easily seen that the set $c(X)$ of all converging elements in $\ell_\infty(X)$ is a Riesz subspace of $\ell_\infty(X)$ and

$$\psi(t) = \lim(t(x) : x \in X)$$

is a positive linear functional of norm one on $c(X)$. Similarly as in Lemma 39.4 we extend the functional ψ to a positive linear functional Ψ on $\ell_\infty(X)$ such that $\| \Psi \| = 1$ holds.

Theorem 39.6. *The following conditions in the normed Riesz space E are equivalent.*

(i) For every downwards directed set $D \downarrow 0$ in E we have $\varphi(D) \downarrow 0$ for every positive $\varphi \in E^*$.

(ii) For every downwards directed set $D \downarrow 0$ in E we have $(\| u \| : u \in D) \downarrow 0$ (i.e., E has order continuous norm).

Proof. The proof is similar to the proof in the countable case. If D is downwards directed in E^+ and $(\| u \| : u \in D) \downarrow 0$, i.e., if D converges in norm to the zero element, then $D \downarrow 0$ (see Theorem 15.7). Hence, if E has order continuous norm, then $\varphi(D) \downarrow 0$ for every positive $\varphi \in E^*$. For the proof in the converse direction, assume that we have a downwards directed set D in E^+ and $\varphi(D) \downarrow 0$ holds for every positive $\varphi \in E^*$, but $\| u \| \ge \alpha > 0$ holds for all $u \in D$. For each $u \in D$ there exists a positive linear functional φ_u on E such that $\| \varphi_u \| = 1$ and $\varphi_u(u) = \| u \|$. Hence, for $f \in E$ arbitrary, we have $|\varphi_u(f)| \le \| f \|$, and so the set of numbers $(\varphi_u(f) : u \in D)$ is a member of $\ell_\infty(D)$. Let Ψ be the positive linear functional on $\ell_\infty(D)$ as introduced above and define $\varphi_0(f)$ for each $f \in E$ by

$$\varphi_0(f) = \Psi\{\varphi_u(f) : u \in D\}.$$

Then φ_0 is a positive linear functional on E and

$$|\varphi_0(f)| \le \Psi\{|\varphi_u(f)| : u \in D\} \le \Psi\{\| f \| \text{ for each } u \in D\} = \| f \|,$$

where it is used now that $\Psi(t) = \lim(t(u) : u \in D)$ for any $t \in c(D)$. This shows that $\varphi_0 \in E^*$. Now fix an element $u^* \in D$ and observe that for all $u \in D$ satisfying $u \le u^*$ we have

$$\varphi_u(u^*) \ge \varphi_u(u) = \| u \| \ge \alpha.$$

Hence

$$\varphi_0(u^*) = \Psi\{\varphi_u(u^*) : u : D\} \ge \Psi\{t(u) : u \in D\} = \alpha$$

for some $t \in \ell_\infty(D)$ such that $t(u) = \alpha$ for all $u \leq u^*$, and therefore $t \in c(D)$ and $\Psi(t) = \alpha$. It has been shown thus that $\varphi_0(u^*) \geq \alpha > 0$ for all $u^* \in D$, which contradicts the hypothesis that $\varphi_0(D) \downarrow 0$. We may conclude, therefore, that $(\parallel u \parallel : u \in D) \downarrow 0$. ∎

CHAPTER 21
Spectrum, Resolvent Set and the Krein-Rutman Theorem

40. Spectrum and Resolvent Set

In the last two chapters we shall discuss some spectral properties of a (linear) norm bounded operator T in a complex Banach space V; in particular we deal with the case that V is a Banach lattice and the operator T is positive. In the present section we restrict ourselves, however, to the case that V is a Banach space. Not everything will be proved. The reader is assumed to be familiar with (or at least to accept without full proof) some elementary results about compact sets and compact operators in a Banach space. We have to explain first what is meant by spectral properties of an operator T. We assume, therefore, that T is a norm bounded operator mapping V into itself, i.e., $T \in \mathcal{B}(V)$ in the notation introduced in section 18. Recall that $\mathcal{B}(V)$ is a Banach space (see Theorem 18.2). Choose a complex number λ and, for brevity, denote the operator $T - \lambda I(I$ the identity operator) by T_λ. Assume now that T_λ maps V onto the whole of V in a one-one way, so that therefore the inverse operator $(T_\lambda)^{-1}$ exists with domain V, and assume also that $(T_\lambda)^{-1}$ is norm bounded, so $(T_\lambda)^{-1} \in \mathcal{B}(V)$. The set of complex numbers λ for which this situation occurs is called the *resolvent set* of T. We shall denote this set by $\rho(T)$. The complementary set (that is to say, the set of all complex λ for which $T - \lambda I$ does not have a norm bounded inverse with domain V) is called the *spectrum* of T and denoted by $\sigma(T)$. Operator spectra have been extensively investigated for many different classes of operators. In the sections following after this one we mainly restrict our attention to positive operators in a Banach lattice. Here, however, it will be proved first that if V is a Banach space and T is norm bounded, then the spectrum $\sigma(T)$ is a non-empty bounded set in the complex plane. Recall the notion of the limes superior (limsup) of a sequence $(a_n : n = 1, 2, \ldots)$ of non-negative numbers. Let $b_n = \sup(a_n, a_{n+1}, \ldots)$ for $n = 1, 2, \ldots$, where it is possible that $b_n = \infty$. The sequence $(b_1, b_2, \ldots)$ is non-increasing, so $s = \lim b_n$ (as $n \to \infty$) exists, where it is possible that $b_n = \infty$ for all n, so $s = \infty$ in this case. The number s is denoted by $s = \limsup a_n$. It is important to note that if t is a real number satisfying $t > s$ (this can happen only if s is finite), then $a_n \geq t$ holds for only finitely many n. On the other hand, if t is a real number satisfying $t < s$, then $a_n \geq t$ holds for infinitely many n. As well-known, any series $\Sigma^\infty a_n z^n$ (all a_n and z complex) is called a power series. More generally, we shall consider now series $\Sigma a_n z^n$, where z is complex and the coefficients a_n are members of a given Banach space V. Given the coefficients, one may ask for which z the series converges in V (i.e., converges in norm). Observe first that if Σf_n is a series with all $f_n \in V$ and $\Sigma \parallel f_n \parallel$ converges, then Σf_n converges (because

$$\parallel \Sigma_m^n f_k \parallel \le \Sigma_m^n \parallel f_k \parallel < \epsilon$$

for any finite sum with m sufficiently large). In other words, if a series in V converges absolutely, then the series converges (in norm).

Theorem 40.1. *Let $\Sigma a_n z^n$ be a power series in the Banach space V, i.e., z is complex and the coefficients a_n are elements of V. Define the number $R(0 \le R \le \infty)$ by*

$$R^{-1} = \limsup \parallel a_n \parallel^{1/n} .$$

Then the following holds.

(i) For $\mid z \mid < R$ the series converges absolutely.

(ii) For $\mid z \mid > R$ the sequence $(\parallel a_n z^n \parallel : n = 0, 1, 2, \ldots)$ is unbounded.

As in the familiar case that the coefficients are complex numbers, the number R is called the radius of convergence of $\Sigma a_n z^n$.

Proof. (i) For $\mid z \mid < R$ there exists a number r such that $\mid z \mid < r < R$. Hence $r^{-1} > R^{-1}$, so there are only finitely many n satisfying $\parallel a_n \parallel^{1/n} > r^{-1}$. In other words, there exists a natural number N such that $\parallel a_n \parallel^{1/n} \le r^{-1}$ for all $n \ge N$. It follows that $\parallel a_n \parallel \le r^{-n}$ and so

$$\parallel a_n z^n \parallel \le (\mid z \mid /r)^n$$

for all $n \ge N$. Since the geometric series $\Sigma (\mid z \mid /r)^n$ is converging (because $\mid z \mid < r$), this shows that $\Sigma \parallel a_n z^n \parallel$ converges.

(ii) For $\mid z \mid > R$ there exists a number r such that $\mid z \mid > r > R$. Hence $r^{-1} < R^{-1}$, so there are infinitely many n satisfying $\parallel a_n \parallel^{1/n} > r^{-1}$. It follows that

$$\parallel a_n z^n \parallel > (\mid z \mid /r)^n$$

for these n. Since $\mid z \mid > r$, we have $(\mid z \mid /r)^n \to \infty$ for the subsequence of these n, so the sequence $(\parallel a_n z^n \parallel : n = 0, 1, 2, \ldots)$ is unbounded.

Note that for $\mid z \mid \le R_1 < R$ the convergence of $\Sigma \parallel a_n z^n \parallel$, and therefore of $\Sigma a_n z^n$, is uniform, i.e., for any $\epsilon > 0$ there exists $N_0 = N_0(\epsilon, R_1)$ such that

$$\Sigma^\infty \parallel a_n z^n \parallel - \Sigma^N \parallel a_n z^n \parallel < \epsilon$$

for all $n \ge N_0$ and all z satisfying $\mid z \mid \le R_1$.   ■

Assume now that $T \in \mathcal{B}(V)$, i.e., T is a norm bounded operator in the Banach space V. Then $T^n \in \mathcal{B}(V)$ for $n = 1, 2, \ldots$ (and also for $n = 0$, if we agree that $T^n = I$ for $n = 0$). Since $\| T_1 T_2 \| \leq \| T_1 \| \cdot \| T_2 \|$ holds for all T_1 and T_2 in $\mathcal{B}(V)$, we see that $\| T^n \| \leq \| T \|^n$ for $n = 1, 2, \ldots$. It follows that $\| T^n \|^{1/n} \leq \| T \|$, which implies that

$$\limsup_n \| T^n \|^{1/n} \leq \| T \| \,.$$

We introduce the series

$$-\lambda^{-1} I - \lambda^{-2} T - \lambda^{-3} T^2 - \cdots,$$

where $\lambda \neq 0$ is complex and where, as before, I is the identity operator in V. This is a power series in λ^{-1} with coefficients in the Banach space $\mathcal{B}(V)$. The formula for the radius of convergence shows that the series converges for $| \lambda | > \limsup \| T^n \|^{1/n}$ and not for $| \lambda | < \limsup \| T^n \|^{1/n}$. Therefore, the series converges certainly for $| \lambda | > \| T \|$. Denoting the sum of the series by S_λ for these values of λ, the operator S_λ is a member of $\mathcal{B}(V)$. A simple calculation shows that

$$S_\lambda (T - \lambda I) = (T - \lambda I) S_\lambda = I.$$

Hence, for $| \lambda | > \limsup \| T^n \|^{1/n}$, the operator $T - \lambda I$ has an inverse $(T - \lambda I)^{-1}$ in $\mathcal{B}(V)$, given by

$$(T - \lambda I)^{-1} = -\lambda^{-1} I - \lambda^{-2} T - \lambda^{-3} T^2 - \cdots.$$

All λ satisfying $| \lambda | > \limsup \| T^n \|^{1/n}$ are, therefore, members of the resolvent set $\rho(T)$ of T. The spectrum $\sigma(T)$ of T (i.e., the complement of the resolvent set) is then contained in the circular set of all λ for which $| \lambda | \leq \limsup \| T^n \|^{1/n}$, so the spectrum is a bounded set in the complex plane. In general, there will also be points λ in the resolvent set of T satisfying $| \lambda | \leq \limsup \| T^n \|^{1/n}$. It will be proved now first that the resolvent set $\rho(T)$ is an open set in the complex plane. For convenience, denote $(T - \lambda I)^{-1}$ by R_λ for all λ in the resolvent set of T.

Theorem 40.2. *If $\lambda_0 \in \rho(T)$ and $| \lambda - \lambda_0 | < \| R_{\lambda_0} \|^{-1}$, then $\lambda \in \rho(T)$ and*

$$R_\lambda = R_{\lambda_0} + (\lambda - \lambda_0)(R_{\lambda_0})^2 + (\lambda - \lambda_0)^2 (R_{\lambda_0})^3 + \cdots$$

holds in $\mathcal{B}(V)$. This shows that $\rho(T)$ is an open set in the complex plane and R_λ is a continuous function of λ on $\rho(T)$.

Proof. The convergence of the series follows by observing that

$$\Sigma \mid \lambda - \lambda_0 \mid^n \cdot \parallel (R_{\lambda_0})^{n+1} \parallel \le \Sigma \mid \lambda - \lambda_0 \mid^n \cdot \parallel R_{\lambda_0} \parallel^{n+1} < \infty.$$

Writing $S = \Sigma(\lambda - \lambda_0)^n (R_{\lambda_0})^{n+1}$, we have

$$\{I - (\lambda - \lambda_0)R_{\lambda_0}\}S =$$

$$\Sigma_0^\infty (\lambda - \lambda_0)^n (R_{\lambda_0})^{n+1} - \Sigma_1^\infty (\lambda - \lambda_0)^n (R_{\lambda_0})^{n+1} = R_{\lambda_0},$$

and hence

$$(T - \lambda_0 I)\{I - (\lambda - \lambda_0)R_{\lambda_0}\}S = I,$$

Corollary 40.3. *If $\lambda_0 \in \sigma(T)$ and if there exists a sequence $(\lambda_n : n = 1, 2, \ldots)$ of complex numbers in $\rho(T)$ such that $\lambda_n \to \lambda_0$, then $\parallel R_{\lambda_n} \parallel \to \infty$ as $n \to \infty$.*

Proof. If $\parallel R_{\lambda_n} \parallel \to \infty$ does not hold, we may assume (passing to a subsequence if necessary) that $\parallel R_{\lambda_n} \parallel \le M$ for some $M > 0$ and all n. Then $\mid \lambda_n - \lambda_0 \mid < M^{-1}$ for n sufficiently large, which implies that $\mid \lambda_0 - \lambda_n \mid < \parallel R_{\lambda_n} \parallel^{-1}$. It follows now from the theorem above that $\lambda_0 \in \rho(T)$, contrary to the hypothesis. ∎

Before discussing further properties of the spectrum it is of some interest to observe that $\lim \parallel T^n \parallel^{1/n}$ (as $n \to \infty$) exists (i.e., the limsup is actually a limit), so that therefore the series

$$-\lambda^{-1}I - \lambda^{-2}T - \lambda^{-3}T^2 - \cdots \tag{1}$$

converges for $\mid \lambda \mid > \lim \parallel T^n \parallel^{1/n}$ and diverges for $\mid \lambda \mid < \lim \parallel T^n \parallel^{1/n}$. The existence of the limit follows by observing that the series (1) converges certainly for $\mid \lambda \mid > \parallel T \parallel$, and so (if n is any natural number)

$$-\lambda^{-n}I - \lambda^{-2n}T^n - \lambda^{-3n}T^{2n} - \cdots \tag{2}$$

converges certainly for all complex λ satisfying $\mid \lambda^n \mid > \parallel T^n \parallel$, i.e., $\mid \lambda \mid > \parallel T^n \parallel^{\frac{1}{n}}$. Multiplying the series (2) by

$$T^{n-1} + \lambda T^{n-2} + \cdots + \lambda^{n-1}I,$$

we obtain the series (1) and this does not change the convergence, so (1) converges certainly for $\mid \lambda \mid > \parallel T^n \parallel^{1/n}$. Because this holds for all $n = 1, 2, \ldots$, we see that (1) converges for $\mid \lambda \mid > \inf_n \parallel T^n \parallel^{1/n}$. Comparing this with the result for the radius of convergence for (1), it follows that

$$\limsup_n \parallel T^n \parallel^{1/n} \le \inf_n \parallel T^n \parallel^{1/n} .$$

This shows that $\lim_n \parallel T^n \parallel^{1/n}$ exists and

$$\lim_n \| T^n \|^{1/n} = \inf_n \| T^n \|^{1/n} .$$

It is evident from what has been proved so far that the spectrum $\sigma(T)$ of T is a closed and bounded set in the complex plane $\mathbb{C}$. The radius of the smallest circle having as its centre the origin and containing the spectrum is called the *spectral radius* of T and is denoted by $r(T)$. Hence

$$r(T) = \sup(| \lambda |: \lambda \in \sigma(T)) = \max(| \lambda |: \lambda \in \sigma(T)).$$

We shall indicate how to prove that $\sigma(T)$ is not empty. Hence, if $r(T) = 0$, then $\sigma(T)$ consists of the number zero only. We shall also indicate (but this is a little more difficult) how to prove that the spectral radius satisfies

$$r(T) = \lim_n \| T^n \|^{1/n} .$$

Before doing so, we first prove a simple lemma dealing with the behaviour of $R_\lambda = (T - \lambda I)^{-1}$.

Lemma 40.4.
(i)   *If the complex numbers λ and μ are in the resolvent set $\rho(T)$ of T, then*

$$R_\lambda - R_\mu = (\lambda - \mu)R_\lambda R_\mu = (\lambda - \mu)R_\mu R_\lambda.$$

(ii) $\| R_\lambda \| \to 0$ as $| \lambda | \to \infty$.

Proof. (i) We have $R_\lambda = R_\lambda(T - \mu I)R_\mu$ and $R_\mu = R_\lambda(T - \lambda I)R_\mu$, so

$$R_\lambda - R_\mu = R_\lambda \{ (T - \mu I) - (T - \lambda I) \} R_\mu =$$

$$R_\lambda(\lambda I - \mu I)R_\mu = (\lambda - \mu)R_\lambda R_\mu.$$

Similarly $R_\lambda = R_\mu(T - \mu I)R_\lambda$ and $R_\mu = R_\mu(T - \lambda I)R_\lambda$, so

$$R_\lambda - R_\mu = (\lambda - \mu)R_\mu R_\lambda.$$

(ii) For $| \lambda |$ sufficiently large (certainly for $| \lambda | > \| T \|$) we have $\lambda \in \rho(T)$, and so it follows for these λ from

$$\| (T - \lambda I)f \| = \| \lambda f - Tf \| \geq (| \lambda | - \| T \|) \cdot \| f \|$$

for every $f \in V$ that

$$\| g \| \geq (| \lambda | - \| T \|) \cdot \| R_\lambda g \|$$

for every $g \in V$. This shows that $\| R_\lambda \| \leq (| \lambda | - \| T \|)^{-1}$ for $| \lambda | > \| T \|$. Hence $\| R_\lambda \| \to 0$ as $| \lambda | \to \infty$. ■

As promised above, we turn our attention to the spectrum and the spectral radius of T. For this purpose it is necessary to introduce the integral of a function assuming its values in a Banach space. Let F be a function, defined for $a \leq t \leq b$ and with values in the Banach space V. If F is continuous at each $t \in [a, b]$, then F is uniformly continuous on $[a, b]$ exactly as in the familiar case that F assumes real or complex values, i.e., for each $\epsilon > 0$ there exists $\delta > 0$ such that for all t_1, t_2 in $[a, b]$ satisfying $\mid t_1 - t_2 \mid < \delta$ we have $\parallel F(t_1) - F(t_2) \parallel \leq \epsilon$. Now, let

$$(a = t_0 < t_1 < t_2 < \cdots < t_n = b)$$

be a partition of $[a, b]$. If $\max_k(t_k - t_{k-1}) \leq \delta$ for $k = 1, \ldots, n$, we shall say for convenience that we have a δ-partition. If moreover F is continuous on $[a, b]$ and ξ_k is a point in $[t_{k-1}, t_k]$ for $k = 1, \ldots, n$, we shall say that

$$F(\xi_1)(t_1 - t_0) + F(\xi_2)(t_2 - t_1) + \cdots + F(\xi_n)(t_n - t_{n-1})$$

is a δ-sum for F.

Lemma 40.5. *If F is continuous on $[a, b]$ and $c > 0$ is given, there exists a number $\delta > 0$ such that the difference between any two δ-sums is at most equal to ϵ in norm.*

Proof. Given $\epsilon > 0$, let $\delta > 0$ be such that $\mid t_1 - t_2 \mid \leq \delta$ implies

$$\parallel F(t_1) - F(t_2) \parallel \leq \frac{\epsilon}{[2(b - a)]}.$$

If S_1 and S_2 are now δ-sums for F such that one of the corresponding δ-partitions is a refinement of the other one, then

$$\parallel S_1 - S_2 \parallel \leq \{\frac{\epsilon}{[2(b - a)]}\}\Sigma(t_k - t_{k-1}) = \frac{\epsilon}{2}.$$

Hence, if S_1 and S_2 are arbitrary δ-sums and S_3 is their common refinement, then

$$\parallel S_1 - S_2 \parallel \leq \parallel S_1 - S_3 \parallel + \parallel S_3 - S_2 \parallel \leq \epsilon. \qquad \blacksquare$$

Therefore, the following definition may be given.

Definition 40.6. If F, with values in the Banach space V, is continuous on $[a, b]$, the δ-sums for F tend to a limit depending only on F as $\delta \to 0$. The limit is called the integral of F over $[a, b]$ and is denoted by $\int_a^b F(t)dt$. Note that the existence of the limit depends on the hypothesis that V is a Banach space, i.e., each Cauchy sequence in V has a limit.

It is easily seen that the integral has the usual linearity properties. Also, if $F = \Sigma F_n$ on $[a, b]$ and the series converges uniformly on $[a, b]$, then $\int F dt = \Sigma \int F_n dt$. If $F = Gf$, where G is complexvalued and continuous and $f \in V$, then $\int F dt = (\int G dt) \cdot f$. If F is constant on $[a, b]$, i.e., $F(t) = f$ for all $t \in [a, b]$, then $\int F dt = (b - a)f$. Finally, it follows from

$$\| \Sigma F(\xi_k)(t_k - t_{k-1}) \| \le \Sigma \| F(\xi_k) \| (t_k - t_{k-1})$$

that $\| \int F(t) dt \| \le \int \| F(t) \| dt$.

Lemma 40.7. *As before, let T be a bounded operator in the Banach space V and let $\mu \in \rho(T)$, so that therefore all $\lambda = \mu + re^{i\varphi}$ (r fixed with $0 < r < \| R_\mu \|^{-1}, \varphi$ real) are in $\rho(T)$, and*

$$R_\varphi = R_\lambda = R_\mu + \Sigma_1^\infty r^n e^{in\varphi}(R_\mu)^{n+1}$$

holds for these λ (see Theorem 40.2 above). Then

$$R_\mu = (2\pi)^{-1} \int_0^{2\pi} R_\varphi d\varphi.$$

Proof. Follows from the uniform convergence of the series for R_φ on $[0, 2\pi]$ (so that termwise integration is possible) and from $\int_0^{2\pi} e^{in\varphi} d\varphi = 0$ which holds for $n = 1, 2, \ldots$. ∎

Corollary 40.8. *The function $\| R_\lambda \|$ on the resolvent set $\rho(T)$ has no proper maximum on $\rho(T)$. Precisely, if $\mu \in \rho(T)$ and the sequence $(\lambda_n : n = 1, 2, \ldots)$, converging to μ, has the property that $\| R_{\lambda_n} \| < \| R_\mu \|$ for all $n = 1, 2, \ldots$, then there exists a sequence $(\nu_n : n = 1, 2, \ldots)$, also converging to μ, such that $\| R_{\nu_n} \| > \| R_\mu \|$ for all $n = 1, 2, \ldots$.*

Proof. Follows from

$$\| R_\mu \| \le (2\pi)^{-1} \int_0^{2\pi} \| R_\varphi \| d\varphi,$$

where the integration is over a circle with centre μ and radius $| \lambda_n - \mu |$. ∎

Theorem 40.9. *The spectrum of T is not empty.*

Proof. Assume that T has an empty spectrum, i.e., every complex number is in the resolvent set $\rho(T)$ of T. In particular $\lambda = 0$ is in $\rho(T)$, i.e., T has a bounded inverse. Hence, T^{-1} exists and this is then R_λ for $\lambda = 0$, i.e., $R_0 = T^{-1}$ maps V onto V. Let $\| R_0 \| = \alpha$, so $\alpha > 0$. Since $\| R_\lambda \| \to 0$ as $\lambda \to \infty$, there exists a number $C > 0$ such that $\| R_\lambda \| < \alpha/2$ for $| \lambda | \ge C$. The function $\| R_\lambda \|$ is continuous on $(\lambda : | \lambda | \le C)$, so $\| R_\lambda \|$ has a finite maximum β on $(\lambda : | \lambda | \le C)$ which is at least equal to α (since $\| R_0 \| = \alpha$). The points at which $\| R_\lambda \|$ attains its maximum β are interior points of $(\lambda : | \lambda | \le C)$ since $\| R_\lambda \| < \alpha/2$ for $| \lambda | \ge C$. It follows that there exists at least one point λ_0 in $(\lambda : | \lambda | < C)$ such that $\| R_{\lambda_0} \| = \beta$ and $\| R_{\lambda_n} \| < \beta$ for all λ_n in an appropriate sequence $(\lambda_n : n = 1, 2, \ldots)$ converging to λ_0. Then, in view of the corollary above, there also exists a sequence $(\nu_n : n = 1, 2, \ldots)$ converging to λ_0 such that $\| R_{\nu_n} \| > \beta$ for all n. This is not possible since β is the maximal value of $\| R_\lambda \|$. Hence, the spectrum of T is not empty.   ∎

We proceed with a discussion of integrals of Banach space valued functions along curves in the complex plane. By definition, a continuous curve (or arc)γ in the complex λ-plane $\mathbb{C}$ is a point set $\gamma : \gamma = (\lambda(t) : a \leq t \leq b)$ in $\mathbb{C}$, where $\lambda(t)$ is continuous on $[a, b]$. We shall assume here even that $\lambda(t)$ has a continuous derivative $\lambda'(t)$ on $[a, b]$. Then, if $F = F(\lambda)$, with values in the Banach space V, is defined and continuous on γ, the function $F\{\lambda(t)\}\lambda'(t)$ is continuous on $[a, b]$. We define the integral of F along γ by

$$\int_\gamma F(\lambda)d\lambda = \int_a^b F\{\lambda(t)\}\lambda'(t)dt.$$

The definition may immediately be extended to the case that the derivative $\lambda'(t)$ is only piecewise continuous. Example: γ is the perimeter of a rectangle. This type of integral is related to the integral of F with respect to arc length, which is by definition the integral

$$\int_\gamma F ds = \int_\gamma F(\lambda) \mid d\lambda \mid = \int_a^b F\{\lambda(t)\} \mid \lambda'(t) \mid dt.$$

For the case that $V = \mathbb{C}$ and $F(\lambda) = 1$ for all $\lambda \in \gamma$, the integral

$$\int_\gamma F ds = \int_\gamma ds = \int \mid \lambda'(t) \mid dt$$

is the arc length of γ. Let us mention as a well-known example that the length of the circle $(\lambda : \lambda = \lambda_0 + re^{i\varphi})$, for $0 \leq \varphi \leq 2\pi$ and $r \geq 0$ fixed, is

$$\int_0^{2\pi} \mid ire^{i\varphi} \mid d\varphi = \int_0^{2\pi} rd\varphi = 2\pi r.$$

We recall that the subset Ω of $\mathbb{C}$ is said to be a *region* if Ω is open and *connected*, i.e., Ω is open and each pair of points in Ω is connected by a continuous arc contained in Ω. Let F, with values in the Banach space V, be defined on the region Ω. As in the familiar case that V is the complex plane $\mathbb{C}$, the function F is said to be *differentiable* at the point $\lambda_0 \in \Omega$ if the limit

$$F'(\lambda_0) = \lim_{\lambda \to \lambda_0} [\{F(\lambda) - F(\lambda_0)\}/(\lambda - \lambda_0)]$$

exists (i.e., exists in norm) and $F'(\lambda_0)$ is then called the derivative of F at λ_0. If F is differentiable at each point of the region Ω, then F is said to be *analytic* on Ω. As an example, let T be a bounded operator in the Banach space V and let F be the function $R_\lambda = (T - \lambda I)^{-1}$ in the resolvent set $\rho(T)$ of T. The set $(\lambda :\mid \lambda \mid > r(T))$ is a region Ω contained in the resolvent set

$\rho(T)$ and R_λ is analytic on Ω. To see this, note that if λ and μ are points in Ω, then

$$(R_\mu - R_\lambda)/(\mu - \lambda) = R_\mu R_\lambda$$

by Lemma 40.4(i), so the limit as $\mu \to \lambda$ exists and is equal to $(R_\lambda)^2$, i.e., $(R_\lambda)' = (R_\lambda)^2$.

Returning to the general case, let F be analytic on the region Ω and let γ be the perimeter of a rectangle such that γ and its interior are contained in Ω. Then $\int_\gamma F d\lambda = 0$. The proof is almost exactly the same as in the case that $V = \mathbb{C}$. This can be extended (as in the case again that $V = \mathbb{C}$) to the situation that γ is a more general closed curve, in particular to the situation that γ is a circle such that both γ and its interior are contained in Ω. Now let λ_0 be a point in Ω and let G be analytic on Ω except perhaps at λ_0. Assume that γ is a circle such that γ and its interior are contained in Ω and such that λ_0 is in the interior of γ. Also, let γ_1 be another circle with centre λ_0 and radius so small that γ_1 is contained in the interior of γ. It follows in a familiar manner that $\int_\gamma G d\lambda = \int_{\gamma_1} G d\lambda$. We apply this result to the case that $G(\lambda) = F(\lambda)/(\lambda - \lambda_0)$, where F is analytic on Ω. It is easy to see that if we let the radius of γ_1 tend to zero, then

$$\int_{\gamma_1} [F(\lambda)/(\lambda - \lambda_0)] d\lambda \to 2\pi i F(\lambda_0).$$

Hence, it follows from $\int_\gamma G d\lambda = \int_{\gamma_1} G d\lambda$ that

$$2\pi i F(\lambda_0) = \int_\gamma G(\lambda) d\lambda = \int_\gamma [F(\lambda)/(\lambda - \lambda_0)] d\lambda.$$

This is, therefore, the extension of Cauchy's integral formula from complex-valued functions to Banach space valued functions. The formula will be used now to prove the following theorem.

Theorem 40.10. *Let F (with values in the Banach space V) be analytic in the interior of a circle γ with centre the origin and radius r. Then there exists a power series $\Sigma_0^\infty a_n \lambda^n$ (with coefficients $a_n \in V$) such that $F(\lambda) = \Sigma_0^\infty a_n \lambda^n$ for all λ in the interior of γ (i.e., for all λ satisfying $\mid \lambda \mid < r$). The coefficients are uniquely determined.*

Proof. Choose a point λ satisfying $\mid \lambda \mid = \rho < r$ and keep λ fixed. Then choose a number ρ_1 such that $\rho < \rho_1 < r$. Let γ_1 be the circle with centre the origin and radius ρ_1. For any point τ on γ_1 we have

$$(\tau - \lambda)^{-1} = \tau^{-1}\{1 - (\lambda/\tau)\}^{-1} =$$

$$= \tau^{-1}\{1 + (\lambda/\tau) + (\lambda/\tau)^2 + \cdots\} = \Sigma_0^\infty (\lambda^n/\tau^{n+1}),$$

where the series on the right converges uniformly on γ_1 (because $|\lambda/\tau| < 1$). Then, since $\| F \|$ is bounded on γ_1, the series

$$F(\tau)/(\tau - \lambda) = \Sigma_0^\infty \{F(\tau)/\tau^{n+1}\}\lambda^n$$

likewise converges uniformly on γ_1. Hence, by means of Cauchy's integral formula and by termwise integration, it follows that

$$F(\lambda) = (2\pi i)^{-1} \int_{\gamma_1} \{F(\tau)/(\tau - \lambda)\}d\tau =$$

$$= \Sigma_0^\infty [(2\pi i)^{-1} \int_{\gamma_1} \{F(\tau)/\tau^{n+1}\}d\tau]\lambda^n = \Sigma_0^\infty a_n \lambda^n,$$

where a_n is the expression between square brackets. The uniqueness of the coefficients is a consequence of the following simple lemma. ∎

Lemma 40.11. *If $\Sigma_0^\infty a_n \lambda^n$ and $\Sigma_0^\infty b_n \lambda^n$ both have a positive radius of convergence and their sumfunctions, say F and G respectively, are equal in an (open) neighbourhood of the origin, then $a_n = b_n$ for all n.*

Proof. The proof is by induction. Observe first that it follows from the uniform convergence of $\Sigma a_n \lambda^n$ in any (closed) circular disc with radius smaller than the radius of convergence (see the remark at the end of the proof of Theorem 40.1) that F is continuous. Since $F(0) = G(0)$ it is evident that $a_0 = b_0$. Assume now that it has been proved that $a_k = b_k$ for $k = 0, 1, \ldots, n-1$. Then

$$a_n + a_{n+1}\lambda + a_{n+2}\lambda^2 + \cdots = b_n + b_{n+1}\lambda + b_{n+2}\lambda^2 + \cdots$$

for all λ in a neighbourhood of the origin, except perhaps the origin itself. Letting $\lambda \to 0$ through an appropriate sequence $\lambda_n \neq 0$, it follows by continuity that $a_n = b_n$. ∎

The proof that the spectral radius $r(T)$ of a norm bounded operator T in the Banach space V satisfies $r(T) = \lim_n \| T^n \|^{1/n}$ is now relatively easy. As before we denote $(T - \lambda I)^{-1}$ for λ in the resolvent set of T by R_λ and, for convenience of notation, we shall denote $\lim_n \| T^n \|^{1/n}$ by $\alpha(T)$. It was proved that

$$R_\lambda = -\lambda^{-1}I - \lambda^{-2}T - \lambda^{-3}T^2 - \cdots$$

for $|\lambda| > \alpha(T)$. Precisely, the power series in λ^{-1} on the right has radius of convergence $\{\alpha(T)\}^{-1}$. Furthermore, $R_\lambda - R_\mu = (\lambda - \mu)R_\lambda R_\mu$ for λ and μ in the resolvent set of T, so the derivative $(R_\lambda)'$ exists and satisfies $(R_\lambda)' = (R_\lambda)^2$. It follows that R_λ is analytic in the region $\{\lambda : |\lambda| > \alpha(T)\}$. Therefore, since there is at least one point λ_0 in the spectrum of T such that $|\lambda_0| = r(T)$, it is evident that $r(T) \leq \alpha(T)$. It remains to be proved that $\alpha(T) \leq r(T)$.

Let $\tau = \lambda^{-1}, \sigma = \mu^{-1}$ and, for convenience again, denote R_λ and R_μ by S_τ and S_σ respectively. Then

$$S_\tau = -\tau I - \tau^2 T - \tau^3 T^2 - \cdots, \tag{3}$$

$$S_\tau - S_\sigma = \{-(\tau - \sigma)S_\tau S_\sigma\}/(\tau\sigma) \tag{4}$$

for $|\tau|$ and $|\sigma|$ not zero and less than $\{\alpha(T)\}^{-1}$. It is evident from (4) that S_τ (as a function of τ) is analytic in the region $(\tau : 0 <| \tau |< \{\alpha(T)\}^{-1})$. Defining now S_τ for $\tau = 0$ by $S_0 = 0$ (i.e., S_0 is the null operator), it follows immediately that S_τ is differentiable also at $\tau = 0$ (note from (3) that $(S_\tau - S_0)/\tau = S_\tau/\tau$ tends to $-I$ as $\tau \to 0$), so S_τ is analytic and equal to the sum of the series (3) in the interior of the circle of convergence with radius $\{\alpha(T)\}^{-1}$ in the τ-plane. On the other hand, since R_λ is differentiable at each point λ in the resolvent set and so in particular for $|\lambda|> r(T)$, it follows that S_τ is analytic in $(\tau :| \tau |< \{r(T)\}^{-1})$. Hence, by Theorem 40.10 above, S_τ is the sum of a power series in τ converging in the interior of a circle with centre the origin and radius at least $\{r(T)\}^{-1}$. By the uniqueness in Theorem 40.10 this power series must be the same series as the one in (3) with convergence radius $\{\alpha(T)\}^{-1}$. It follows that $\{r(T)\}^{-1} \leq \{\alpha(T)\}^{-1}$, i.e., $\alpha(T) \leq r(T)$. This is the desired result which we now state as a theorem.

Theorem 40.12. *The spectral radius $r(T)$ of the norm bounded operator T in the Banach space V satisfies*

$$r(T) = \lim_n \| T^n \|^{1/n} .$$

We mention some terminology. As well-known, the operator T in the Banach space V is called *nilpotent* if there exists a natural number n such that T^n is the null operator (equivalently, $\| T^n \|= 0$ if T is bounded). The bounded operator T is said to be *quasi-nilpotent* if $\lim \| T^n \|^{1/n}= 0$ as $n \to \infty$, i.e., if $r(T) = 0$.

To proceed with the present section we recall some elementary facts from Banach space theory. Let F be a subset of the Banach space V. According to Lemma 14.1 the set F is closed (i.e., norm closed) if and only if it follows from $f_n \in F(n = 1, 2, \ldots)$ and $f_n \to f$ in norm that $f \in F$. Now, if $F \subseteq V$ is arbitrary, we define F^- to be the set of all points in V that are the norm limit of some sequence in F. Then every point of F is (trivially) a point of F^-, but F^- may contain points not in F unless F is closed (because it is obvious from Lemma 14.1 as cited above that $F = F^-$ if F is closed). The set F^- is closed. To see this, let $f_n^\sim \in F^-(n = 1, 2, \ldots)$ and $f_n^\sim \to f^\sim$ in norm. For each n there exists $f_n \in F$ such that $\| f_n^\sim - f_n \|< n^{-1}$, so (triangle inequality) $f_n \to f^\sim$ in norm. This shows that $f^\sim \in F^-$, and so F^- is closed (by Lemma 14.1 again). We may conclude from this that if $F = F^-$, then F is closed. Hence, we have $F = F^-$ if and only if F is closed. The set F^- is called the *closure* of F. Evidently, F^- is the smallest closed set containing F. If $F^- = V$, then F is said to be *norm dense* in V.

We mention some further properties which a subset F of V may possess. If $F \subseteq \cup_\tau O_\tau$ (all O_τ open), then the collection of all O_τ is said to be an *open cover* of F.

Definition 40.13. (i) The subset F of the Banach space V is called compact if every open cover of F has a finite subcover.

(ii) The subset F of the Banach space V is called sequentially compact if every sequence in F has a subsequence converging to a point of F.

(iii) The subset F of the Banach space V is called relatively compact (relatively sequentially compact) if the closure F^- is compact (sequentially compact).

It can be proved that F is compact if and only if F is sequentially compact. Therefore, F is relatively compact if and only if F is relatively sequentially compact. We recall here only the proof that compactness of F implies sequential compactness of F. Assume that F is compact but not sequentially compact. Then there exists a sequence $(f_n : n = 1, 2, \ldots)$ in F without any subsequence converging to a point of F. This implies that for any $g \in F$ there exists an open ball B_g with centre g such that B_g contains only finitely many of the points f_n. The union $\cup_{g \in F} B_g$ is an open cover of F, so there exists a finite subcover, say $A = B_{g_1} \cup \cdots \cup B_{g_k}$. Then $A \supseteq F$, so A contains all f_n. On the other hand A contains only finitely many f_n. Contradiction. It follows immediately that relative compactness implies relative sequential compactness.

It is easy to see (by means of Lemma 14.1) that any sequentially compact set is closed. Hence, any compact set is closed. We prove that any closed subset F_1 of a compact set F is compact. Let $\cup_\tau O_\tau$ be an open cover of F_1. Then

$$(\cup_\tau O_\tau) \cup (V \backslash F_1)$$

is an open cover of F (it is even an open cover of V). There is a finite subcover which of course also covers F_1. The set $V \backslash F_1$ does not contribute to the covering of F_1, so $\cup_\tau O_\tau$ has a finite subcover for F_1. Hence, F_1 is compact. It is easy to see that every closed subset of a sequentially compact set is sequentially compact. It follows that any arbitrary subset of a relatively compact set (relatively sequentially compact set) is relatively compact (relatively sequentially compact). Note that any sequentially compact set F in V is norm bounded (if not, there would exist a sequence $(f_n : n = 1, 2, \ldots)$ in F satisfying $\| f_n \| \to \infty$, contradicting the sequential compactness of F). Hence, any compact set in F is norm bounded. If V is of finite dimension, then conversely every norm bounded set in V is relatively compact (for a simple proof we refer to Exercise 40.15 below).

Finally, we shall make some remarks about the spectrum of a norm bounded operator in the Banach space V, in particular about the spectrum of a compact operator in V. As defined above, the resolvent set $\rho(T)$ of T consists of all complex λ for which $T - \lambda I$ maps V in a one-one manner onto V such that the inverse $(T - \lambda I)^{-1}$ is norm bounded. All complex λ not in $\rho(T)$ form the spectrum of T. The spectrum of T can be decomposed into several parts. If there exists an element $f \neq 0$ in V such that $(T - \lambda I)f = 0$, i.e., if $Tf = \lambda f$, then λ is called an *eigenvalue* of T and f is a corresponding *eigenelement*. The set of all f satisfying $Tf = \lambda f$ is a closed subspace of V, called the *eigenspace* corresponding to λ. The dimension of the eigenspace (finite or not) is called the *(geometrical) multiplicity* of λ. There exist more possibilities for λ. If the operator $T - \lambda I$ is one-one with range norm dense in V but the inverse mapping (defined on the range) is not norm bounded, then λ is in the *continuous spectrum* of T. Finally, if the range of $T - \lambda I$ is not norm dense in V, then λ is said to be in the *residual spectrum* of T.

Definition 40.14. The operator T in the Banach space V is called a compact operator if T transforms the unit ball $(f : \| f \| \leq 1)$ in V into a relatively compact set.

Evidently, T is compact if and only if T transforms any norm bounded set in V into a relatively compact set. Since every relatively compact set is norm bounded, it follows that any compact operator is norm bounded. Also, since compactness and sequential compactness in V are identical, the operator T in V is compact if and only if the image under T of any norm bounded sequence is a sequence possessing a Cauchy subsequence.

It can be proved (but we do not include the proofs here; for these we refer to any book on functional analysis) that any $\lambda \neq 0$ in the spectrum of the compact operator T is an eigenvalue of finite multiplicity. The number of these eigenvalues is finite or countable and, if not finite, they can be written in a sequence $(\lambda_n : n = 1, 2, \ldots)$ such that $\lambda_n \to 0$. It follows that each λ_n is an isolated point in the spectrum, i.e., λ_n is the centre of an open circular disc containing only points in the resolvent set except λ_n itself.

Exercise 40.15. Let F be a norm bounded set in the normed space V of finite dimension. Hence, if $(e_1, e_2, \ldots, e_n)$ is a basis in V, each $f \in V$ can be written uniquely as $f = \alpha_1 e_1 + \cdots + \alpha_n e_n$ with $\alpha_1, \ldots, \alpha_n$ complex. For abbreviation we write $A(f) = | \alpha_1 | + \cdots + | \alpha_n |$. Prove that $\| f \| \leq M \cdot A(f)$ for $M = \max(\| e_1 \|, \ldots, \| e_n \|)$. Hence, $A(f_k) \to 0$ implies that $\| f_k \| \to 0$. There also exists a number $m > 0$ such that $A(f) \leq m \| f \|$ for every f. If not, there would exist a sequence $(f_1, f_2, \ldots)$ such that $\| f_k \| \to 0$ but $A(f_k) = 1$ for all k. Show that there exists then a subsequence $(f_{k_1}, f_{k_2}, \ldots)$ such that each coordinate converges, say

$$\alpha_p^{(k_j)} \to \alpha_p \text{ for } p = 1, \ldots, n \text{ as } j \to \infty.$$

Let $g = \Sigma_1^n \alpha_p e_p$, so $A(g) = 1$. On the other hand we have $A(g - f_{k_j}) \to 0$, so $\| g - f_{k_j} \| \to 0$ as proved above. But $\| f_{k_j} \| \to 0$, so $g = 0$ which implies $A(g) = 0$. Contradiction. It follows that if F is a norm bounded set, then each coordinate separately is bounded. Hence, by the Bolzano-Weierstrass theorem, each sequence in F has a converging subsequence, i.e., F is relatively sequentially compact.

Exercise 40.16. (i) Let V be the Banach space ℓ_1 of all points $f = (x_1, x_2, \ldots)$ with $\| f \| = \Sigma_1^\infty | x_k |$ finite. Denote, for $k = 1, 2, \ldots$, by e_k the element in V having the k-th coordinate equal to one and all other coordinates zero, so

$$f = (x_1, x_2, \ldots) = \Sigma_1^\infty x_k e_k,$$

and let T be the operator in V defined by

$$T(\Sigma x_k e_k) = \Sigma k^{-1} x_k e_k.$$

Show that the range of T is norm dense in V but not equal to the whole of V. Show that for each $k = 1, 2, \ldots$ the element e_k is an eigenelement with eigenvalue k^{-1}. Show that if $\lambda \neq 0$ and $\lambda \neq k^{-1}$ for $k = 1, 2, \ldots$, then λ is in the resolvent set of T. Finally, show that $\lambda = 0$ is in the continuous spectrum of T.

(ii) Let V and $e_k (k = 1, 2, \ldots)$ be the same as in part (i) and let T be the operator in V defined by

$$T(\Sigma_1^\infty x_k e_k) = \Sigma_1^\infty x_k e_{k+1}.$$

Show that $\lambda = 0$ is in the residual spectrum of T.

We finally present the following theorem in which some further properties of spectra and spectral radii are discussed.

Theorem 40.17. *All operators mentioned here are norm bounded operators in the (complex) Banach space V.*

(i) If the operators S and T commute (i.e., $ST = TS$) and T^{-1} exists (as a bounded operator in V), then S and T^{-1} commute. Hence, if S^{-1} exists as well, then S, T, S^{-1} and T^{-1} all mutually commute and

$$(ST)^{-1} = S^{-1}T^{-1} = T^{-1}S^{-1}.$$

As a simple example, if the complex numbers λ and μ are in the resolvent set $\rho(T)$ of T, then $T, (T - \lambda I)^{-1}$ and $(T - \mu I)^{-1}$ commute.

(ii) Let $T = T_1 T_2 \cdots T_n$, where $T_1, \ldots, T_n$ commute. If T_k^{-1} exists for $k = 1, 2, \ldots$, then T^{-1} exists and satisfies

$$T^{-1} = T_1^{-1} T_2^{-1} \cdots T_n^{-1}.$$

Conversely, if T^{-1} exists, then T_k^{-1} exists for all k and

$$T_k^{-1} = T^{-1}(T_1 \cdots T_{k-1} T_{k+1} \cdots T_n).$$

(iii) Let $p(\lambda)$ be the polynomial

$$p(\lambda) = \alpha_0 + \alpha_1 \lambda + \cdots + \alpha_n \lambda^n$$

of degree $n \geq 1$ (complex coefficients). For the operator T, let

$$p(T) = \alpha_0 I + \alpha_1 T + \cdots + \alpha_n T^n.$$

Then the complex number λ_0 is in the resolvent set $\rho\{p(T)\}$ of $p(T)$ if and only if all roots of the equation $p(\lambda) = \lambda_0$ are in the resolvent set $\rho(T)$ of T. In other words, if $\lambda^{\sim}$ is in the spectrum $\sigma(T)$ of T, then $p(\lambda^{\sim})$ is in the spectrum $\sigma\{p(T)\}$ of $p(T)$ and, conversely, if λ_0 is in the spectrum of $p(T)$, then one at least of the roots of the equation $p(\lambda) = \lambda_0$ is in the spectrum of T. Hence

$$\sigma\{p(T)\} = \{p(\lambda) : \lambda \in \sigma(T)\}.$$

In particular $\sigma(T^k) = \{\lambda^k : \lambda \in \sigma(T)\}$ for $k = 1, 2, \ldots$. It follows that the spectral radii satisfy $r(T^k) = \{r(T)\}^k$.

(iv) As before, the Banach adjoint of the operator T in V is denoted by T^*. We recall that T^* is a bounded operator in the adjoint space V^*. The adjoint of the identity operator I in V is the identity operator in V^* (for convenience denoted by I as well). If T^{-1} exists, then $(T^*)^{-1}$ exists and $(T^*)^{-1} = (T^{-1})^*$. Hence, if $\lambda \in \rho(T)$, i.e., if $(T - \lambda I)^{-1}$ exists, then $(T^* - \lambda I)^{-1}$ exists, so $\lambda \in \rho(T^*)$. It follows that the spectral radii satisfy $r(T^*) \leq r(T)$.

(v) If S and T commute, then $r(ST) \leq r(S) \cdot r(T)$.

(vi) We have $r(ST) = r(TS)$, also if S and T do not commute.

(vii) If S is norm bounded and T is compact, then ST and TS are compact.

Proof. (i) We have

$$T^{-1}S = T^{-1}(ST)T^{-1} = T^{-1}(TS)T^{-1} = ST^{-1}.$$

(ii) The first part is evident from (i). Now assume that T^{-1} exists. Since T commutes with each of $T_1, \ldots, T_n$, the same holds for T^{-1}. Hence, denoting

$$T^{-1}(T_1 \cdots T_{k-1} T_{k+1} \cdots T_n)$$

by S_k, it follows that $S_k T_k = T_k S_k = I$, so $S_k = T_k^{-1}$.

(iii) Let λ_0 be given and denote by $\lambda_1, \ldots, \lambda_n$ the roots of the equation $p(\lambda) = \lambda_0$. Then

$$p(\lambda) - \lambda_0 = (\lambda - \lambda_1) \cdots (\lambda - \lambda_n),$$

so

$$p(T) - \lambda_0 I = (T - \lambda_1 I) \cdots (T - \lambda_n I).$$

Hence, in view of (ii), we have $\lambda_0 \in \rho\{p(T)\}$ if and only if $\lambda_k \in \rho(T)$ for $k = 1, \ldots, n$.

(iv) Assume that T^{-1} exists. For all $f \in V$ and $\varphi \in V^*$ we have

$$\varphi(f) = \varphi(T^{-1}Tf) = \{(T^{-1})^*\varphi\}(Tf) = \{T^*(T^{-1})^*\varphi\}(f).$$

This holds for all $f \in V$, so $\varphi = \{T^*(T^{-1})^*\}(\varphi)$. This holds for all $\varphi \in V^*$, and so $T^*(T^{-1})^*$ is the identity operator in V^*. Similarly

$$\varphi(f) = \varphi(TT^{-1}f) = (T^*\varphi)(T^{-1}f) = \{(T^{-1})^*T^*\varphi\}(f)$$

for all $f \in V$, so $(T^{-1})^*T^*$ is the identity operator in V^*. It follows that

$$T^*(T^{-1})^* = (T^{-1})^*T^* = I.$$

In other words, $(T^*)^{-1}$ exists and is equal to $(T^{-1})^*$.

(v) For $ST = TS$ we have $(ST)^n = S^n T^n$ for $n = 1, 2, \ldots$, so

$$\| (ST)^n \| \leq \| S^n \| \cdot \| T^n \| .$$

It follows that

$$r(ST) = \inf_n \| (ST)^n \|^{1/n} \leq \| S^n \|^{1/n} \cdot \| T^n \|^{1/n}$$

for $n = 1, 2, \ldots$, which implies ($n \to \infty$) that $r(ST) \leq r(S) \cdot r(T)$.

(vi) We may assume that neither S nor T is the null operator, so $\| S \| > 0$ and $\| T \| > 0$. From $(ST)^n = S(TS)^{n-1}T$ for $n = 2, 3, \ldots$ it follows that

$$\| (ST)^n \| \leq \| S \| \cdot \| (TS)^{n-1} \| \cdot \| T \| .$$

Note now that $\| S \|^{1/n}$ and $\| T \|^{1/n}$ tend to one as $n \to \infty$ and $\| (TS)^{n-1} \|^{1/n}$ tends to $r(TS)$. It follows that $r(ST) \leq r(TS)$. By symmetry we have $r(TS) \leq r(ST)$ as well, and so $r(ST) = r(TS)$.

(vii) The proof is easy if one uses the fact that compactness and sequential compactness in V are the same.   ∎

41. The Krein-Rutman Theorem

Assume that E is a complex Banach lattice. For any norm bounded operator T in E the notations $\rho(T), \sigma(T)$ and $r(T)$ have the same meaning as before. Note that any positive operator in E is norm bounded (Theorem 18.4).

Lemma 41.1. *Let T be a positive operator in E. Then the following holds.*

(i) If $\mid \lambda \mid > r = r(T)$, then $\parallel (T - \lambda I)^{-1} f \parallel \leq \parallel (T - \mid \lambda \mid I)^{-1}(\mid f \mid) \parallel$ for any $f \in E$. It follows that $\parallel (T - \lambda I)^{-1} \parallel \leq \parallel (T - \mid \lambda \mid I)^{-1} \parallel$.

(ii) The number $r = r(T)$ is in the spectrum of T, i.e., $r(T) \in \sigma(T)$.

Proof. (i) Writing

$$S_k(\lambda) = -(\lambda^{-1} I + \lambda^{-2} T + \cdots + \lambda^{-(k+1)} T^k)$$

for abbreviation, we have $S_k(\lambda) f \to (T - \lambda I)^{-1} f$ in norm as $k \to \infty$ for every $f \in E$. Furthermore

$$\mid S_k(\lambda) f \mid \leq \mid \lambda^{-1} f \mid + \mid \lambda^{-2} T f \mid + \cdots + \mid \lambda^{-(k+1)} T^k f \mid \leq$$

$$\leq (\mid \lambda \mid^{-1} I + \mid \lambda \mid^{-2} T + \cdots \mid \lambda \mid^{-(k+1)} T^k)(\mid f \mid) = -S_k(\mid \lambda \mid)(\mid f \mid).$$

It follows that $\parallel S_k(\lambda) f \parallel \leq \parallel S_k(\mid \lambda \mid)(\mid f \mid) \parallel$. Letting $k \to \infty$, we see that

$$\parallel (T - \lambda I)^{-1} f \parallel \leq \parallel (T - \mid \lambda \mid I)^{-1}(\mid f \mid) \parallel$$

Since this holds for every $f \in E$, the norms satisfy

$$\parallel (T - \lambda I)^{-1} \parallel \leq \parallel (T - \mid \lambda \mid I)^{-1} \parallel .$$

(ii) Since the spectrum $\sigma(T)$ is a closed set in the complex plane there is at least one point λ_0 in $\sigma(T)$ satisfying $\mid \lambda_0 \mid = r$. If the spectrum is finite this is evident. If the spectrum is infinite the spectrum contains a sequence $r_n e^{i\theta_n}$ with $r_n \uparrow r$, these $r_n e^{i\theta_n}$ not necessarily mutually different. The sequence has a subsequence for which θ_n converges to some θ_0. It follows that $r e^{i\theta_0}$ is in the spectrum. Let $\lambda_n \to \lambda_0$ such that $\mid \lambda_n \mid > r$ for all n and $\mid \lambda_n \mid \downarrow r$. All λ_n are, therefore, in the resolvent set $\rho(T)$ and, on account of Corollary 40.3, we have $\parallel (T - \lambda_n I)^{-1} \parallel \to \infty$. Then also $\parallel (T - \mid \lambda_n \mid I)^{-1} \parallel \to \infty$ in view of part (i) above. Since $\mid \lambda_n \mid \downarrow$ it is impossible now that r is in the resolvent set (because if r were in the resolvent set then $\parallel (T - \lambda I)^{-1} \parallel$ would be continuous, and therefore bounded, in some circular neighbourhood of r, see Theorem 40.2). Hence $r \in \sigma(T)$. $\blacksquare$

The theorem about positive compact operators which follows was published in 1948; the theorem is due to M.G. Krein and M.A. Rutman.

Theorem 41.2. *(Krein-Rutman theorem). Let T be a positive and compact operator in the Banach lattice E having a strictly positive spectral radius, so $r = r(T) > 0$. Then there exists a positive element u in E such that $Tu = ru$, i.e., $u > 0(u \neq 0)$ and $Tu = ru$. In other words, r is an eigenvalue of T with a corresponding positive eigenelement.*

Proof. According to the last lemma the number r is in the spectrum of T. Let $\lambda_n \downarrow r$. Then $\| (T - \lambda_n I)^{-1} \| \to \infty$ by Lemma 40.3. It follows that there exists a sequence $(f_n : n = 1, 2, \ldots)$ in E such that $\| f_n \| \leq 1$ for all n and $\| (T - \lambda_n I)^{-1} f_n \| \to \infty$. Observe now that

$$-(T - \lambda_n I)^{-1} = \lambda_n^{-1} I + \lambda_n^{-2} T + \cdots + \lambda_n^{-k} T^{k-1} + \cdots$$

is a positive operator, so

$$\| (T - \lambda_n I)^{-1} f_n \| \leq \| (T - \lambda_n I)^{-1}(| f_n |) \|$$

for all n. Writing $v_n = | f_n |$, we have $v_n \geq 0$ and $\| v_n \| \leq 1$ for all n and also $\| (T - \lambda_n I)^{-1} v_n \| \to \infty$. Let now

$$u_n = \{-(T - \lambda_n I)^{-1} v_n\} / \| (T - \lambda_n I)^{-1} v_n \|,$$

so $u_n \geq 0$ and $\| u_n \| = 1$ for all n. Furthermore

$$Tu_n - ru_n = (T - \lambda_n I)u_n + (\lambda_n - r)u_n =$$

$$-v_n / \| (T - \lambda_n I)^{-1} v_n \| + (\lambda_n - r)u_n,$$

which tends to zero in norm. Note next that since T is (sequentially) compact there exists a subsequence $(Tu_{n_k} : k = 1, 2, \ldots)$ converging in norm. Passing to the subsequence we may assume that Tu_n converges to some $u \in E$. All Tu_n are positive, so u is positive, i.e., $u \geq 0$. Since $Tu_n - ru_n \to 0$ and $Tu_n \to u$, we have $ru_n \to u$, and so $rTu_n \to Tu$. On the other hand it follows from $Tu_n \to u$ that $rTu_n \to ru$. This shows that rTu_n converges to Tu as well as to ru. Hence $Tu = ru$. Furthermore

$$\| u \| = \lim \| ru_n \| = r \cdot \lim \| u_n \| = r.$$

Since $r > 0$ by hypothesis it follows that $\| u \| > 0$, so $u \neq 0$. This completes the proof. The method of proof as presented here is due to F.F. Bonsall (1958). ∎

For an understanding of the example of a compact operator which will follow in Theorem 41.6 it is necessary to recall some facts about measure and integration.

The (non-negative and σ-additive) measure μ in the point set X is said to be *separable* if there exists a finite or countable collection Z of μ-measurable sets of finite measure such that for each set A of finite measure and for each $\epsilon > 0$ there exists a set B in the collection Z satisfying

$$\mu(A\backslash B) + \mu(B\backslash A) < \epsilon,$$

i.e., the measure of the symmetric difference

$$\Delta(A, B) = (A\backslash B) \cup (B\backslash A)$$

is less than ϵ.

Lebesgue measure in $\mathbb{R}^n$ is separable. To see this, consider the collection $S^\sim$ of all finite or countable unions of half-open intervals in $\mathbb{R}^n$ (open on the left, closed on the right). For each measurable set A of finite measure and each $\epsilon > 0$ there exists a set $C = \cup_1^\infty B_n$ in $S^\sim$ (each B_n a half-open interval) such that $A \subseteq C$ and $\mu(C\backslash A) < \epsilon/2$, where μ is now Lebesgue measure. The same holds if we restrict ourselves to the collection S of all finite or countable unions of half-open intervals with rational endpoints. If, with the same notations, we then write $B = \cup_1^k B_n$, then $B \subseteq C$ and $\mu(C\backslash B) < \epsilon/2$ for k sufficiently large. We thus have

$$\mu(A\backslash B) \leq \mu(C\backslash B) < \epsilon/2,$$

$$\mu(B\backslash A) \leq \mu(C\backslash A) < \epsilon/2,$$

so $\mu\{\Delta(A, B)\} < \epsilon$. Observe now that the collection Z of all finite unions of (half-open) intervals with rational endpoints is a countable collection and the set B is a member of Z. The collection Z satisfies, therefore, the conditions for separability of μ.

Let μ be a separable measure in the arbitrary point set X and let Z be the finite or countable collection of subsets of X as introduced above. Any finite sum $\Sigma_1^n \alpha_k \chi_{B_k}$, where χ_{B_k} is the characteristic function of the set $B_k \in Z$ for all k and all α_k are rational complex (the real and imaginary parts of α_k are therefore rational) is called a rational step function on Z. The collection of all these rational step functions on Z is countable. For $1 \leq p \leq \infty$ each of these step functions is (evidently) a member of the space $L_p(X, \mu)$.

The normed space V is said to be separable if there exists a countable subset Y of V which is norm dense in V, i.e., to each $f \in V$ and each $\epsilon > 0$ there exists an element $g \in Y$ such that $\| f - g \| < \epsilon$.

Theorem 41.3. *If the measure μ in X is σ-finite and separable and if $1 \leq p < \infty$, then $L_p(X, \mu)$ is separable.*

Proof. We shall prove that the countable collection Y of all rational step functions on Z (where Z is the same collection of sets as above) is dense in L_p. Let $f \in L_p$ and $\epsilon > 0$ be given. Since there exists an increasing sequence $X_n^\sim \uparrow X$, each $X_n^\sim$ of finite measure, we have $| f - f\chi_{X_n^\sim} | \downarrow 0$. Hence, there exists a set X_1 of finite measure ($X_1 = X_n^\sim$ for an appropriate n) such that if we define $f_1 = f\chi_{X_1}$, then $\| f - f_1 \| < \epsilon/3$.

The function f_1 can be written as $f = (g_1 - g_2) + i(h_1 - h_2)$, where g_1, g_2, h_1, h_2 are non-negative members of L_p vanishing outside X_1. For g_1 there exists a sequence $g^{(n)}$ of rational step functions, non-negative and vanishing outside X_1, such that $g^{(n)} \uparrow g_1$. Then $\mid g_1 - g^{(n)} \mid^p \downarrow 0$, so $\| g_1 - g^{(n)} \| < \epsilon/12$ for n sufficiently large. Similarly for g_2, h_1 and h_2. By linear combination we obtain a rational step function f_2, vanishing outside X_1 and such that $\| f_1 - f_2 \| < \epsilon/3$.

The function f_2 can be written as $f_2 = \Sigma_1^k \alpha_j \chi_{A_j}$, where all α_j are rational complex, all A_j are mutually disjoint and $\cup_1^k A_j = X_1$. For each A_j there exists a set $B_j \in Z$ such that

$$\mu\{\Delta(A_j, B_j)\} < (\epsilon/3k \mid \alpha_j \mid)^p,$$

so

$$\| \chi_{A_j} - \chi_{B_j} \| = (\mu\{\Delta(A_j, B_j)\})^{1/p} < \epsilon/(3k \mid \alpha_j \mid).$$

Hence, writing $f_3 = \Sigma_1^k \alpha_j \chi_{B_j}$, we have

$$\| f_2 - f_3 \| < \Sigma_1^k \mid \alpha_j \mid \cdot \| \chi_{A_j} - \chi_{B_j} \| < \epsilon/3.$$

It follows that $\| f - f_3 \| < \epsilon$. Since f_3 is a member of the countable collection Y of rational step functions on Z this shows that L_p is separable.   ∎

Exercise 41.4. In many cases separability of the measure μ implies σ-finiteness. Show that if μ is separable and has the property that any set of (strictly) positive measure has a subset of finite and strictly positive measure, then μ is σ-finite.

Let μ be a (positive) σ-finite separable measure in X and let p be a number satisfying $1 < p < \infty$. Then, if $p^{-1} + q^{-1} = 1$, we have $1 < q < \infty$ and both $L_p = L_p(X, \mu)$ and $L_q = L_q(X, \mu)$ are separable Banach spaces. Furthermore, for any $f \in L_p$ and $g \in L_q$, the products fg and $\mid fg \mid$ are μ-summable over X (by Hölder's inequality, see Exercise 15.15).

Lemma 41.5. *Let μ be a separable measure in X, let $1 < p < \infty$ and $p^{-1} + q^{-1} = 1$. Furthermore, let A be a norm bounded set of functions in $L_p = L_p(X, \mu)$. Then there exists a sequence $(f_n : n = 1, 2, \ldots)$ in A and a function $f_0 \in L_p$ such that*

$$\lim_n \int f_n g \, d\mu = \int f_0 g \, d\mu$$

for every $g \in L_q = L_q(X, \mu)$.

Proof. We may assume that the set A is a norm bounded sequence $(h_n : n = 1, 2, \ldots)$ such that $\| h_n \| \le \alpha$ for some $\alpha > 0$ and all n. The spaces L_p and L_q are separable Banach spaces; let $(g_k : k = 1, 2, \ldots)$ be norm dense in L_q. On account of $| \int h_n g_1 d\mu | \le \alpha \| g_1 \|$ the sequence $(\int h_n g_1 d\mu : n = 1, 2, \ldots)$ is a bounded sequence of complex numbers; it follows that there exists a subsequence (h_{n1}) of (h_n) such that $\int h_{n1} g_1 d\mu$ converges. Similarly it is seen that there exists a subsequence (h_{n2}) of (h_{n1}) such that also $\int h_{n2} g_2 d\mu$ converges, and so on. Then, writing $f_n = h_{nn}$ for all n, the sequence $(\int f_n g_k d\mu : n = 1, 2, \ldots)$ converges for every g_k. We shall prove that $(\int f_n g d\mu : n = 1, 2, \ldots)$ converges for every $g \in L_q$. For this purpose, let $g \in L_q$ and $\epsilon > 0$ be given. Choose g^* from the sequence $(g_k : k = 1, 2, \ldots)$ such that $\| g^* - g \| < \epsilon/3\alpha$. Then

$$| \int f_n g d\mu - \int f_m g d\mu | \le | \int f_n g^* d\mu - \int f_m g^* d\mu | +$$

$$+ | \int f_n (g - g^*) d\mu | + | \int f_m (g - g^*) d\mu | \, .$$

The last two terms are each less than $\epsilon/3$, and the first term on the right is also less than $\epsilon/3$ for m and n sufficiently large. Hence, the sequence $(\int f_n g d\mu : n = 1, 2, \ldots)$ converges for every $g \in L_q$. Denote the limit by $\varphi(g)$. Since $| \int f_n g d\mu | \le \alpha \| g \|$ for all n and all $g \in L_q$, we see that $| \varphi(g) | \le \alpha \| g \|$ for all $g \in L_q$, i.e., φ is a norm bounded linear functional on L_q. As shown in Example 29.4, part (ii), there exists a (unique) function $f_0 \in L_p$ such that $\varphi(g) = \int f_0 g d\mu$ for every $g \in L_q$. Hence

$$\lim_n \int f_n g d\mu = \int f_0 g d\mu \text{ for every } g \in L_q. \qquad \blacksquare$$

The moment has come now to introduce the compact operator that was promised above. Let μ be a σ-finite (positive) measure in X with $\mu \times \mu$ the product measure in $X \times X$. If for example μ is Lebesgue measure in $\mathbb{R}$, then $\mu \times \mu$ is the Lebesgue measure in $\mathbb{R}^2$. Furthermore, let $1 < p < \infty$ and $p^{-1} + q^{-1} = 1$. For convenience, the norms in $L_p(X, \mu)$ and $L_q(X, \mu)$ will be denoted by $\| \cdot \|_p$ and $\| \cdot \|_q$ respectively. Assume now that the complex $(\mu \times \mu)$-measurable function $T(x, y)$ on $X \times X$ has the property that, for almost every $x \in X$, the function $T(x, y)$, as a function of y, is a member of L_q, i.e., $\| T(x, \cdot) \|_q$ exists for almost every $x \in X$ as a finite number. Write $t(x) = \| T(x, \cdot) \|_q$ and assume, finally, that the thus defined function t is a member of L_p, so $\| t \|_p < \infty$ or, written out fully,

$$\| T \|_{pq} = [\int \{\int | T(x, y) |^q \, dy\}^{p/q} dx]^{1/p} < \infty,$$

where we have written dx and dy shortly for $d\mu_x$ and $d\mu_y$.

It is easy to see that the operator T, defined by

$$(Tf)(x) = \int_X T(x,y)f(y)dy,$$

maps L_p into L_p because for almost every x and every $f \in L_p$ we have (by Hölder's inequality) that

$$\left| \int T(x,y)f(y)dy \right| \leq \| T(x,\cdot) \|_q \cdot \| f \|_p \in L_p,$$

and so $\int T(x,y)f(y)dy \in L_p$. The thus defined operator T in L_p is an example of an *integral operator* (also called *kernel operator*) and $T(x,y)$ is said to be the *kernel* of T. The number $| T |_{pq}$ is sometimes called the *double norm* of T (precisely, the pq-double norm) and T itself is sometimes called a *Hille-Tamarkin operator* because this particular example of operators was introduced by E. Hille and J.D. Tamarkin, 1934.

Note that for $p = 2$ we have $q = 2$, and so $| T |_{pq}$ is now

$$| T |_{22} = [\int_X \{ \int_X | T(x,y) |^2 \, dy \} dx]^{1/2} = [\int_{X \times X} | T(x,y) |^2 \, dxdy]^{1/2},$$

In this case T maps L_2 into itself and $T(x,y)$ is now called a *Hilbert-Schmidt kernel*.

Theorem 41.6. *Let* $1 < p < \infty$. *If* T *is a Hille-Tamarkin operator in* $L_p(X,\mu)$ *and* μ *is separable, then* T *is compact.*

Proof. We have to show that the image under T of any norm bounded set A in L_p contains a norm convergent subsequence. Assume that $\| f \|_p \leq \alpha$ for all f in A. According to the last lemma there is a sequence $(f_n : n = 1, 2, \ldots)$ in A and a function $f_0 \in L_p$ such that

$$\lim_n \int f_n g d\mu = \int f_0 g d\mu$$

for every $g \in L_q$ (where q is defined by $p^{-1} + q^{-1} = 1$). Since $T(x,\cdot) \in L_q$ for almost every x, we may choose $g(x) = T(x,\cdot)$ for these values of x. Hence

$$\lim (Tf_n)(x) = \lim \int T(x,y)f_n(y)dy = \int T(x,y)f_0(y)dy = (Tf_0)(x),$$

which shows that Tf_n converges pointwise almost everywhere to Tf_0. We prove that the convergence is with respect to the L_p-norm as well. Note that

$$| (Tf_n)(x) - (Tf_0)(x) | \leq \| f_n - f_0 \|_p \cdot \| T(x,\cdot) \|_q \leq \alpha_1 \| T(x,\cdot) \|_q,$$

where $\alpha_1 = \alpha + \| f_0 \|_p$. Since $\alpha_1 \| T(x,\cdot) \|_q \in L_p$, we see that the sequence $| (Tf_n)(x) - (Tf_0)(x) |^p$ converges pointwise (almost everywhere) to zero and is dominated in absolute value by the μ-summable function $(\alpha_1 \| T(x,\cdot) \|_q)^p$.

In view of the dominated convergence theorem in integration theory it follows that

$$\lim_n \int |\, (Tf_n)(x) - (Tf_0)(x)\,|^p \, d\mu = 0,$$

i.e., Tf_n converges in the L_p -norm to Tf_0. This is the desired result.   ∎

Assume that T is a Hille-Tamarkin operator in $L_p = L_p(X,\mu)$, where μ is a σ-finite measure in X and p is a number satisfying $1 < p < \infty$. It is evident that if the kernel $T(x,y)$ of T is non-negative for almost every $(x,y) \in X \times X$, then T is a positive operator. As before, let q be defined by $p^{-1} + q^{-1} = 1$. Since L_p is a Banach lattice the order adjoint space $L_p^\sim$ and the Banach adjoint space L_p^* are identical, i.e., $L_p^\sim = L_p^*$ (see Theorem 25.8). The norm in L_p is order continuous which implies that every order bounded linear functional in L_p is order continuous, i.e., $L_p^\sim = (L_p)_n^\sim$. Hence $L_p^* = L_p^\sim = (L_p)_n^\sim$, see Theorem 25.10. According to Example 28.4 we have $\varphi \in L_p^* = (L_p)_n^\sim$ if and only if there exists a function $g \in L_q$ such that $\varphi(f) = \int f g d\mu$ holds for all $f \in L_p$ and, in this case, g is uniquely determined and $\| \varphi \| = \| g \|_q$. The space L_p^* may be identified thus (with respect to the ordering as well) with the space L_q. It follows that the norm adjoint $T^* : L_q \to L_q$ of the given operator T may be identified with the order adjoint $T^\sim$. We determine T^*.

Theorem 41.7. *Let T be a Hille-Tamarkin operator in $L_p = L_p(X,\mu)$, where $1 < p < \infty$. Then, for $p^{-1} + q^{-1} = 1$, the adjoint operator T^* in L_q is given by*

$$(T^*g)(y) = \int T(x,y)g(x)d\mu_x$$

for every $g \in L_q$.

Proof. For $f \in L_p$ and $g \in L_q$ we have

$$\int \int |\, T(x,y)f(y)g(x)\,|\, d(\mu \times \mu) < \infty,$$

so the function

$$h(y) = \int |\, T(x,y)g(x)\,|\, d\mu_x$$

is a μ-measurable function on X with the property that $\int h(y) \,|\, f(y)\,|\, d\mu_y$ is finite for every $f \in L_p$. Writing

$$\varphi(f) = \int h(y)f(y)d\mu_y \text{ for } f \in L_p,$$

it is evident that φ is a positive linear functional on L_p. It follows therefore from what has been shown in Example 28.4(ii) that $h \in L_q$. Then, writing $T^*(x,y) = T(y,x)$, the function

$$\int T^*(y,x)g(x)d\mu_x = \int T(x,y)g(x)d\mu_x$$

is likewise a member of L_q for every $g \in L_q$, i.e., $T^*(x,y)$ is the kernel of an integral operator T^* in L_q. We show that $T^* = T^*$. For this purpose, choose $f \in L_p$ and $g \in L_q$, and observe that

$$(T^*g)(f) = g(Tf) = \int \left(\int T(x,y)f(y)d\mu_y \right)g(x)d\mu_x =$$

$$= \int \left(\int T(x,y)g(x)d\mu_x \right)f(y)d\mu_y = (T^\star g)(f),$$

so $T^*g = T^\star g$. This holds for every $g \in L_q$, so $T^* = T^*$.   ∎

CHAPTER 22
Spectral Theory of Positive Operators

42. Irreducible Operators

We present a definition.

Definition 42.1. The order bounded operator T in the Archimedean Riesz space E is called *band irreducible* if T does not leave any band invariant except E itself and the band $\{0\}$ consisting of the null element, i.e., any band B in E such that $Tf \in B$ for every $f \in B$ satisfies either $B = E$ or $B = \{0\}$.

If E is Dedekind complete, the set of all order bounded operators in E is a Dedekind complete Riesz space $\mathcal{L}_b(E)$, see Theorem 20.2. In this case T leaves the band B invariant if and only if T^+ and T^- do so. In other words, T leaves B invariant if and only if $\mid T \mid$ does so. For the proof, assume first that T leaves B invariant. For any $u \geq 0$ in B we have $T^+u = \sup(Tv : 0 \leq v \leq u)$. All v are in B, so all Tv are in B. Therefore, $T^+u \in B$ (since B is a band). It follows that T^+ leaves B invariant. The same holds then for T^- and $\mid T \mid$. Conversely, if $\mid T \mid$ leaves the band B invariant, it follows immediately from $0 \leq T^+ \leq \mid T \mid$ and $0 \leq T^- \leq \mid T \mid$ that T^+ and T^- (and, therefore, also T) leave B invariant.

For Banach lattices there exists the similar notion of an ideal irreducible operator.

Definition 42.2. The positive operator T in the Banach lattice E is called *ideal irreducible* if T does not leave any norm closed ideal invariant except E itself and $\{0\}$.

Since any band in a Banach lattice is norm closed it is evident that if the positive operator T in E is ideal irreducible, then T is band irreducible. If E has order continuous norm it follows for any upwards directed set D in E^+ satisfying $\sup(u : u \in D) = u_0$ that $(\parallel u_0 - u \parallel : u \in D) \downarrow 0$ which implies that any norm closed ideal in E is now a band. In this case, therefore, any band irreducible positive operator is also ideal irreducible.

If the norm in E is not order continuous there may exist a band irreducible operator that is not ideal irreducible, as the following example in the space ℓ_∞ of all bounded sequences shows. Let $T : \ell_\infty \to \ell_\infty$ be given by

$$T(x_1, x_2, \ldots) = (y_1, y_2, \ldots),$$

where $y_n = n^{-1}\Sigma_1^\infty k^{-2}x_k$ for $n = 1, 2, \ldots$. Note that T is positive and if $x \geq 0$ has at least one of its coordinates strictly positive, then $y = Tx$ has all of its coordinates strictly positive. It follows that ℓ_∞ and $\{0\}$ are the only T-invariant bands, so T is band irreducible. For each $x \in \ell_\infty$ the image $y = Tx$ is a null sequence, i.e., $Tx \in (c_0)$. Therefore, the norm closed ideal c_0 in ℓ_∞ is T-invariant, and so T is not ideal irreducible. We continue with a necessary and sufficient condition for an integral operator to be irreducible.

Example 42.3. As in earlier examples, let μ be a σ-finite (positive) measure in the point set X and let E be an ideal in the Riesz space L_0 of all μ-measurable (real or complex) functions on X. Furthermore, let the $\mu \times \mu$-measurable and non-negative function $T(x,y)$ on $X \times X$ have the property that

$$\int_X T(x,y)f(y)d\mu_y$$

is a member of E for any $f \in E$. Then the operator $T : E \to E$, defined by

$$(Tf)(x) = \int_X T(x,y)f(y)d\mu_y$$

for $f \in E$, is a positive integral operator in E. Note that E is not necessarily equipped with a norm here. Without loss of generality it may be assumed that X is the carrier of E.

The thus defined operator T is band irreducible if and only if

$$\int_{X\setminus Y}\left\{\int_Y T(x,y)d\mu_y\right\}d\mu_x > 0 \tag{1}$$

holds for every subset Y of X satisfying $\mu(Y) > 0$ and $\mu(X\setminus Y) > 0$, where it is possible that the repeated integral assumes the value $+\infty$.

For the proof, observe first that T is order continuous because if D is a set in E^+ such that $(f : f \in D) \downarrow 0$, then D contains a sequence $f_n \downarrow 0$ (because L_0 is super Dedekind complete, see Theorem 17.6 and the remark attached to this theorem). This implies that $\int T(x,y)f_n(y)d\mu_y \downarrow 0$. Assume now that T is band irreducible and at the same time there exists a subset Y of X such that $\mu(Y) > 0$ as well as $\mu(X\setminus Y) > 0$ and the integral (1) above is equal to zero. Then, for any subset Z of Y such that $\chi_Z \in E$, we have

$$\int_{X\setminus Y}(T\chi_Z)(x)d\mu_x = 0,$$

which implies that $(T\chi_Z)(x) = 0$ for almost every $x \in X\setminus Y$. Since any non-negative $f \in E$ vanishing on $X\setminus Y$ can be approximated (pointwise) from below by finite linear combinations of such functions χ_Z, it follows now from the order continuity of T that $(Tf)(x) = 0$ almost everywhere on $X\backslash Y$ for any $f \in E$ satisfying $f(x) = 0$ on $X\backslash Y$. This shows that the band B of all $f \in E$ vanishing on $X\backslash Y$ is T-invariant. But $\mu(Y)$ and $\mu(X\backslash Y)$ are strictly positive, so $B \neq E$ as well as $B \neq \{0\}$. Contradiction since T is band irreducible.

Conversely, assume that the integral in (1) is strictly positive for all Y satisfying $\mu(Y) > 0$ and $\mu(X\backslash Y) > 0$. Let B be a T-invariant band and let $X\backslash Y$ be the carrier of B, so any $f \in B$ vanishes almost everywhere on Y. Since $Tf \in B$ for any $f \in B$, it follows that Tf vanishes almost everywhere on Y. Now, let $Z_n \uparrow X\backslash Y$ such that $\chi_{Z_n} \in B$ for all n (see Theorem 29.1 and the remarks preceding it to understand that the characteristic function of $X\backslash Y$ is not necessarily a member of B). Then $T\chi_{Z_n} \in B$ for all n, so $T\chi_{Z_n}$ vanishes almost everywhere on Y, i.e., $\int_{Z_n} T(x,y)d\mu_y = 0$ almost everywhere on Y for every n. Since $Z_n \uparrow X\backslash Y$, it follows that $\int_{X\backslash Y} T(x,y)d\mu_y = 0$, so

$$\int_Y \{ \int_{X\backslash Y} T(x,y)d\mu_y \} d\mu_x = 0.$$

Hence, by our present assunption, we must have either $\mu(Y) = 0$ or $\mu(X\backslash Y) = 0$, i.e., $B = E$ or $B = \{0\}$. But then T is band irreducible.

As a simple special case note that if $T(x,y) > 0$ almost everywhere on $X \times X$, then T is band irreducible.

Exercise 42.4. As above, let T with kernel $T(x,y) \geq 0$ be a positive integral operator in the ideal E of L_0. We may assume again that X is the carrier of E. Let X be the disjoint union of sets $X_k(k = 1, 2, \ldots, n)$ of positive measure such that $T(x,y) = 0$ on $X_k \times X_k(k = 1, 2, \ldots, n)$ and $T(x,y) > 0$ elsewhere on $X \times X$. Show that T is band irreducible. Show also that if $T(x,y) > 0$ on each $X_k \times X_k(k = 1, 2, \ldots, n)$ and zero elsewhere, then for each $k = 1, \ldots, n$ the band having X_k as carrier is T-invariant.

Exercise 42.5. In the Riesz space $\mathbb{R}^n(n \geq 2)$ with coordinatewise ordering let $e_k(k = 1, \ldots, n)$ be the element with the k-th coordinate one and all other coordinates zero. Choose a subset from $e_1, \ldots, e_n$ and after renumbering denote the subset by $d_1, \ldots, d_p$. Show that the set of all linear combinations of $d_1, \ldots, d_p$ is a band in $\mathbb{R}^n$ and, conversely, each band in $\mathbb{R}^n$ is of this type. Note that ideals and bands are the same. Assume now that, for example, the band generated by $e_1, \ldots, e_q(1 \leq q < n)$ is invariant under the operator T. Show that in the matrix of T the first q columns have the first q numbers possibly nonzero and the last $(n - q)$ numbers zero.

Note that if T_1 and T_2 are positive operators in the Archimedean Riesz space E such that $T_1 \leq T_2$ and T_1 is band irreducible, then so is T_2 (because any T_2-invariant band will be T_1-invariant as well). Furthermore, if T is positive (not the null operator), σ-order continuous and band irreducible, then $f_0 > 0$ (i.e., $f_0 \geq 0, f_0 \neq 0$) implies $Tf_0 > 0$. To see this, assume that $f_0 > 0$, but $Tf_0 = 0$. Then $Tf = 0$ for any f in the ideal generatrd by f_0. If $g \geq 0$ is an element in the band $[f_0]$ generated by f_0, then (as shown in Corollary 11.6) we have

$$(g \wedge nf_0 : n = 1, 2, \ldots) \uparrow g,$$

and so $Tg = 0$ (by the σ-order continuity of T). It follows that the band $[f_0]$ is T- invariant. Then $[f_0] = \{0\}$ or $[f_0] = E$. The first is impossible since $f_0 > 0$. Hence $[f_0] = E$, which implies that T is the null operator. Contradiction. Hence, $f_0 > 0$ implies $Tf_0 > 0$.

Similarly, if T is positive (not the null operator) and ideal irreducible in the Banach lattice E, then $f_0 > 0$ implies $Tf_0 > 0$.

In our next example we assume again that T is positive (not the null operator), σ-order continuous and band irreducible. Let $f_0 > 0$ and let the band $[Tf_0]$ be a subset of the band $[f_0]$. Note that $Tf_0 > 0$ (as proved above), so $[Tf_0]$ does not consist of the null element only. As above it follows that $g \in [f_0]$ implies $Tg \in [Tf_0]$. In particular, since $[Tf_0]$ is a subset of $[f_0]$, it follows from $g \in [Tf_0]$ that $Tg \in [Tf_0]$. The band $[Tf_0]$ is, therefore, T-invariant. Then $[Tf_0] = E$, and so $[f_0] = E$. In other words, $[Tf_0] \subseteq [f_0]$ occurs if and only if f_0 is a weak order unit in E, and in this case Tf_0 is likewise a weak order unit.

Similarly, if T is positive (not the null operator) and ideal irreducible in the Banach lattice E and if $f_0 > 0$, then the norm closed ideal $\{Tf_0\}$ generated by Tf_0 is a subset of the norm closed ideal $\{f_0\}$ if and only if $\{Tf_0\} = E$, and so $\{f_0\} = E$ holds then as well in this case.

If E is a Banach lattice and there exists a positive irreducible operator in E, then there also exist operators in E having the stronger property of transforming any positive non-zero element in E into a weak order unit. Details are contained in the following theorem.

Theorem 42.6. *(i) Let E be a (real or complex) Banach lattice and let T be a positive and σ-order continuous operator in E (not the null operator). Furthermore, let $\lambda > r(T)$, where $r(T)$ is the spectral radius of T. Then T is band irreducible if and only if $S = (\lambda I - T)^{-1}$ is band irreducible and in this case Sf is a weak unit in E for any $f > 0$ in E.*

(ii) Similarly, if T is positive (not the null operator), then T is ideal irreducible if and only if $S = (\lambda I - T)^{-1}$ is ideal irreducible and in this case, for any $f > 0$ in E, the norm closed ideal generated by Sf is the space E itself.

Proof. (i) Assume that T is band irreducible. Choose an arbitrary $f > 0$ in E. Then $Tf > 0$, as proved above. For abbreviation, denote $Sf = (\lambda I - T)^{-1} f$ by g. Then

$$g = (\lambda I - T)^{-1} f = (\lambda^{-1} I + \lambda^{-2} T + \lambda^{-3} T^2 + \cdots) f,$$

as proved in the first part of section 40. The series converges in norm and all terms are positive; the partial sums form therefore an increasing sequence. The norm limit is then also the order limit (see Theorem 15.3), i.e., $g = Sf$ is the supremum of the sequence $(s_n : n = 1, 2, \ldots)$ of partial sums, that is to say, we have $s_n \uparrow g$. Note that $g = Sf \geq \lambda^{-1} f$, so $g > 0$. Furthermore, $T s_n$ converges in norm to Tg (since T, being positive, is norm bounded, see Theorem 18.4). It follows similarly as for $s_n \uparrow g$ that $T s_n \uparrow Tg$, i.e.,

$$Tg = \sup_n (\lambda^{-1} T + \lambda^{-2} T^2 + \cdots + \lambda^{-n} T^n) f =$$

$$= \lambda \sup_n (\lambda^{-2} T + \lambda^{-3} T^2 + \cdots + \lambda^{-n} T^{n-1}) f \leq \lambda Sf = \lambda g.$$

This shows that Tg is a member of the band $[g]$. Since T is σ-order continuous, it follows easily that $Th \in [g]$ for any $h \in [g]$, i.e., $[g]$ is T-invariant. We have now that $g > 0$ and T is band irreducible, so we must have $[g] = E$, i.e., $g = Sf$ is a weak order unit in E. It is trivial to conclude then that S is band irreducible.

Conversely, assume that $S = (\lambda I - T)^{-1}$ is band irreducible. From the formula for S it follows that any T-invariant band B is S-invariant. Then $B = E$ or $B = \{0\}$, and so T is band irreducible.

(ii) The proof is similar. Note that σ-order continuity of T is not needed here. ∎

For abbreviation we shall say that a positive operator T with the property that Tf is a weak order unit for any $f > 0$ is *strongly irreducible*. For an integral operator T we have now the following necessary and sufficient condition to be strongly irreducible.

Theorem 42.7. *As before, let μ be a σ-finite measure in the point set X and let L_0 be the Riesz space of all μ-measurable (real or complex) functions on X. Furthermore, let E be an ideal in L_0 and let $T(x,y)$ be the kernel of a positive integral operator in E. By Δ we shall denote the set of all points (x,y) for which $T(x,y) = 0$. The following conditions are now equivalent.*

(i) Δ does not contain a rectangle of positive measure, i.e., Δ does not have a subset $A \times B$, where A and B are subsets of X of positive measure (all this modulo sets of measure zero).

(ii) T is strongly band irreducible.

Observe already that the operator in Exercise 42.4 is irreducible but not strongly irreducible.

Proof. (i)$\Rightarrow$ (ii) Assume that (i) holds and that for some $f > 0$ in E the image Tf fails to be a weak unit. Then $A = (x : (Tf)(x) = 0)$ is of positive measure. The set $B = (y : f(y) > 0)$ is also of positive measure. For any $x \in A$ we have

$$0 = (Tf)(x) = \int_X T(x,y)f(y)d\mu_y = \int_B T(x,y)f(y)d\mu_y.$$

This shows that $T(x,y) = 0$ for almost every $y \in B$, i.e., for every $x \in A$ we have $T(x,y) = 0$ for almost every $y \in B$. Hence $T(x,y) = 0$ almost everywhere on $A \times B$. This contradicts our hypothesis (i).

(ii)$\Rightarrow$(i) Assume that T is strongly band irreducible and that A and B are subsets of X of positive measure such that $A \times B$ is a subset of Δ (except for a set of measure zero). The set B has a subset S of positive measure such that $\chi_S \in E$. Therefore, $T\chi_S$ is a weak unit by hypothesis. On the other hand, for $x \in A$ we have

$$(T\chi_S)(x) = \int_S T(x,y)d\mu_y \leq \int_B T(x,y)d\mu_y = 0,$$

which shows that $T\chi_S$ vanishes on a set of positive measure. Contradiction. $\blacksquare$

As a particular case we get the result that if $T(x,y) > 0$ on $X \times X$, then T is strongly irreducible. It can be proved (for example if μ is Lebesgue measure) that it is possible that the set Δ on which $T(x,y) = 0$ is of positive measure but Δ does not contain a rectangle of positive measure.

Let us say now a few words about products of irreducible operators. The first point to observe is that the product of two positive irreducible operators is not necessarily irreducible. As an example, let $E = \mathbb{R}^2$ with coordinatewise ordering and let $e_1 = (1,0), e_2 = (0,1)$. The operator T in E, defined by $Te_1 = e_2$ and $Te_2 = e_1$, is irreducible and $T^2 = I$ (the identity operator) is not irreducible. It is true, however, in any Archimedean Riesz space, that the product of a finite number of (positive and σ-order continuous) band irreducible operators, one at least of which is strongly irreducible, is strongly irreducible. To see this, recall that if T is band irreducible and $f > 0$, then $Tf > 0$ and if f is a weak order unit, then Tf is a weak order unit.

The question arises whether the adjoint operator of a positive irreducible operator is likewise irreducible. For an answer we recall several definitions and notations. Let E, F be Riesz spaces with E Archimedean and F Dedekind complete and let $E^{\sim}, F^{\sim}$ be their order duals, i.e., $E^{\sim}$ is the Dedekind complete Riesz space of all order bounded linear functionals on E (and similarly for $F^{\sim}$), see section 25. The set $E_n^{\sim}$ of all order continuous linear functionals on E is a band in $E^{\sim}$. Similarly for $F_n^{\sim}$. The set $\mathcal{L}_b(E,F)$ of all order bounded operators from E into F is a Dedekind complete Riesz space and for any $T \in \mathcal{L}_b(E,F)$ the order adjoint $T^{\sim} : F^{\sim} \to E^{\sim}$ is defined by

$$(T^\sim\psi)(f) = \psi(Tf) \text{ for all } f \in E \text{ and all } \psi \in F^\sim,$$

see section 26. The operator $T^\sim$ is called the order adjoint of T and it was proved in Theorem 26.1 that $T^\sim$ is order bounded and order continuous. The restriction of $T^\sim$ to $F_n^\sim$ is denoted by T'. If T is order continuous, then T' maps $F_n^\sim$ into $E_n^\sim$, i.e., T' is then order continuous and satisfies $T' : F_n^\sim \to E_n^\sim$ (see Theorem 26.6). These last remarks are of interest only if $F_n^\sim \neq \{0\}$, i.e., if there exist order continuous linear functionals on F which are not the null functional. This will hold in particular if ${}^o(F_n^\sim) = \{0\}$, that is to say, if $F_n^\sim$ separates the points of F, i.e., $\psi(g) = 0$ for all $\psi \in F_n^\sim$ implies $g = 0$ in F. Note that if ${}^o(F_n^\sim) = \{0\}$ holds and T is not the null operator, then T' is not the null operator. Indeed, if $T'\psi = 0$ for all ψ, then $\psi(Tf) = (T'\psi)(f) = 0$ for all ψ, f, so $Tf = 0$ for all f (on account of ${}^o(F_n^\sim) = \{0\}$), which shows that $T = 0$.

Lemma 42.8. *Let E and F be Riesz spaces such that E is Archimedean, F is Dedekind complete and ${}^o(F_n^\sim) = \{0\}$.*

(i) If D is an upwards directed set of positive and order continuous operators from E into F such that $(T : T \in D) \uparrow T_0$, then T_0 is order continuous. Furthermore, the corresponding set of adjoints $(T' : T \in D)$ is upwards directed and satisfies $(T' : T \in D) \uparrow T_0'$. Similarly for σ-order continuous operators.

(ii) If $T : E \to F$ is positive and order continuous and T transforms every $u > 0$ in E into a weak order unit Tu in F, then $T'\psi$ is a weak order unit in $E_n^\sim$ for every $\psi > 0$ in $F_n^\sim$.

Proof. (i) As observed above, the set $\mathcal{L}_b(E, F)$ of all order bounded operators from E into F is a Dedekind complete Riesz space. The order continuous operators in $\mathcal{L}_b(E, F)$ form a band, see Theorem 22.2. Hence, if D is an upwards directed set of positive and order continuous operators from E into F such that $(T : T \in D) \uparrow T_0$, then T_0 is order continuous.

To see that $(T' : T \in D) \uparrow T_0'$, choose $0 \le \psi \in F_n^\sim$ and $0 \le u \in E$ arbitrarily and observe that

$$\{(T'\psi)(u) : T \in D\} = \{\psi(Tu) : T \in D\} \uparrow \psi(T_0 u) = (T_0'\psi)(u).$$

(ii) Assume that $\psi(Tu) = 0$ for some $0 < u \in E$ and some $0 \le \psi \in F_n^\sim$. Then $\psi(g) = 0$ for all g in the ideal generated by Tu. Since ψ is order continuous it is even so that $\psi(g) = 0$ for all g in the band generated by Tu. By our special hypothesis Tu is a weak unit in F, so the band generated by Tu is the space F itself. Hence $\psi(g) = 0$ for all $g \in F$, i.e., $\psi = 0$. It has been shown thus that $\psi(Tu) = 0$ only if $\psi = 0$. It follows that if $0 < u \in E$ and $0 < \psi \in F_n^\sim$ are given, then $\psi(Tu) > 0$, i.e., $(T'\psi)(u) > 0$. In other words, the linear functional $\varphi = T'\psi$ in $E_n^\sim$ is strictly positive, i.e., the null ideal

$N_\varphi = (f \in E : \varphi(|\ f\ |) = 0)$ satisfies $N_\varphi = \{0\}$. Therefore, the carrier $C_\varphi = (N_\varphi)^d$ of φ satisfies $C_\varphi = E$. Now recall that for φ and φ_1 in $E_n^\sim$ it was proved in Theorem 24.6 that $\varphi \perp \varphi_1$ if and only $C_\varphi \perp C_{\varphi_1}$. In the present situation, therefore, it follows from $\varphi \perp \varphi_1$ for some $\varphi_1 \in E_n^\sim$ that $C_{\varphi_1} \perp C_\varphi$. But $C_\varphi = E$, so $C_{\varphi_1} = \{0\}$. This implies that the null ideal N_{φ_1} satisfies $N_{\varphi_1} = E$, i.e., $\varphi_1 = 0$. We have proved thus that in $E_n^\sim$ the only element disjoint to $\varphi = T'\psi$ is the null element. Therefore, $\varphi = T'\psi$ is a weak order unit in $E_n^\sim$.   ∎

We specialize to the case that $F = E$. Assume, therefore, that E is a Dedekind complete Riesz space such that $E_n^\sim \neq \{0\}$ and let T be a positive and order continuous operator in E. Then T' is a positive and order continuous operator in $E_n^\sim$. The powers $T^n (n = 1, 2, \ldots)$ are positive order continuous operators in E, so for every n the operator $(T^n)'$ is positive and order continuous in $E_n^\sim$. Since it is easy to see that $(T_1 T_2)' = T_2' T_1'$ holds generally, it is true in particular that $(T^n)' = (T')^n$ for $n = 1, 2, \ldots$.

Theorem 42.9. *Let E be a Dedekind complete Banach lattice such that $^\circ(E_n^\sim) = \{0\}$ and let T be a positive and order continuous operator in E, not the null operator. Then, if T is band irreducible, T' is also band irreducible.*

Proof. Assume that T is band irreducible. Choose a positive number $\lambda > r(T)$, where $r(T)$ is the spectral radius of T. Then in view of Theorem 42.6 in the present section, $S = (\lambda I - T)^{-1}$ is strongly band irreducible. Writing

$$S_n = \lambda^{-1} I + \lambda^{-2} T + \cdots + \lambda^{-n} T^{n-1}$$

for $n = 1, 2, \ldots$, we have $S_n f \uparrow Sf$ for any $f > 0$, i.e., $S_n \uparrow S$, and all S_n are positive and order continuous. Therefore, $S_n' \uparrow S'$ by part (i) of the preceding lemma. Using now that $(T^k)' = (T')^k$ for $k = 1, 2, \ldots$, we see that

$$S_n' = \lambda^{-1} I + \lambda^{-2} T' + \cdots + \lambda^{-n} (T')^{n-1}.$$

Note also that the series $\Sigma \lambda^{-k} (T')^{k-1}$ converges in norm since the terms are majorized in norm by the terms of $\Sigma \lambda^{-k} \|\ T\ \|^{k-1}$ (in view of $\|\ T'\ \| \leq \|\ T\ \| < \lambda$). Hence, S_n' converges to S' in norm as well as in order, so

$$S' = \Sigma_1^\infty \lambda^{-n} (T')^{n-1} = (\lambda I - T')^{-1}.$$

Since, as shown above, S is strongly band irreducible, it follows from part (ii) of the preceding lemma that $S' = (\lambda I - T')^{-1}$ is strongly band irreducible, and therefore (from Theorem 42.6 again) that T' is band irreducible.   ∎

As observed earlier, ideal irreducibility of a positive operator T in a Banach lattice implies band irreducibility, but the converse does not always hold. If T is positive, σ-order continuous and compact, then band irreducibility of T nearly implies ideal irreducibility in the sense explained in the following theorem.

Theorem 42.10. *(i) Let T be a positive, σ-order continuous and band irreducible operator in the Banach lattice E, not the null operator. Then every norm closed T-invariant ideal $A \neq \{0\}$ contains a weak order unit of E.*

(ii) Assume in addition that the operator T in part (i) is compact and denote by M the norm closed ideal generated by the range $T(E)$ of T. Then the restriction T_M of T to M is ideal irreducible in M.

Proof. (i) This is a variant upon the result in part (i) of Theorem 42.6. Let $A \neq \{0\}$ be a norm closed T-invariant ideal. Choose $f > 0$ in A and choose $\lambda > r(T)$, where $r(T)$ is the spectral radius of T. Then

$$g = (\lambda I - T)^{-1} f = \Sigma \lambda^{-n} T^{n-1} f$$

converges in norm as well as in order, so $g \in A$. As in Theorem 42.6 it follows that the band generated by g is equal to E, i.e., g is a weak order unit contained in A. Note that the band generated by A is now also equal to E, i.e., A is order dense in E (see Theorem 23.6).

(ii) We have to prove that any norm closed T_M-invariant ideal $A \neq \{0\}$ in M is equal to M. Let $A \neq \{0\}$ be a norm closed and T_M-invariant ideal in M. Then A is a T-invariant norm closed ideal in E, so by part (i) A contains a weak order unit $g > 0$ of E. Now choose $0 \leq u \in E$ arbitrarily and denote $u \wedge ng$ for $n = 1, 2, \ldots$ by u_n. We have $u_n \uparrow u$, which implies that $Tu_n \uparrow Tu$ (since T is σ-order continuous). Since T is compact, the sequence $(Tu_n : n = 1, 2, \ldots)$ has a subsequence converging in norm. The sequence is monotone, so the sequence itself converges, i.e., $Tu_n \rightarrow Tu$ in norm. All u_n are in A, so all Tu_n are in A, and therefore $Tu \in A$ (since A is norm closed). Since u is arbitrary, this shows that the range $T(E)$ is a subset of A, and so the norm closed ideal M generated by $T(E)$ is a subset of A. Conversely, A is a subset of M by definition. Hence $A = M$.   ∎

Remark 42.11. Let M and T_M be as in part (ii) above. Then the following holds.

(i) M is a norm closed ideal in E, so M is a Banach lattice by itself.

(ii) Since the range $T(E)$ is a subset of M, the set $T(E)$ is also a subset of the band $[M]$ generated by M, i.e., $Tf \in [M]$ for every $f \in E$. Then $Tf \in [M]$ for every $f \in [M]$, so $[M]$ is a T-invariant band in E. Since $M \neq \{0\}$ and T is band irreducible, we must have $[M] = E$, i.e., M is order dense in E.

(iii) The operator T_M in the Banach lattice M is positive, σ-order continuous, compact and ideal irreducible.

43. The Spectrum of a Compact Irreducible Operator

One of the main results in the present section will be the theorem that if E is a Dedekind σ-complete Banach lattice and T is a positive ideal irreducible compact operator in E (not the null operator), then the spectral radius $r(T)$ of T satisfies $r(T) > 0$. The Dedekind σ-completeness of E may be omitted from the hypotheses, but we shall restrict ourselves here to the result as stated above. The theorem is due to B. de Pagter (1986), extending earlier less general results. We first prove some simple lemmas.

The operator T in the Archimedean Riesz space E is said to be *band preserving* if for each band B in E it is true that $f \in B$ implies $Tf \in B$ (see the proof of Lemma 34.2 where such a band preserving operator already occurs). Similarly, T is *ideal preserving* if for every ideal A in E it is true that $f \in A$ implies $Tf \in A$.

Lemma 43.1. *For an operator T in an Archimedean Riesz space E the following conditions are equivalent.*

(i) T is band preserving.
(ii) $u \wedge v = 0$ in E implies $Tu \perp v$.
(iii) $f \perp g$ in E implies $Tf \perp g$.
(iv) $Tf \in \{f\}^{dd}$ holds for every $f \in E$.

If these conditions are satisfied, then $\mid Tf \mid = \mid T(\mid f \mid) \mid$ holds for every $f \in E$.

Proof. $(i) \Rightarrow (ii)$ If $u \wedge v = 0$, then $u \in \{v\}^d$. Hence $Tu \in \{v\}^d$ since $\{v\}^d$ is a band and T is band preserving. This shows that $Tu \perp v$.

$(ii) \Rightarrow (iii)$ From $f \perp g$ it follows that $f^+ \wedge \mid g \mid = 0$ and $f^- \wedge \mid g \mid = 0$. Then $Tf^+ \perp g$ and $Tf^- \perp g$, and so $Tf = (Tf^+ - Tf^-) \perp g$.

$(iii) \Rightarrow (iv)$ If $f \in E$ is arbitrary and g is any element in $\{f\}^d$, then $f \perp g$ and so $Tf \perp g$. This shows that $Tf \perp \{f\}^d$, i.e., $Tf \in \{f\}^{dd}$.

$(iv) \Rightarrow (i)$ Let B be a band in E. For any $f \in B$ we have $Tf \in \{f\}^{dd} \subseteq B$. This shows that T is band preserving.

Finally, assume that (i) until (iv) are satisfied and let $f \in E$ be given. It follows from $f^+ \perp f^-$ that $Tf^+ \perp f^-$, and so $Tf^+ \perp Tf^-$. Then

$$\mid Tf \mid = \mid Tf^+ - Tf^- \mid = \mid Tf^+ + Tf^- \mid = \mid T(\mid f \mid) \mid . \qquad \blacksquare$$

Any order bounded operator in E which is band preserving is called an *orthomorphism*. Since positive operators are order bounded it is evident that if π is a positive operator in E, then π is an orthomorphism in E if and only if $u \wedge v = 0$ in E implies $\pi u \wedge v = 0$. The set of all orthomorphisms in E is a linear subspace of the space $\mathcal{L}_b(E)$ of all order bounded operators in E. With respect to the ordering inherited from $\mathcal{L}_b(E)$ the space of all orthomorphisms is an ordered vector space. Note that if $0 \leq \pi_1 \leq \pi_2$ in $\mathcal{L}_b(E)$ and π_2 is an orthomorphism, then so is π_1. In the case that E is Dedekind complete, the space $\mathcal{L}_b(E)$ is a Dedekind complete Riesz space (Theorem 20.2). In this case the order bounded operator π in E is an orthomorphism if and only if π^+ and π^- are so. To see that π^+ is band preserving if π is so, observe that

$$\pi^+ u = \sup(\pi w : 0 \leq w \leq u) \text{ for } u \geq 0.$$

It follows now that π is an orthomorphism if and only if $\mid \pi \mid$ is so. Hence, the space of orthomorphisms is an ideal in $\mathcal{L}_b(E)$; it is even a band in $\mathcal{L}_b(E)$.

If π is an operator in E for which there exists a number $\lambda > 0$ such that $\mid \pi f \mid \leq \lambda \mid f \mid$ holds for all $f \in E$, then π is obviously order bounded and ideal preserving, so π is an orthomorphism. The set $\mathcal{H}$ of all orthomorphisms π possessing this special property (λ depending on π but not on f) is a linear subspace of the space of orthomorphisms. Let $\pi \in \mathcal{H}$, so $\mid \pi f \mid \leq \lambda \mid f \mid$ for all $f \in E$. Then, if $f \geq 0$, we have $\pi f \leq \mid \pi f \mid \leq \lambda f$, and similarly $\pi f \geq -\lambda f$, so

$$-\lambda f \leq \pi f \leq \lambda f, \text{ i.e., } -\lambda I \leq \pi \leq \lambda I,$$

where I is the identity operator in E. In the special case that π is positive, we have, therefore, that $0 \leq \pi \leq \lambda I$. In the converse direction, if it is given that $-\lambda I \leq \pi \leq \lambda I$, then $\mid \pi f \mid \leq \lambda f$ for $f \geq 0$, and so for f arbitrary we have

$$\mid \pi f \mid \leq \mid \pi f^+ \mid + \mid \pi f^- \mid \leq \lambda(f^+ + f^-) = \lambda \mid f \mid .$$

Any $\pi \in \mathcal{H}$ can be written as $\pi = \pi_1 - \pi_2$ with $0 \leq \pi_1 \in \mathcal{H}$ and $0 \leq \pi_2 \in \mathcal{H}$. To see this, observe that if $-\lambda I \leq \pi \leq \lambda I$, then $\pi = \lambda I - (\lambda I - \pi)$ with both λI and $(\lambda I - \pi)$ positive.

We assume now further that E has the principal projection property (so that, therefore, E is Archimedean). Let $B_1, \ldots, B_n$ be disjoint projection bands in E with corresponding band projections $P_1, \ldots, P_n$. Then the algebraic sum $B_1 + \cdots + B_n$ is a direct sum $B_1 \oplus \cdots \oplus B_n$ and this is likewise a projection band with band projection $P_1 + \cdots + P_n$ (see Theorem 32.3). We assume that $B_1 \oplus \cdots \oplus B_n = E$, so $P_1 + \cdots P_n = I$, the identity operator. Concerning this, observe that if $B_1 \oplus \cdots \oplus B_n \neq E$, then the disjoint complement $B_{n+1} = (B_1 \oplus \cdots \oplus B_n)^d$ is likewise a projection band such that $B_1 \oplus \cdots \oplus B_{n+1} = E$ holds. Hence, in view of what follows, it is no restriction of the generality to assume that $B_1 \oplus \cdots \oplus B_n = E$ holds. Given the real numbers $\alpha_1 \leq \alpha_2 \leq \cdots \leq \alpha_n$, let $\pi = \Sigma_1^n \alpha_k P_k$. Then $\pi \leq \Sigma_1^n \alpha_n P_k = \alpha_n I$, and similarly $\pi \geq \alpha_1 I$. Hence $-\lambda I \leq \pi \leq \lambda I$ for $\lambda = \max(\mid \alpha_1 \mid, \mid \alpha_n \mid)$. This shows that $\pi \in \mathcal{H}$ and $\mid \pi f \mid \leq \lambda \mid f \mid$ for every $f \in E$.

We have seen orthomorphisms of the special form $\Sigma \alpha_k P_k$ before, where in the sections 32 and 33 Freudenthal's spectral theorem was discussed. Given $e \in E^+$, any element πe, where $\pi = \Sigma \alpha_k P_k$ with disjoint band projections

$P_k(k = 1, \ldots, n)$ is called an e-step element (see section 32); it was proved in section 33 that any f in the ideal generated by e can be approximated e-uniformly by a sequence of e-step elements $\pi_j e (j = 1, 2, \ldots)$, see the proof of Freudenthal's spectral theorem in Theorem 33.2. For each of these π_j it may be assumed that the sum of the in π_j occurring P_k satisfies $\Sigma P_k = I$ by adding if necessary an extra term with coefficient zero. If $0 \le f \le e$ and $\pi_j e \to f$ holds e-uniformly, then (for j sufficiently large) all coefficients α_k in the approximating π_j may be assumed to satisfy $0 \le \alpha_k \le 1$, so $0 \le \pi_j \le I$. Finally, if one looks at how the terms $\alpha_k P_k$ in the approximating e-step elements in the spectral theorem are defined, then one sees that if $g \perp e$, then $\alpha_k P_k g = 0$ for all k, and so $\pi_j g = 0$ for all the approximating operators π_j.

In the main theorem which follows later we shall be interested in the situation that $0 \le u \le\mid f \mid\le e$, where e is a weak unit. Then $f = f^+ - f^-$ and $\mid f \mid = f^+ + f^-$ with f^+ and f^- positive and disjoint. It follows that u can be written as $u = u_1 + u_2$ with $0 \le u_1 \le f^+$ and $0 \le u_2 \le f^-$, so $u_1 \perp u_2$. As seen above, there exists a sequence $\pi_{1n}(n = 1, 2, \ldots)$ with $0 \le \pi_{1n} \in \mathcal{H}$ such that $\pi_{1n} f^+ \to u_1$ holds f^+-uniformly, and so $\pi_{1n} f^+ \to u_1$ holds e-uniformly (since $f^+ \le e$). Note that $\pi_{1n} f^- = 0$ for all n (because $f^- \perp f^+$). Similarly, there exists a sequence $\pi_{2n}(n = 1, 2, \ldots)$ with $0 \le \pi_{2n} \in \mathcal{H}$ such that $\pi_{2n} f^- \to u_2$ holds e-uniformly and $\pi_{2n} f^+ = 0$ for all n. It follows that $\pi_{1n} f \to u_1$ and $\pi_{2n} f \to -u_2$ hold e-uniformly. Writing now $\pi_n = \pi_{1n} - \pi_{2n}$, we have $\pi_n \in \mathcal{H}$ for all n and

$$\pi_n f = \pi_{1n} f - \pi_{2n} f \to u_1 + u_2 = u \text{ as } n \to \infty$$

holds e-uniformly.

We add two more observations. The first is that if E is a normed Riesz space then the e-uniform convergence of $\pi_n f$ to u implies norm convergence of $\pi_n f$ to u. Secondly, if E is normed and $e \in E^+$ has the property that the norm closure $(A_e)^-$ of the ideal A_e generated by e satisfies $(A_e)^- = E$, then the band B_e generated by e is also equal to E(since B_e is norm closed), i.e., e is now a weak order unit. Collecting all these results, we obtain the following lemma.

Lemma 43.2. *Let E be a normed Riesz space with the principal projection property and let $e \in E^+$ have the property that the norm closure $(A_e)^-$ of the ideal A_e generated by e satisfies $(A_e)^- = E$. Then, if $0 \le u \le\mid f \mid$ in E, there exists a sequence $(\pi_n : n = 1, 2, \ldots)$ of orthomorphisms such that $\pi_n \in \mathcal{H}$ for all n and $\pi_n f \to u$ in norm.*

Proof. We may assume that $\mid f \mid\le e$ since if necessary we may replace e by $\mid f \mid \vee e$. $\blacksquare$

We turn to the theorem that if E is a Dedekind σ-complete Banach lattice and T is a positive operator in E (not the null operator) such that T is compact and ideal irreducible, then $r(T) > 0$. The proof follows immediately from the following theorem which is of independent interest. Actually, the theorems are reformulations of each other.

Theorem 43.3. *If E is a (complex) Dedekind σ-complete Banach lattice and T is a positive operator in E (not the null operator) which is compact and has spectral radius equal to zero, then there exists a norm closed T-invariant non-trivial ideal in E (non-trivial means that the ideal is neither equal to E nor to $\{0\}$).*

Proof. (i) For each $n = 1, 2, \ldots$ the value of $\| T^n \|$ depends only on the behaviour of T on the real part of the space E (see Lemma 36.2(i)). For the proof we may assume, therefore, that the space E is real and T is a non-zero compact positive operator in E with $\| T^n \|^{1/n} \to 0$ as $n \to \infty$. Choose an arbitrary $u > 0$ in E and define e by

$$e = \Sigma_0^\infty 2^{-n} \| T \|^{-n} T^n u,$$

where the series obviously converges in norm (and, therefore, e is the supremum of the partial sums). It is evident that the ideal A_e generated by e is T-invariant, and so the norm closure $(A_e)^-$ is T-invariant. If $(A_e)^- \neq E$, then $(A_e)^-$ is a norm closed non-trivial T-invariant ideal, so in this case the theorem has been proved. We now assume, therefore, that $(A_e)^- = E$. This will enable us to apply the preceding lemma somewhat later in the proof.

(ii) Some notations will be useful. Let $\mathcal{L}(E)$ be the space of all norm bounded operators in E and note that every positive operator in E is a member of $\mathcal{L}(E)$, see Theorem 18.4. Furthermore, let

$$\mathcal{C}^+ = \{R : 0 \leq R \in \mathcal{L}(E) : RT = TR\},$$

$$\mathcal{S}^+ = \{S : 0 \leq S \leq R \text{ for some } R \in \mathcal{C}^+\},$$

$$\mathcal{S} = \{S = S_1 - S_2 \text{ with } S_1 \in \mathcal{S}^+, S_2 \in \mathcal{S}^+\}.$$

It is easy to see that $\mathcal{S}$ is a linear subspace of $\mathcal{L}(E)$. Furthermore $TR \in \mathcal{C}^+$ for every $R \in \mathcal{C}^+$, and so $TS \in \mathcal{S}^+$ for every $S \in \mathcal{S}^+$. It follows that $TS \in \mathcal{S}$ for every $S \in \mathcal{S}$. Note that $\mathcal{C}^+$ is a subset of $\mathcal{S}^+$ and the identity operator I is a member of $\mathcal{C}^+$. Finally we show that if $S \in \mathcal{S}$, then $\pi S \in \mathcal{S}$ for any orthomorphism π such that $\pi \in \mathcal{H}$. For the proof, assume that $S = S_1 - S_2$ with

$$0 \leq S_1 \leq R_1 \in \mathcal{C}^+ \text{ and } 0 \leq S_2 \leq R_2 \in \mathcal{C}^+,$$

$$\pi = \pi_1 - \pi_2 \text{ with } 0 \leq \pi_1 \leq \lambda I, 0 \leq \pi_2 \leq \lambda I.$$

Then $\pi S = (\pi_1 S_1 + \pi_2 S_2) - (\pi_1 S_2 + \pi_2 S_1)$ with each bracket positive and majorized by $\lambda(R_1 + R_2) \in \mathcal{C}^+$. Hence $\pi S \in \mathcal{S}$.

We fix $f \in E$ temporarily and define

$$\mathcal{S}(f) = (Sf : S \in \mathcal{S}).$$

By what has been observed above, $\mathcal{S}(f)$ is a linear subspace of E such that $f \in \mathcal{S}(f)$. We shall prove now that the norm closure $(\mathcal{S}(f))^-$ is an ideal in E. First, let $0 \le u \le \mid Sf \mid$ for some $S \in \mathcal{S}$. By the preceding lemma there exists a sequence of orthomorphisms $(\pi_n : n = 1, 2, \ldots)$ such that $\pi_n \in \mathcal{H}$ for all n and $\pi_n(Sf) \to u$ in norm. We have $\pi_n S \in \mathcal{S}$, and so $\pi_n Sf \in \mathcal{S}(f)$ for all n. It follows that $u \in (\mathcal{S}(f))^-$. Now let $\{\mathcal{S}(f)\}$ be the ideal generated by $\mathcal{S}(f)$ and choose $v \ge 0$ in this ideal. There exist $S_1, \ldots S_n$ in $\mathcal{S}$ such that $0 \le v \le \Sigma_1^n \mid S_k f \mid$, so $v = \Sigma_1^n v_k$ with $0 \le v_k \le \mid S_k f \mid$ for $k = 1, \ldots, n$. It follows that $v_k \in (\mathcal{S}(f))^-$ for each k, so $v \in (\mathcal{S}(f))^-$. It has been shown thus that

$$\mathcal{S}(f) \subseteq \{\mathcal{S}(f)\} \subseteq (\mathcal{S}(f))^-,$$

which implies that $(\mathcal{S}(f))^- = \{\mathcal{S}(f)\}^-$. Since the norm closure of an ideal is an ideal (see Theorem 15.19), we see now that $(\mathcal{S}(f))^-$ is an ideal. As seen above, we have $TS \in \mathcal{S}$ for $S \in \mathcal{S}$, so $\mathcal{S}(f)$ is T-invariant. Then $(\mathcal{S}(f))^-$ is T-invariant. If $(\mathcal{S}(f_0))^- \ne E$ for at least one $f_0 \ne 0$ in E, then $(\mathcal{S}(f_0))^-$ is a norm closed and non-trivial T-invariant ideal in E, so the proof is then complete. It will be sufficient, therefore, to show that the assumption that $(\mathcal{S}(f))^- = E$ for all $f \ne 0$ in E leads to a contradiction.

(iii) The compactness of T will now be of major importance. Assume that $(\mathcal{S}(f))^- = E$ for all $f \ne 0$ in E and choose $u > 0$ in E such that $Tu > 0$. Let $\mathcal{B}$ be an open ball in E with centre u such that the null element is neither contained in the closure $\mathcal{B}^-$ nor in $(T(\mathcal{B}))^-$, where $T(\mathcal{B}) = (Tf : f \in \mathcal{B})$ is the image of $\mathcal{B}$ under T. Every $f \in (T(\mathcal{B}))^-$ satisfies $f \ne 0$ so that, by hypothesis, $(\mathcal{S}(f))^- = E$. In particular, $u \in (\mathcal{S}(f))^-$, so there exists an operator $S_f \in \mathcal{S}$ such that $S_f f \in \mathcal{B}$. It is even so that, by continuity, there exists an open neighbourhood $\mathcal{U}_f$ of f such that $S_f(\mathcal{U}_f) \subseteq \mathcal{B}$. Observe now that the sets $\{\mathcal{U}_f : f \in (T(\mathcal{B}))^-\}$ form an open cover of the compact set $(T(\mathcal{B}))^-$, so there exist $f_1, \ldots, f_n$ in $(T(\mathcal{B}))^-$ with corresponding $S_j = S_{f_j}$ and $\mathcal{U}_j = \mathcal{U}_{f_j} (j = 1, \ldots, n)$ such that

$$(T(\mathcal{B}))^- \subseteq \mathcal{U}_1 \cup \ldots \cup \mathcal{U}_n.$$

Since $Tu \in (T(\mathcal{B}))^-$, there exists an index $j_1 \in \{1, \ldots, n\}$ such that $Tu \in \mathcal{U}_{j_1}$, so $S_{j_1} Tu \in \mathcal{B}$ which implies that $TS_{j_1} Tu \in (T(\mathcal{B}))^-$. Repeating the argument it follows that there exists a sequence $(j_m : m = 1, 2, \ldots)$ in $\{1, \ldots, n\}$ such that

$$g_m = S_{j_m} T S_{j_{m-1}} T \cdots T S_{j_1} Tu \in \mathcal{B}$$

for all $m = 1, 2, \ldots$. We shall now estimate the norm $\parallel g_m \parallel$ of g_m. To do this, we first make a general remark concerning an element of the form $h = Sg$, where $g \in E$ is arbitrary and $S = S_1 - S_2$ with S_1 and S_2 positive. Then

$$| h |=| S_1g - S_2g |\leq| S_1g | + | S_2g |=$$

$$| S_1g^+ - S_1g^- | + | S_2g^+ - S_2g^- |\leq$$

$$S_1g^+ + S_1g^- + S_2g^+ + S_2g^- = (S_1 + S_2)(| g |).$$

Now, in the formula for g_m above, let $S_j = S_{j,1} - S_{j,2}$ with

$$0 \leq S_{j,1} \leq R_{j,1} \in \mathcal{C}^+ \text{ and } 0 \leq S_{j,2} \leq R_{j,2} \in \mathcal{C}^+$$

for $j = 1, \ldots, n$ and

$$\alpha = \max(\| R_{j,1} \|, \| R_{j,2} \|; j = 1, \ldots, n).$$

Then

$$| g_m |=| S_{j_m}T \cdots S_{j_1}Tu |\leq$$

$$(R_{j_m} + R_{j_m})T \cdots (R_{j_1} + R_{j_1})Tu =$$

$$(R_{j_m} + R_{j_m}) \cdots (R_{j_1} + R_{j_1})T^m u,$$

so $\| g_m \|\leq (2\alpha)^m \| T^m \| \cdot \| u \|$, which implies that

$$\| g_m \|^{1/m}\leq 2\alpha\cdot \| T^m \|^{1/m} \cdot \| u \|^{1/m} .$$

For $m \to \infty$ we have $\| T^m \|^{1/m}\to 0$ since the spectral radius of T is zero. Hence $\| g_m \|^{1/m}\to 0$ which easily implies that $\| g_m \|\to 0$. All g_m are contained in $\mathcal{B}$, so it follows from $\| g_m \|\to 0$ that the null element is in $\mathcal{B}^-$. This contradicts our choice of $\mathcal{B}$. The final result is, therefore, that it is not so that $(\mathcal{S}(f))^- = E$ holds for all $f \in E$. In other words, there exists an element $f \in E$ such that the norm closed (and non-zero) ideal $(\mathcal{S}(f))^-$ is T-invariant and non-trivial. ∎

Theorem 43.4. *If E is a complex Dedekind σ-complete Banach lattice and T is a positive ideal irreducible and compact operator in E (not the null operator), then the spectral radius $r(T)$ of T satisfies $r(T) > 0$.*

Proof. If $r(T)$ would be zero, then (by the preceding theorem) there would exist a non-trivial T-invariant norm closed ideal. This is not so here since T is ideal irreducible. Hence $r(T) > 0$. ∎

As observed above, the theorem is due to B. de Pagter (1986). The last part of the proof (part (iii) in the proof of Theorem 43.3) is similar to one of the proofs of the theorem that if T is a compact operator in the Banach space V, then there exists a non-trivial T-invariant norm closed subspace of V.

There is an analogous theorem for band irreducible operators, as follows.

Theorem 43.5. *Let E be a Dedekind σ–complete complex Banach lattice and let T be a positive operator in E (not the null operator) such that T is σ–order continuous, band irreducible and compact. Then the spectral radius $r(T)$ of T satisfies $r(T) > 0$.*

Proof. We may assume that E is a real space. As in Theorem 42.10, let M be the norm closed ideal generated by the range $T(E)$ of T. Denote the restriction of T to M by T_M. Observe first that M is a Dedekind σ-complete Banach lattice by itself and obviously T_M is a positive operator mapping M into itself. Since $f > 0$ in E implies $Tf > 0$ (because T is band irreducible; see the remarks that precede Theorem 42.6), it follows that $T_M(Tf) = T(Tf) > 0$, which shows that T_M is not the null operator. Furthermore, in view of Theorem 42.10(ii), T_M is ideal irreducible. We can now apply the preceding theorem to conclude that $r(T_M) > 0$. The operator T_M is a restriction of T, so $\| (T_M)^n \| \leq \| T^n \|$ for $n = 1, 2, \ldots$, which implies that $r(T_M) \leq r(T)$. Hence $r(T) > 0$. ∎

There is also a result about the adjoint operator of T.

Theorem 43.6. *Let E be a Dedekind complete Banach lattice such that $\,^{\circ}(E_n^\sim) = \{0\}$ and let T be an order bounded and order continuous operator in E. The restriction of $T^\sim = T^*$ to $E_n^\sim = E_n^*$ is denoted by T' (as before). Then T' is order continuous and maps $E_n^\sim$ into itself (see Theorem 26.6). If T is not the null operator, then T' is not the null operator (see the remark preceding Lemma 42.8). If T is positive (not the null operator), order continuous, compact and band irreducible, then T' is positive (not the null operator), compact and band irreducible. Furthermore $r(T) > 0$ as well as $r(T') > 0$.*

Proof. It was proved in the preceding theorem that $r(T) > 0$. It is easy to see that T' is a positive operator and, as observed already, T' is not the null operator. The compactness of T implies the compactness of T^* (this is a general theorem in Banach space theory). Since T' is a restriction of T^*, the operator T' is compact as well. It follows from Theorem 42.9 that T' is band irreducible. Applying the preceding theorem now to T', we see that $r(T') > 0$. ∎

Under the same hypotheses for E and T we can prove more about the spectrum of T.

Theorem 43.7. *(generalized Perron-Jentzsch theorem). Let E be a Dedekind complete Banach lattice such that $\,^{\circ}(E_n^\sim) = \{0\}$ and let T be a positive operator in E (not the null operator) such that T is order continuous, compact and band irreducible. Then the spectral radius $r(T)$ is an eigenvalue of T of multiplicity one, i.e., the corresponding eigenspace is of dimension one.*

*Furthermore, each element $\neq 0$ in the eigenspace is a weak order unit in E.
Finally, we have $r(T') = r(T)$.*

Proof. As proved above, we have $r(T) > 0$ and $r(T') > 0$. Hence, since T is
positive and compact with $r(T) > 0$, the Krein-Rutman theorem (Theorem
41.2) holds for T. Thus, $r(T)$ is an eigenvalue of T and there exists an element
$u > 0$ such that $Tu = r(T)u$. Evidently $Tu \in \{u\}^{dd}$ and so, for any f in the
ideal generated by u, the image Tf is an element in $\{u\}^{dd}$. Since T is order
continuous, the same holds for any f in the band generated by u, i.e., it
follows from $f \in \{u\}^{dd}$ that $Tf \in \{u\}^{dd}$. This shows that the band $\{u\}^{dd}$ is
T- invariant. The operator T is band irreducible, so it follows that $\{u\}^{dd} = E$,
i.e., u is a weak order unit in E.

Let $f \neq 0$ be any other eigenelement belonging to the eigenvalue $r(T)$, so
$Tf = r(T)f$. If $f = g + ih(g$ and h real), then $Tg = r(T)g$ and $Th = r(T)h$.
For the proof that f is a real or complex multiple of u we may assume,
therefore, that f is real. From $Tf = r(T)f$ it follows that

$$r(T)\cdot \mid f \mid = \mid Tf \mid \leq T(\mid f \mid);$$

we shall prove now first that there is equality in this inequality. To this end
we apply the Krein-Rutman theorem to T' to conclude that there exists a
positive weak order unit φ in $E_n^{\sim}$ satisfying $T'\varphi = r(T')\varphi$. Since every $g \geq 0$
in E acts as an order continuous linear functional on $E_n^{\sim}$ (see the remarks
which precede Lemma 31.2), it follows from $\varphi(g) = 0$ for some $g \geq 0$ that
$\psi(g) = 0$ for all ψ in the band generated by φ in $E_n^{\sim}$. This band is equal to
$E_n^{\sim}$ (since φ is a weak order unit). Hence, $\varphi(g) = 0$ for some $g \geq 0$ implies
$g = 0$. In other words, $g > 0$ implies that $\varphi(g) > 0$, i.e., φ is a strictly
positive linear functional. Observing now that $r(T') \leq r(T)$ and returning to
the eigenelement f, we see that

$$0 \leq \varphi\{T(\mid f \mid) - r(T)\cdot \mid f \mid\} = (T'\varphi)(\mid f \mid) - r(T) \cdot \varphi(\mid f \mid) =$$

$$\{r(T') - r(T)\}\varphi(\mid f \mid) \leq 0.$$

Hence, φ being strictly positive, it follows that $r(T') = r(T)$ and $T(\mid f \mid) =
r(T)\cdot \mid f \mid$. It has been proved thus that the eigenspace E_0 of T belonging to
the eigenvalue $r(T)$ has the property that $f \in E_0$ implies $\mid f \mid \in E_0$. Hence, if
f and g are in E_0, then

$$f \vee g = (f + g + \mid f - g \mid)/2$$

is in E_0. Similarly for $f \wedge g$. This shows that E_0 is a Riesz subspace of E.
Furthermore, E_0 has the property that any $v > 0$ in E_0 is a weak order unit
in E, which implies that if v_1 and v_2 are elements in E_0^+ such that $v_1 \wedge v_2 = 0$,
then $v_1 = 0$ or $v_2 = 0$ (or both). Now let f_1 and f_2 be given in E_0. Then, for
$m = f_1 \wedge f_2$, we have

$$(f_1 - m) \wedge (f_2 - m) = 0,$$

so $f_1 - m = 0$ or $f_2 - m = 0$. In other words, $f_1 \leq f_2$ or $f_2 \leq f_1$, i.e., E_0 is linearly ordered. For the proof that E_0 is of dimension one it remains to be shown that if $0 < v \leq u$ in E_0, then $v = \alpha_0 u$ for some positive constant α_0. Since $\alpha u \downarrow 0$ as $\alpha \downarrow 0$, it is impossible that $v \leq \alpha u$ for all $\alpha > 0$. Let $\alpha_0 = \inf(\alpha : v \leq \alpha u)$, so $\alpha_0 > 0$ and $v \leq \alpha_0 u$. For any α satisfying $0 < \alpha < \alpha_0$ we have $0 < (v - \alpha u)^+ \in E_0$, so $(v - \alpha u)^+$ is a weak unit in E. Then $(v - \alpha u)^- = 0$, so $v \geq \alpha u$. Letting now $\alpha \uparrow \alpha_0$ it follows that $v \geq \alpha_0 u$. Hence $v = \alpha_0 u$. This concludes the proof. ∎

The thus proved theorem has applications in the special case that we have to deal with integral operators. Let μ be a σ-finite (positive) measure in the point set X(such as for example Lebesgue measure in $\mathbb{R}$) and let E be an ideal in the Riesz space L_0 of all μ-measurable complex functions on X. Without loss of generality we may assume that the carrier of E is X itself. Assume furthermore that E is equipped with a Riesz norm such that E is a Banach lattice with respect to this norm and such that $\mathscr{C}(E_n^\sim) = \mathscr{C}(E_n^*) = \{0\}$. Let $T : E \to E$ be a positive integral operator with kernel $T(x, y) \geq 0$ such that T is compact and band irreducible. For an operator satisfying these conditions the following theorem, known as Jentzsch's theorem, holds.

Theorem 43.8. *(Jentzsch's theorem). Let $T : E \to E$ be an integral operator with kernel $T(x, y) \geq 0$ satisfying the conditions mentioned above. Then the spectral radius $r(T)$ of T is positive and $r(T)$ is an eigenvalue of T with corresponding eigenfunction $u(x)$ such that $u(x) > 0$ almost everywhere on X. Any other eigenfunction belonging to $r(T)$ is a constant multiple of $u(x)$.*

The theorem goes back to R. Jentzsch (1910) for the case that μ is Lebesgue measure in an interval $X = [a, b]$ with $T(x, y)$ continuous and strictly positive on $X \times X$. The theorem generalizes a result about operators in $\mathbb{R}^n$ having matrices with strictly positive entries, due to O. Perron (1907), see Exercise 43.9 at the end of the present section. Some final remarks.

(i) The assumption that $\mathscr{C}(E_n^\sim) = \{0\}$ is equivalent to assuming that the carrier of $E_n^\sim$ is X; see the remarks which follow after Theorem 31.3.

(ii) The assumption that T is band irreducible is equivalent to assuming that

$$\int_{X \setminus Y} \left\{ \int_Y T(x, y) d\mu_y \right\} d\mu_x > 0$$

for every subset Y of X satisfying $\mu(Y) > 0$ and $\mu(X \setminus Y) > 0$, see Example 42.3. This holds in particular if $T(x, y) > 0$ almost everywhere on $X \times X$.

(iii) If T is a Hille-Tamarkin operator in $L_p(X, \mu)$ as in Theorem 42.6 with μ separable and $T(x, y) \geq 0$ and if T satisfies the irreducibility condition as in (ii), then Jentzsch's theorem holds for T.

Exercise 43.9. Let the point set X consist of n points, each point of measure one. The Riesz space E consists of all complex functions on X with pointwise ordering for the real functions, i.e., $E = \mathbb{C}^n = \mathbb{R}^n + i\mathbb{R}^n$. The subset A of $\mathbb{C}^n$ is an ideal if and only $A = A_r + iA_r$, where A_r is an ideal in $\mathbb{R}^n$ (see Theorem 13.8). As in the real case, ideals and bands are the same. Let $(t_{jk} : j, k = 1, \ldots, n)$ be the matrix of the operator T in E. Observe that T is positive if and only if all t_{jk} are non-negative. Show that the positive operator T is irreducible if and only if

$$\Sigma_{j \in S_1}(\Sigma_{k \in S_2} t_{jk}) > 0$$

for every decomposition of X into non-empty subsets S_1 and S_2 such that $S_1 \cup S_2 = X$. Furthermore, show that if T is positive and irreducible, then $r(T) > 0$ and $r(T)$ is an eigenvalue of T with an eigenvector u having all its coordinates strictly positive. Each further eigenvector belonging to $r(T)$ is a constant multiple of u. For all $t_{jk} > 0$ this is Perron's theorem.

44. The Peripheral Spectrum of a Positive Operator

In Theorem 13.4 it was proved that if the Riesz space E is Archimedean and uniformly complete, then

$$\mid h \mid = \sup(f \cos\theta + g\sin\theta : 0 \le \theta \le 2\pi) =$$

$$\sup(\mid f\cos\theta + g\sin\theta \mid : 0 \le \theta \le 2\pi)$$

exists in E for all f and g in E. This supremum is then by definition the absolute value $\mid h \mid$ of the element $h = f + ig$ in the complexification $E + iE$ of E. We shall prove now first that in any Archimedean Riesz space E the infimum

$$\inf(\mid f\sin\theta - g\cos\theta \mid : 0 \le \theta \le 2\pi)$$

exists for all f, g in E and is equal to the null element. This will lead then to the result that if E is Dedekind σ-complete with a multiplication having a strong order unit as multiplicative unit, then $\mid h \mid^2 = f^2 + g^2$ for every $h = f + ig$ in $E + iE$, and so

$$(f + ig)(f - ig) = f^2 + g^2 = \mid h \mid^2 .$$

We shall use this result in the further investigation of the spectrum of a positive, compact and irreducible operator.

Lemma 44.1. *Let $f_1, \ldots, f_n$ be a finite sequence in the Riesz space E and let $u \in E^+$ be such that*

(i) $|f_k - f_{k+1}| \le u$ for $k = 1, \ldots, n-1$,
(ii) $\inf(f_1, \ldots, f_n) \le u$ and $\inf(-f_1, \ldots, -f_n) \le u$.

   Then $\inf(|f_1|, \ldots, |f_n|) \le u$.

Proof. By making use of the distributive laws repeatedly, we can write $\inf(|f_1|, \ldots, |f_n|)$ as the supremum of 2^n terms, one of which is $\inf(f_1, \ldots, f_n)$ and another one is $\inf(-f_1, \ldots, -f_n)$. All other terms have the form

$$f_{j_1} \wedge \cdots \wedge f_{j_k} \wedge (-f_{j_{k+1}}) \wedge \cdots \wedge (-f_{j_n}). \tag{1}$$

It is sufficient to show that each of the terms is majorized by u. This holds for $\inf(f_1, \ldots, f_n)$ and for $\inf(-f_1, \ldots, -f_n)$ by hypothesis. For the term in (1) there exist two consecutive indices in $(1, \ldots, n)$ such that one of these is contained in $(j_1, \ldots, j_k)$ and the other one in $(j_{k+1}, \ldots, j_n)$. For example, let $p \in (j_1, \ldots, j_k)$ and $p + 1 \in (j_{k+1}, \ldots, j_n)$. Then, denoting the term in (1) by t, we have

$$t \le f_p \wedge (-f_{p+1}) \le (f_p - f_{p+1})/2 \le u/2 \le u.$$

Similarly if $p + 1 \in (j_1, \ldots, j_k)$ and $p \in (j_{k+1}, \ldots, j_n)$.   ∎

Theorem 44.2. *For all f and g in the Archimedean space E we have*

$$\inf(|f \sin\theta - g \cos\theta| : 0 \le \theta \le 2\pi) = 0.$$

Proof. Choose a natural number $n \ge 2$ and define

$$h_k = f \sin(\pi k/n) - g \cos(\pi k/n)$$

for $k = 0, 1, \ldots, 2n$. Since

$$|\sin\{\pi(k+1)/n\} - \sin(\pi k/n)| \le \pi/n$$

by the mean value theorem in calculus, and since we have a similar inequality for the cosine, it follows that

$$|h_k - h_{k+1}| \le (\pi/n)(|f| + |g|) \text{ for } k = 0, 1, \ldots, 2n-1.$$

Furthermore

$$h_0 \wedge \cdots \wedge h_{2n} \le h_0 \wedge h_n = -|g| \le (\pi/n)(|f| + |g|),$$

and similarly for $(-h_0) \wedge \cdots \wedge (-h_{2n})$. The conditions of the preceding lemma are satisfied, therefore. It follows that if we define w_n for $n = 1, 2, \ldots$ by $w_n = \inf(|h_1|, \ldots, |h_{2n}|)$, then

$$0 \leq w_n \leq (\pi/n)(\mid f \mid + \mid g \mid).$$

Since $(\pi/n)(\mid f \mid + \mid g \mid) \downarrow 0$ as $n \to \infty$ because E is Archimedean, we have $\inf_n w_n = 0$, and so

$$\inf(\mid f \sin\theta - g \cos\theta \mid : 0 \leq \theta \leq 2\pi) = 0. \qquad \blacksquare$$

Assume now that E is Dedekind σ-complete. Let $e > 0$ be given in E. Then it is possible to introduce a commutative multiplication in the principal ideal A_e having e as multiplicative unit, so $fg = gf$ and $fe = ef = f$ for all f, g in A_e (see section 35). We recall that if f, g and h are in A_e, then $gh = 0$ if and only $g \perp h$, so $f^2 = \mid f \mid^2$ (since $f^+ f^- = 0$) and $f^2 = 0$ if and only $f = 0$.

Lemma 44.3. *(i) If u and v are elements in A_e^+, then $u^2 \leq v^2$ if and only if $u \leq v$. Hence $u^2 = v^2$ if and only $u = v$. It follows that if $0 \leq p \in A_e$ and p has a positive square root q, i.e., $q \geq 0$ and $q^2 = p$, then this square root is unique.*

(ii) If $0 \leq u_0 \in A_e$ and D is a subset of A_e^+, then $\sup(u : u \in D) = u_0$ if and only if $\sup(u^2 : u \in D) = u_0^2$. Similarly for the infimum.

Proof. (i) If $0 \leq u \leq v$, then $0 \leq u^2 \leq uv \leq v^2$. Conversely, let $u^2 \leq v^2$. Then

$$0 \leq v^2 - u^2 = (v + u)(v - u) =$$

$$(v + u)(v - u)^+ - (v + u)(v - u)^-,$$

where the last two terms are positive and disjoint. Hence, by Theorem 5.6, $(v + u)(v - u)^-$ is the negative part of $v^2 - u^2$. But $v^2 - u^2$ is positive, so $(v + u)(v - u)^- = 0$. Then $(v - u)^- \perp (v + u)$, i.e. $(u - v)^+ \perp (u + v)$. This shows that $(u - v)^+$ is disjoint to as well as majorized by $(u + v)$. It follows that $(u - v)^+ = 0$, i.e., $u \leq v$.

(ii) Assume first that $\sup(u : u \in D) = u_0$, i.e., $\inf(u_0 - u : u \in D) = 0$. For any $u \in D$ we have

$$0 \leq u_0^2 - u^2 = (u_0 + u)(u_0 - u) \leq 2u_0(u_0 - u).$$

But $2u_0 \leq \alpha e$ for some number $\alpha > 0$, so

$$2u_0(u_0 - u) \leq \alpha e(u_0 - u) = \alpha(u_0 - u).$$

It follows that $\inf(u_0^2 - u^2) = 0$, so $\sup(u^2 : u \in D) = u_0^2$.

Conversely, let $\sup(u^2 : u \in D) = u_0^2$. From $u^2 \leq u_0^2$ it follows first that $u \leq u_0$ for all $u \in D$. Assume now that $u_0 - u \geq w$ for some $w \geq 0$ and all

$u \in D$. Then $u + w \leq u_0$, so $u^2 + w^2 \leq (u+w)^2 \leq u_0^2$, i.e., $w^2 \leq u_0^2 - u^2$ for all $u \in D$. It follows that

$$w^2 \leq \inf(u_0^2 - u^2 : u \in D) = 0.$$

Hence $w^2 = 0$, so $w = 0$. Then $\sup(u : u \in D) = u_0$. ∎

Theorem 44.4. *Once more, let $e > 0$ be given in the Dedekind σ-complete Riesz space E and let the commutative multiplication in A_e with e as unit element be introduced as explained above. The multiplication is extended to the complexification $A_e + iA_e$ in the natural manner, i.e.,*

$$(f + ig)(p + iq) = (fp - gq) + i(fq + gp).$$

Then, for any $h = f + ig \in A_e + iA_e$ we have

$$\mid h \mid^2 = f^2 + g^2,$$

i.c., $\mid h \mid$ is the (unique) positive square root of $f^2 + g^2$.

Proof. Combining the preceding theorem and lemma, we see that

$$\inf\{(f \sin\theta - g\cos\theta)^2 : 0 \leq \theta \leq 2\pi\} = 0$$

for all f and g in A_e. Note now that

$$p - \inf(q : q \in D) = \sup(p - q : u \in D)$$

holds generally for arbitrary p, q and D. Choose

$$p = f^2 + g^2 \text{ and } q = (f \sin\theta - g\cos\theta)^2.$$

This gives

$$f^2 + g^2 = (f^2 + g^2) - \inf\{f \sin\theta - g\cos\theta)^2 : 0 \leq \theta \leq 2\pi\} =$$
$$\sup\{f^2 + g^2 - (f\sin\theta - g\cos\theta)^2 : 0 \leq \theta \leq 2\pi\} =$$
$$\sup\{(f\cos\theta + g\sin\theta)^2 : 0 \leq \theta \leq 2\pi\} =$$
$$\sup\{\mid f\cos\theta + g\sin\theta \mid^2 : 0 \leq \theta \leq 2\pi\} = \mid h \mid^2,$$

where $\sup(\mid f\cos\theta + g\sin\theta \mid : 0 \leq \theta \leq 2\pi)$ is the absolute value $\mid h \mid$ of $h = f + ig$ according to the definition in Chapter 13. ∎

Corollary 44.5. *The multiplication in $A_e + iA_e$ is commutative with e as unit element and with the following further properties.*

(i) $\mid h_1 h_2 \mid = \mid h_1 \mid \cdot \mid h_2 \mid$ *for all h_1 and h_2 in $A_e + iA_e$.*

(ii) $h_1 h_2 = 0$ *if and only if $\mid h_1 \mid \perp \mid h_2 \mid$.*

(iii) $\mid h_1 \mid \perp \mid h_2 \mid$ *implies that $\mid h_3 h_1 \mid \perp \mid h_2 \mid$ for any $h_3 \in A_e + iA_e$.*

(iv) $h^k = 0$ *for some natural number k implies that $h = 0$.*

(v)   *If $(h_n : n = 1, 2, \ldots)$ and k are in $A_e + iA_e$ and h_n converges to h_0 in order, then $kh_n \to kh_0$ in order.*

Proof. Note first that for $h = f + ig$ we have $\mid f \mid \leq \mid h \mid$ and $\mid g \mid \leq \mid h \mid$ by Theorem 13.5. Hence, if $\mid h \mid = 0$, then $\mid f \mid = \mid g \mid = 0$, which implies that $f = g = 0$ and so $h = 0$.

(i) It is easy to see by means of the last theorem that

$$\mid h_1 h_2 \mid^2 = \mid h_1 \mid^2 \cdot \mid h_2 \mid^2 = (\mid h_1 \mid \cdot \mid h_2 \mid)^2,$$

so $\mid h_1 h_2 \mid = \mid h_1 \mid \cdot \mid h_2 \mid$ by the uniqueness of the square root.

(ii) $h_1 h_2 = 0$ if and only if $\mid h_1 \mid \cdot \mid h_2 \mid = \mid h_1 h_2 \mid = 0$, i.e., if and only if $\mid h_1 \mid \perp \mid h_2 \mid$.

(iii) From $\mid h_1 \mid \perp \mid h_2 \mid$ it follows that $(\mid h_3 \mid \cdot \mid h_1 \mid) \perp \mid h_2 \mid$, i.e., $\mid h_3 h_1 \mid \perp \mid h_2 \mid$.

(iv) If $h^2 = 0$, then $\mid h \mid^2 = \mid h^2 \mid = 0$, so $\mid h \mid \perp \mid h \mid$, i.e., $\mid h \mid = 0$, which implies that $h = 0$. If $h^k = 0$ and $k = 2^n$ for some natural number n, it follows easily that $h = 0$. If $2^{n-1} < k < 2^n$ and $h^k = 0$, then $h^{2^n} = 0$, and so $h = 0$.

(v) We have $\mid k \mid \leq \beta e$ for some $\beta > 0$ and $\mid h_0 - h_n \mid \leq p_n \downarrow 0$ for some sequence $(p_n : n = 1, 2, \ldots)$ in A_e^+. Hence

$$\mid kh_n - kh \mid = \mid k \mid \cdot \mid h_n - h \mid \leq (\beta e)p_n = \beta p_n \downarrow 0. \qquad \blacksquare$$

For $f \in A_e + iA_e$ multiplication by f is a (linear) operator in $A_e + iA_e$. If $f \in A_e^+$, then $0 \leq f \leq \alpha e$ for some number $\alpha \geq 0$, so $0 \leq fg \leq \alpha eg = \alpha g$ for all $g \geq 0$ in A_e. This shows that multiplication by f is a positive and order continuous operator in $A_e + iA_e$. Now denote by B_e the principal band generated by e in E. If $0 \leq g \in B_e$ is given and $g_n = g \wedge ne$ for $n = 1, 2, \ldots$, we have $g_n \in A_e$ for all n and $0 \leq g_n \uparrow g$. Let $0 \leq f \in A_e$ be as above, so multiplication by f is an order continuous positive operator in A_e majorized by αI, where I is the identity operator. Note that $fg_n \leq \alpha I g_n = \alpha g_n \leq \alpha g$ for all n. Hence, since E is Dedekind σ-complete, the sequence $(fg_n : n = 1, 2, \ldots)$ has an order limit (the supremum of the sequence) which we shall denote by fg. It is easy to see that the limit depends on f and g but not on the approximating sequence. The domain of the multiplication operator has thus been extended to B_e^+ and it is obvious how to extend the domain to $B_e + iB_e$. More generally, proceeding step by step, we define hk for any $h \in A_e + iA_e$

and any $k \in B_e + iB_e$. The extended multiplication, as defined in this manner, is order continuous and coincides on $A_e + iA_e$ (i.e., both h and k in $A_e + iA_e$) with the already existing multiplication.

Lemma 44.6. *Properties (i) and (ii) in the preceding corollary remain true for the extended multiplication, i.e., if $h \in A_e + iA_e$ and $k \in B_e + iB_e$, then $| hk | = | h | \cdot | k |$ and $hk = 0$ if and only if $| h | \perp | k |$.*

Proof. (i) It was observed in section 21 (in connection with Definition 21.5) that if in a (real) Riesz space the positive operator T is σ-order continuous, then $f_n \to 0$(in order) implies $Tf_n \to 0$ (in order). Of course, the same holds if T is a difference of two σ-order continuous positive operators. The extension to the complex case is evident. In the present case multiplication in $B_e + iB_e$ by an element $h \in A_e + iA_e$ is σ-order continuous. Let now $h \in A_e + iA_e$ and $k \in B_e + iB_e$ be given. There exists a sequence $(k_n : n = 1, 2, \ldots)$ in $A_e + iA_e$ converging in order to k. Then hk_n converges to hk in order, and so $| hk_n |$ converges to $| hk |$ in order. But $| hk_n | = | h | \cdot | k_n |$ by part (i) in the preceding corollary, so $| h | \cdot | k_n | \to | hk |$ in order. From $k_n \to k$ (in order) we derive that $| k_n | \to | k |$ in order, and so $| h | \cdot | k_n | \to | h | \cdot | k |$ in order. It has thus been shown that $| h | \cdot | k_n |$ converges in order to $| hk |$ as well as to $| h | \cdot | k |$. Hence $| hk | = | h | \cdot | k |$.

(ii) Let $h \in A_e + iA_e$ and $k \in B_e + iB_e$ be given and let $(k_n : n = 1, 2, \ldots)$ be a sequence in A_e^+ such that $0 \leq k_n \uparrow | k |$. Then

$$| hk | = 0 \Leftrightarrow | h | \cdot | k | = 0 \Leftrightarrow | h | \cdot k_n = 0 \text{ for all } n \Leftrightarrow$$

$$| h | \perp k_n \text{ for all } n \Leftrightarrow | h | \perp | k | . \qquad \blacksquare$$

Every order bounded operator T in the Dedekind complete space E has an absolute value, defined for any $u \geq 0$ by

$$| T | (u) = \sup(| Tf | : | f | \leq u).$$

This holds if E is real (see Theorem 20.4) as well as if E is complex (see Theoem 36.4). We shall be interested now in the case that $e > 0$ is a weak order unit in the Dedekind complete space E. Then the ideal A_e is, therefore, order dense in E and the band B_e is the space E itself. Any $h \in A_e + iA_e$ acts as an order bounded (σ-order continuous) multiplication operator in $B_e + iB_e = E + iE$. To make a distinction between the element h and the corresponding multiplication operator we shall denote the operator by π_h. The element h and the operator π_h have the absolute values $| h |$ and $| \pi_h |$ respectively. Are these the same? The answer is affirmative, i.e., $| \pi_h | (k) = | h | \cdot k$ for every $k \in E + iE$. It is sufficient to prove this for the case that $k = u \geq 0$ in E. Then

$$| \pi_h | (u) = \sup(| \pi_h f | : | f | \leq u) = \sup(| hf | : | f | \leq u) =$$

$$\sup(\mid h \mid \cdot \mid f \mid : \mid f \mid \le u) = \mid h \mid \cdot (u),$$

which shows that $\mid \pi_h \mid$ is multiplication by $\mid h \mid$.

Assume once more that e is a weak unit in the Dedekind complete space E and let $h = f + ig \in E + iE$ satisfy $\mid h \mid = e$. This implies that $h \in A_e + iA_e$, where (as above) A_e is the ideal generated in E by e. The conjugate complex element $f - ig$ will be denoted by h^* and multiplication in $E + iE$ by h and h^* is denoted by π_h and π_h^* respectively. As long as h is fixed we shall simply write π and π^*. Note that both $\pi\pi^*$ and $\pi^*\pi$ are multiplication by

$$hh^* = h^*h = f^2 + g^2 = \mid h \mid^2 = e^2 = e,$$

so $\pi\pi^* = \pi^*\pi = I$. Equivalently, $\pi h^* = \pi^* h = e$. As observed above, the operator $\mid \pi \mid$ is multiplication by $\mid h \mid$. Some further remarks follow.

(i) For any order bounded operator T in $E + iE$ we have $\mid \pi T \mid = \mid T \mid$ and $\mid \pi^* T \mid = \mid T \mid$. This follows from

$$\mid \pi T \mid (u) = \sup(\mid \pi Tf \mid : \mid f \mid \le u) = \sup(\mid hTf \mid : \mid f \mid \le u) =$$

$$\sup(\mid h \mid \cdot \mid Tf \mid : \mid f \mid \le u) = \sup(\mid Tf \mid : \mid f \mid \le u) = \mid T \mid (u)$$

for any $u \ge 0$. Similarly for $\mid \pi^* T \mid$.

(ii) The equalities $\mid T\pi \mid = \mid T \mid$ and $\mid T\pi^* \mid = \mid T \mid$ hold as well. This follows from the observation that for any $u \ge 0$ the sets $(f : \mid f \mid \le u)$ and $(\pi f : \mid f \mid \le u)$ are identical. To see this, note that $\mid f \mid \le u$ implies $\mid \pi f \mid = \mid f \mid \le u$ and conversely, if $\mid g \mid \le u$, then $g = \pi f$ for $f = \pi^* g$ with $\mid f \mid = \mid \pi^* g \mid = \mid g \mid \le u$. It follows that

$$\mid T\pi \mid (u) = \sup(\mid T\pi f \mid : \mid f \mid \le u) = \sup(\mid Tg \mid : \mid g \mid \le u) = \mid T \mid (u).$$

Similarly for $\mid T\pi^* \mid$.

(iii) It follows now from (i) and (ii) that $\mid \pi T\pi^* \mid = \mid \pi^* T\pi \mid = \mid T \mid$.

$$(\pi T\pi^*)^k = \pi T^k \pi^* \quad \text{and} \quad (\pi^* T\pi)^k = \pi^* T^k \pi.$$

Therefore, for λ any complex number, we have

$$(\pi T\pi^* - \lambda I)^k = \{\pi(T - \lambda I)\pi^*\}^k = \pi(T - \lambda I)^k \pi^*$$

for $k = 1, 2, \ldots$. It follows that if E is a Dedekind complete Banach lattice and T is norm bounded as well as order bounded, then $(T - \lambda I)^k$ has a norm bounded inverse if and only if $(\pi T\pi^* - \lambda I)^k$ has a norm bounded inverse. To prove this, note first that π and π^* are norm bounded (simply because $\mid \pi \mid = \mid \pi^* \mid = I$). Furthermore, if the inverse S of $(T - \lambda I)^k$ exists, then $\pi S\pi^*$ is the inverse of $(\pi T\pi^* - \lambda I)^k$. Similarly, if the inverse S_1 of $(\pi T\pi^* - \lambda I)^k$ exists, then $\pi^* S_1 \pi$ is the inverse of $(T - \lambda I)^k$. It follows in particular $(k = 1)$ that T

and $\pi T \pi^*$ have the same spectrum. Hence, if T is compact, the operators T and $\pi T \pi^*$ have the same non-zero eigenvalues, i.e., $\lambda \neq 0$ is an eigenvalue of T if and only if λ is an eigenvalue of $\pi T \pi^*$. Furthermore, if $(f_1, \ldots, f_n)$ is a linearly independent basis for the corresponding eigenspace of T, then $(\pi f_1, \ldots, \pi f_n)$ is a linearly independent basis for the corresponding eigenspace of $\pi T \pi^*$. Conversely, if $(g_1, \ldots, g_n)$ is a linearly independent basis for the eigenspace of $\pi T \pi^*$, then $(\pi^* g_1, \ldots, \pi^* g_n)$ is a linearly independent basis for the eigenspace of T.

Before presenting the main results in the present section we prove another lemma.

Lemma 44.7. *(i) Let E be an Archimedean and uniformly complete Riesz space and let $h = f + ig \in E + iE$ have the property that $\mid h \mid = \mid f \mid$. Then $g = 0$, i.e., $h = f$.*

(ii) Let G be a set of n mutually different complex numbers $(n \geq 1)$, all of these numbers of absolute value one, and let G have the property that if $\alpha \in G$ and $\beta \in G$, then $\alpha \beta^{-1} \in G$. Then G consists of the n roots of the equation $\alpha^n - 1 = 0$.

Proof. (i) It follows from parts (ii) and (iii) in Theorem 13.6 that

$$\mid f \mid = \mid h \mid = \sup(\mid f \mid \cos\theta + \mid g \mid \sin\theta : 0 \leq \theta \leq \pi/2).$$

Hence $\mid g \mid \sin\theta \leq \mid f \mid (1 - \cos\theta)$ for all θ satisfying $0 < \theta \leq \pi/2$. Since

$$\sin\theta = 2\sin(\theta/2)\cos(\theta/2) \text{ and } 1 - \cos\theta = 2\sin^2(\theta/2),$$

it follows that $\mid g \mid \cos(\theta/2) \leq \mid f \mid \sin(\theta/2)$, i.e., $\mid g \mid \leq \mid f \mid tg(\theta/2)$. Letting $\theta \downarrow 0$, it is seen that $g = 0$.

(ii) Choose $\alpha = \beta$ in G. Then $\alpha \beta^{-1} \in G$, i.e., $1 \in G$. Hence, for an arbitrary $\alpha \in G$ and $\beta = 1$ we have $\alpha^{-1} = \beta \alpha^{-1} \in G$. Finally, for any $\alpha \in G$ and $\beta \in G$ we have $\beta^{-1} \in G$, so $\alpha\beta = \alpha(\beta^{-1})^{-1} \in G$. This together shows that G is a *commutative group* with respect to multiplication as the group operation and with the number 1 as unit element. The group G is a *finite group* consisting of n distinct elements. The number n is called the *order* of the group. For $n = 1$ the group consists of the unit element only. It is a fundamental result in elementary group theory that if α is an arbitrary element in a (multiplicative) group of order n, then α^n equals the unit element, so $\alpha^n = 1$ in the present case. For an indication of the proof we refer to Exercise 44.12 at the end of the section. Since G has n different elements and each $\alpha \in G$ satisfies $\alpha^n = 1$, it is evident now that G consists of the n different roots of the equation $\alpha^n - 1 = 0$. ∎

Theorem 44.8. *As in the generalized Perron-Jentzsch theorem (Theorem 43.7), let E be a Dedekind complete Banach lattice such that $\mathcal{N}(E_n^\sim) = \{0\}$ and let T be a positive operator in E such that T is order continuous, compact and band irreducible. Then, as proved already, the spectral radius $r = r(T)$ of T is strictly positive, $r = r(T)$ is an eigenvalue of T of multiplicity one and there exists a corresponding eigenelement which is a weak unit in E. Let λ be any eigenvalue of T satisfying $\mid \lambda \mid = r = r(T)$ and let $h = f + ig \in E + iE$ be a corresponding non-zero eigenelement, so $Th = \lambda h$ with $h \neq 0$. Then $\mid h \mid$ is a weak order unit in E. Assume now that multiplication with $\mid h \mid$ as unit element is introduced (as explained above) and denote by π_h the multiplication by h. For brevity, write π instead of π_h as long as h is kept fixed. Similarly, denote by π_h^* (or π^* for brevity) the multiplication by $h^* = f - ig$. Then*

$$T = \lambda r^{-1} \pi T \pi^*.$$

Proof. From $Th = \lambda h$ it follows that

$$r \mid h \mid = \mid \lambda h \mid = \mid Th \mid \leq T(\mid h \mid).$$

As before (see Theorem 43.6), let T' be the restriction of $T^\sim = T^*$ to $E_n^\sim = E_n^*$. Then T' is a positive operator in $E_n^\sim$ and $r(T') = r(T)$ by the generalized Jentzsch's theorem (Theorem 43.7). As shown in the proof of that theorem, there exists a strictly positive linear functional φ on E such that $T'\varphi = r(T')\varphi$, so $T'\varphi = r\varphi$. Then

$$0 \leq \varphi\{T(\mid h \mid) - r \mid h \mid\} = (T'\varphi - r\varphi) \mid h \mid = 0,$$

which implies that $T(\mid h \mid) = r \mid h \mid$. From the generalized Jentzsch's theorem it follows now that $\mid h \mid$ is a weak order unit in E. Since $\mid h \mid$ is the multiplicative unit, we have $\pi(\mid h \mid) = h \cdot \mid h \mid = h$ and

$$\pi^* h = h^* h = f^2 + g^2 = \mid h \mid^2 = \mid h \mid .$$

Now consider the operator $S = \lambda^* r^{-1} \pi^* T \pi$, where λ^* is the complex conjugate of λ. We have

$$S(\mid h \mid) = \lambda^* r^{-1} \pi^* T \pi(\mid h \mid) = \lambda^* r^{-1} \pi^* Th = \lambda \lambda^* r^{-1} \pi^* h = r \mid h \mid .$$

Writing $S = S_1 + iS_2$ (with S_1 and S_2 real operators), it follows that $S_1(\mid h \mid) = r \mid h \mid$. Since

$$\mid S \mid = \mid \pi^* T \pi \mid = \mid T \mid = T \text{ and } S_1 \leq \mid S_1 \mid \leq \mid S \mid = T,$$

the operator $T - S_1$ is positive and also order continuous (since $0 \leq T - S_1 \leq T + S_1 \leq 2T$). As proved already, we have $T(\mid h \mid) = r \mid h \mid$ as well as $S_1(\mid h \mid) = r \mid h \mid$. It follows that $(T - S_1)(\mid h \mid) = 0$. Since $\mid h \mid$ is a weak unit, this implies similarly as before that $T - S_1$ is the null operator. Hence $S_1 = T = \mid S \mid$, i.e., the real part S_1 of S satisfies $S_1 = \mid S \mid$. Then, by part (i) of Lemma 44.7 above, we have $S_2 = 0$, and so $S = S_1 = T$, i.e., $T = \lambda^* r^{-1} \pi^* T \pi$. It follows that $\pi T \pi^* = \lambda^* r^{-1} T$. Since $\lambda^* = r^2 \lambda^{-1}$, this implies that $\pi T \pi^* = \lambda^{-1} r T$, and so

$$T = \lambda r^{-1} \pi T \pi^*.\qquad\blacksquare$$

We continue with the discussion of the spectral properties of the positive operator introduced in the last theorem. The result which follows is a generalization of results due to G. Frobenius (1912) for the matrix case. The proof is based upon the method of proof of H. Wielandt (1950) for the matrix case.

Theorem 44.9. *(generalized Frobenius theorem). Assume that the Banach lattice E and the operator T in E are the same as in the last theorem. As before, denote by r the spectral radius $r(T)$ of T. If the peripheral spectrum of T (i.e., the set of eigenvalues of T of absolute value r) consists of n different numbers $\lambda_1, \ldots, \lambda_n$, then these numbers are exactly the n roots of the equation $\lambda^n - r^n = 0$. Furthermore, the spectrum $\sigma(T)$ of T is invariant under rotation over an angle $2\pi/n$ (multiplicities of the eigenvalues included). This implies in particular that the eigenvalues in the peripheral spectrum are of multiplicity one.*

Proof. Replacing T by $r^{-1}T$, we may assume without loss of generality that $r = r(T) = 1$. Let α and β be points in the peripheral spectrum of T (so $\mid \alpha \mid = \mid \beta \mid = 1$) and let h_α and h_β be corresponding non-zero eigenelements. Denote the corresponding multiplication operators by π_α and π_β respectively. Applying the last theorem for $\lambda = \alpha$ and $\lambda = \beta$, we see that

$$T = \alpha \pi_\alpha T \pi_\alpha^* \text{ and } T = \beta \pi_\beta T \pi_\beta^*.$$

Since $\beta\beta^* = 1$ (i.e., $\beta^* = \beta^{-1}$) and $\pi_\beta \pi_\beta^* = \pi_\beta^* \pi_\beta = I$, it follows from $T = \beta \pi_\beta T \pi_\beta^*$ that $T = \beta^* \pi_\beta^* T \pi_\beta$. Substituting this in $T = \alpha \pi_\alpha T \pi_\alpha^*$, we get

$$T = \alpha\beta^* \pi_\alpha \pi_\beta^* T \pi_\beta \pi_\alpha^*,$$

which implies that

$$T - \alpha\beta^* I = \alpha\beta^* \pi_\alpha \pi_\beta^* (T - I) \pi_\beta \pi_\alpha^*.$$

As in remark (iv) which precedes Lemma 44.7 it follows that the null spaces of $T - \alpha\beta^* I$ and $T - I$ have the same dimension. Since the null space of $T - I$ is of dimension one (in view of Jentzsch's theorem), the same holds for the null space of $T - \alpha\beta^* I$. In other words, $\alpha\beta^* = \alpha\beta^{-1}$ is an eigenvalue of T of multiplicity one. It has been shown in this manner that if α and β are members of the peripheral spectrum, then $\alpha\beta^{-1}$ is also a member. As proved in part (ii) of Lemma 44.7 it follows that the peripheral spectrum consists exactly of the n roots $\lambda_1, \ldots, \lambda_n$ of the equation $\lambda^n - 1 = 0$. For $\alpha = \lambda_1$ it follows from $T = \alpha \pi_\alpha T \pi_\alpha^*$ that the spectrum satisfies

$$\sigma(T) = \sigma(\alpha \pi_\alpha T \pi_\alpha^*) = \alpha\sigma(\pi_\alpha T \pi_\alpha^*) = \alpha\sigma(T),$$

where we refer again to remark (iv) preceding Lemma 44.7. Hence, the spectrum of T is invariant under rotation by λ_1, i.e., rotation over an angle of $2\pi/n$ (multiplicities included). $\blacksquare$

In some cases the peripheral spectrum of T consists of the number $r(T)$ only. This occurs for example if T is strongly irreducible. We recall (see the remark following upon Theorem 43.7) that a product of irreducible operators, one of which is strongly irreducible, is strongly irreducible.

Theorem 44.10. *Let E and T be as in the preceding theorems. If T is now strongly irreducible, then the only eigenvalue of T having absolute value $r(T)$ is the number $r(T)$ itself. In other words, any non-zero eigenvalue $\lambda \neq r(T)$ satisfies $|\lambda| < r(T)$.*

Proof. Assume that there exists an eigenvalue $\lambda \neq r(T)$ satisfying $|\lambda| = r(T)$. Let f be a corresponding eigenelement, so $Tf = \lambda f$. As shown in the proof of the preceding theorem we have

$$T(|f|) = |\lambda f| = r(T) \cdot |f|.$$

Since $\lambda \neq r(T)$, the elements f and $|f|$ are linearly independent. Once more by the preceding theorem, there exists a natural number n such that $\lambda^n = \{r(T)\}^n$. The positive operator T^n is order continuous, compact and irreducible (even strongly irreducible since T is strongly irreducible), so T^n satisfies all conditions for Jentzsch's theorem. The spectral radius of T^n is $\{r(T)\}^n$ as shown in Theorem 40.15(iii). The eigenspace of T^n belonging to $\{r(T)\}^n$ is, therefore, of dimension one. This contradicts the fact that both f and $|f|$ are elements in this eigenspace. Hence, the only eigenvalue of T having absolute value $r(T)$ is $r(T)$ itself. $\blacksquare$

Exercise 44.11. (i) In $\mathbb{C}^n (n \geq 2)$, let for $k = 1, \ldots, n$ the vector e_k have its k-th coordinare equal to one and all other coordinates zero. Let the positive operator T in $\mathbb{C}^n$ be given by $Te_1 = e_2, Te_2 = e_3, \ldots, Te_n = e_1$. Show that T is irreducible but not strongly irreducible and the eigenvalues of T are the roots $\lambda_1, \ldots, \lambda_n$ of $\lambda^n - 1 = 0$. Show that the eigenspace corresponding to λ_k consists of all (complex) multiples of

$$e_n + \lambda_k e_{n-1} + \lambda_k^2 e_{n-2} + \cdots + \lambda_k^{n-1} e_1.$$

(ii) Show that if in $\mathbb{C}^n (n \geq 2)$ the matrix of T has $t_{jk} = 1$ for all j and k, then T is strongly irreducible and has eigenvalues $\lambda_1 = n$ and $\lambda_2 = \cdots = \lambda_n = 0$ with eigenvector $e_1 + \cdots + e_n$ for $\lambda_1 = n$ and independent eigenvectors $e_1 - e_2, e_2 - e_3, \ldots, e_{n-1} - e_n$ for $\lambda = 0$. Note that $r(T) = n$.

(iii) Show that if in $\mathbb{C}^n (n \geq 2)$ the matrix of T_1 has $t_{jk} = 0$ for $j = k$ and $t_{jk} = 1$ for $j \neq k$, then T_1 is irreducible but not strongly irreducible and T_1 has eigenvalues $\lambda_1 = n - 1$ and $\lambda_2 = \cdots = \lambda_n = -1$ with eigenvector $e_1 + \cdots + e_n$ for $\lambda_1 = n - 1$ and independent eigenvectors $e_1 - e_2, e_2 - e_3, \ldots, e_{n-1} - e_n$ for $\lambda = -1$. Note that $r(T_1) = n - 1$. Consider in particular the case that $n = 2$.

Hint: For (ii) note that the determinant of the matrix of $T - \lambda I$ has the factor $n - \lambda$. To see this, add to the first column all the other columns. After dividing by $n - \lambda$ subtract the remaining first column from the other columns. For (iii) note that $T_1 = T - I$, where T is the operator in (ii).

Exercise 44.12. Assume that G is a non-empty set and τ is a mapping from $G \times G$ into G. Elements of G will be denoted by $a, b, \ldots$ and $\tau(a, b)$ will simply be written as ab. The mapping is called a multiplication and ab is called the product of a and b. We assume here that the multiplication is commutative, i.e., $ab = ba$ for all a and b in G. Some further assumptions are made, as follows.

(i) The multiplication is associative, i.e., $(ab)c = a(bc)$ for all a, b, c in G.

(ii) There exists an element e in G such that $ae = a$ for all $a \in G$. The element e is called a unit element in G. Note that the unit element is unique (if $ae_1 = ae_2$ for all a, then $e_1 = e_1 e_2 = e_2$).

(iii) For any $a \in G$ there exists an element $b \in G$ satisfying $ab = e$. The element b is called an inverse of a. The inverse of a is unique (if $ab_1 = ab_2$, multiply by an inverse c of a, so $(ca)b_1 = (ca)b_2$, i.e., $eb_1 = eb_2$, so $b_1 = b_2$). The inverse of a is denoted by a^{-1}.

The set G with multiplication satisfying (i),(ii),(iii) is called a commutative group. The product of n factors a is written as a^n. It is easy to see that $(ab)^{-1} = b^{-1}a^{-1}$, and so $(a^{-1})^n = (a^n)^{-1}$ for $n = 1, 2, \ldots$. This is written as a^{-n}.

The non-empty subset H of G is called a subgroup if $a \in H, b \in H$ implies $ab \in H$ and $a \in H$ implies $a^{-1} \in H$. For any $a \in G$, the set

$$(\ldots, a^{-2}, a^{-1}, e, a, a^2, \ldots)$$

is a subgroup, called the subgroup generated by a. If G is a finite group, the elements $a, a^2, \ldots$ are not all distinct, so there exist natural numbers $p, q (p > q)$ such that $a^q = a^p$. Then $a^{p-q} = e$. The smallest positive number m (in a finite group) satisfying $a^m = e$ is called the order of a. The number of elements in G is called the order of G.

For any subgroup H and any $a \in G$ the set $aH = (ah : h \in H)$ is called the coset determined by a. Note that $a \in aH$ and $eH = H$. If the order of H is m, then each coset aH consists of m distinct elements ($h_1 \neq h_2$ implies $ah_1 \neq ah_2$ because $ah_1 = ah_2$ implies $a^{-1}ah_1 = a^{-1}ah_2$). If the cosets aH and bH have an element in common, then they coincide. For the proof, assume that $ah_1 = bh_2$. Then $a = b(h_2 h_1^{-1})$, so any $ah_3 \in aH$ satisfies

$ah_3 = b(h_2h_1^{-1}h_3) \in bH$. This shows that $aH \subseteq bH$. Similarly $bH \subseteq aH$. Hence, any two cosets either coincide or they do not have any element in common. It follows that if H is a subgroup of order m in the group G of order n, then G is the disjoint union of H and a finite number of cosets where H and each of these cosets has m elements. This shows that n is an entire multiple of m. In the particular case that H is the subgroup generated by an element $a \in G$ of order m, we have $a^m = e$ (as shown above). The order n of G is an entire multiple of m, so $a^n = e$. This is the result needed in Lemma 44.7(ii).

As promised in the preface we end with a list of some other books on Riesz spaces.

(1) W.A.J. Luxemburg-A.C. Zaanen, Riesz spaces I, 1971, North Holland Publishing Company.

(2) H.H. Schaefer, Banach lattices and positive operators, 1974, Springer-Verlag.

(3) E. de Jonge-A.C.M. van Rooij, Introduction to Riesz spaces, 1977, Mathematical Centre (Amsterdam).

(4) A.C. Zaanen, Riesz spaces II, 1983, North Holland Publishing Company.

(5) H.U. Schwarz, Banach lattices, 1984, Teubner Verlag.

(6) C.D. Aliprantis-O. Burkinshaw, Positive operators, 1985, Academic Press.

(7) P. Meyer-Nieberg, Banach lattices, 1991, Springer-Verlag.

(8) D.H. Fremlin, Topological Riesz spaces and measure theory, 1974, Cambridge University Press.

(9) C.D. Aliprantis - O. Burkinshaw. Locally solid Riesz spaces, 1978, Academic Press.

The books 3 and 5 are (at least in the first half) somewhat more elementary. The books 8 and 9 are more general in the sense that they also deal with topological Riesz spaces.

Index